Tropical Forestry

Tropical Forestry
Volumes Already Published in this Series

Tropical Forest Seed by Schmidt, L. 2007

ISBN: 978-3-540-49028-9

Harvesting Operations in the Tropics by Sessions, J. 2007

ISBN: 3-540-46390-9

Forest Road Operations in the Tropics by Sessions, J. 2007

ISBN: 3-540-46392-5

Tropical Forest Genetics by Finkeldey, R., Hattemer, H. 2007

ISBN: 3-540-37396-9

Sampling Methods, Remote Sensing and GIS Multiresource Forest Inventory by Köhl, M. Magnussen, S., Marchetti, M. 2006

ISBN: 3-540-32571-9

Tropical Forest Ecology - The Basis for Conservation and Management by Montagnini, F., Jordan, C. 2005

ISBN: 3-540-23797-6

Lars Schmidt

Tropical Forest Seed

With 143 Figures and 19 Tables

Springer

Lars Schmidt
Forest Genetic Resources
Forest & Landscape, Denmark
Hoersholm Kongevej 11
DK-2970 Hoersholm
Denmark

ISSN: 1614-9785
ISBN-13: 978-3-540-49028-9
Springer-Verlag Berlin Heidelberg New York

Library of Congress Control Number: 2006938538

Springer-Verlag is a part of Springer Science+Business Media

springer.com

Editor: Dr. Dieter Czeschlik, Heidelberg
Desk Editor: Anette Lindqvist, Heidelberg
Production: SPi
Typesetting: SPi
Cover Design: Design & Production, Heidelberg

Printed on acid-free paper 3/3152-HM 5 4 3 2 1 0

Preface

The role and perception of forests in tropical areas has changed drastically during the last half century. Natural forests, as resources for forest products, are dwindling. Sustainable management of natural forests faces many difficulties in practice, although progress has been made. However, rural people in tropical countries often experience that the forests, which were previously the buffer for agriculture and an important resource, are becoming more and more inaccessible. Remaining forests are to a large extent protected, degraded or so far away from settlement that in practice they are beyond reach. The majority of the world's forest products in the future will come from man-made plantations and cultivated trees. The term 'plantation', usually referring to traditional block plantings of industrial species, is acquiring a wider meaning which includes, for example, smaller woodlots, shelterbelts and various types of agroforestry. Forest seeds are in this context of utmost importance. Not only are seeds the most commonly used propagation material, they are also the carriers of the genetic material from one generation to the next. Forest seed handling is thus an integrated part of selection, management, development and conservation of forest genetic resources in a larger context. With this in mind, and considering how self-evident the matter of seed quality is in agriculture, one can wonder how little attention has been and is given to forest germplasm in many afforestation and plantation programmes. The fact that seeds are small, seemingly ubiquitous and that the result of using good or poor seed will only become apparent far in the future tends to induce low priority or ignorance. The sad observation is that not only are forests degrading and dwindling at an alarming rate, but even the basis for reestablishment, good genetic material, is vanishing. For many species it is getting increasingly difficult to find 'good' seed.

Among potential afforestation or plantation species, relatively few exhibit major seed physiological problems. Yet many are not used because of alleged seed problems, problems that could easily be overcome by a little more careful handling during collection and subsequent procedures. Some tree seeds *are* difficult, or at least appear to be so, because they behave differently from what we expect. Systematic research has shown ways to overcome many practical problems. It has also shown that some features such as desiccation sensitivity and short viability are inert, and we must adapt our practices to these, e.g. using seed quickly if it cannot be stored.

The basic philosophy of this book is that good forestry practice should never be impeded by failure to get access to good-quality seed, and that the solution to possible seed problems is not to use poor seed or 'easy' species, but to improve and develop seed handling practices.

September 2006 Lars Schmidt

Foreword

Danish International Development Assistance (Danida) has a long experience of working with tropical forest seed. Danida Forest Seed Centre (DFSC), now merged with others into the Danish Centre for Forest, Landscape and Planning, has for more than 40 years been involved in research and development of all aspects of tropical forest seed, including tree improvement, seed technology, conservation and seed supply systems. The Centre continues to be a key centre for dissemination of information material via technical notes, lecture notes, seed leaflets, extension material and books. Among the most comprehensive material on seed technology was the book *Guide to Handling of Tropical and Subtropical Forest Tree Seed*, by Lars Schmidt, which was published in 2000. The book has been widely distributed to most tropical countries and has been translated into, for example, Bahasa Indonesia. In 2002 Springer addressed the former DFSC to write a volume on forest seed for the series of books on tropical forestry. The task was agreed after Lars Schmidt returned to the Centre from leave.

Although there are inevitably similarities and some sections have been reused with few changes from the previous publication, the present book is not a mere reissue or revision of the former DFSC publication but rather an independent contribution to the *Tropical Forestry* series of Springer. The book has a slightly different focus: in view of the generally improved access to technical facilities in tropical countries, there is more emphasis on these facilities. In addition, many pieces of new information have been included. The author has, during the years since he wrote the DFSC guide, been working on tree seed projects in Indochina and Indonesia. Experiences from these areas are included in this book.

In the past and present, seed problems have been and are a limiting factor for use of species. Forest seed handling is determined by a combination of knowledge of seed biology, of available technology and of seed demand. Knowledge of seed biology increases with experience and research. It is implicit that experience is primarily directed towards already used species – the more they have been used, the greater the experience. Many research efforts are designed to overcome crucial and limiting bottlenecks for particular problems of particular species. It can thereby make species choice less seed handling dependent. Tremendous progress has been made during the last decade on, for example, desiccation and storage conditions of recalcitrant seed (Sacande et al. 2004). The research has not eliminated the problems of these seeds, but we know far more about the interaction between water content and storage behaviour of desiccation-sensitive seed, in order to optimise seed handling practice. Available technology sets a limit to what can be done in practice. The advantages of cold storage are, for example, of little use in areas without cooling facilities. Fortunately, many tropical countries are also getting access to improved technical facilities.

Seed demand is highly determined by political and economic considerations. Most seed users will select tree species which produce the desired result in the shortest possible time, i.e. good genetic quality, of the best provenance, of the best species. Since planting sites and product demands are diverse, this should imply a much diversified species demand. In reality, however, large afforestation programmes often tend to economise establishment cost by reducing species diversity and chose species which are cheap and easy to propagate and raise. The unfortunate consequence is that many species which could be grown and thereby enrich the environment and provide good return to tree planters in the long term are not used because of short-term economic rationales. Progress in seed technology does not alone overcome the diversity problems, but it helps. Fortunately, the political awareness of diversity and the importance of good seed quality for successful afforestation seem to be improving. With an optimistic view that the discrepancy between political will and practical field implementation will be overcome, there will inevitably be more pronounced focus on handling different tree seeds in the future.

Virtually all trees regenerate from seed and can thus be propagated from seed. Many species may be difficult to propagate at the first attempt. Persistent and systematic trials will usually help in identifying and overcoming the problems. Continuous research is necessary, as there are still many problems to be overcome and methods to be improved. Research and dissemination of research results are cornerstones in building up a better capacity in the supply of forest seeds. The Danish Centre for Forest, Landscape and Planning and the former DFSC have played an active role in research and development of tropical tree seed, with the overall objective of increasing species diversity, and improving seed quality of planted forests in the tropics. It is hoped that this book will be a contribution to this overall objective.

The present book attempts to cover all relevant aspects of practical seed handling, from collection to distribution with inclusion, when deemed necessary for understanding and further development, of relevant physiological or genetic background for the recommended practice. The book is thus primarily addressed to seed practitioners in seed centres or seed enterprises, but can be read by anyone with an interest in seed biology, technology and supply.

July 2006

Niels Elers Koch
Director General, Danish
Centre for Forest,
Landscape and Planning

Acknowledgements

The basis for my knowledge of tropical forest seed handling was set at the former Danida Forest Seed Centre (DFSC), i.e. during the compilation of the former seed handbook, *Guide to Handling of Tropical and Subtropical Forest Tree Seed*. The acknowledgements given there are still valid. My experiences in Indochina and Indonesia have added to my personal understanding about forest seeds, their biology, technology and constraints in seed supply. Some things I thought were complicated appeared to be less so in reality; some things I believed were easy turned out to be more diverse than anticipated. I have been unable to identify the sources of many pieces of practical information, but project colleagues and staff from research institutions helped 'thinking together' to overcome practical problems and willingly shared their knowledge and experience; I owe them thanks for their encouragements and contributions. For this book, Niels Arp Hansen from Levinsen Skovfrø, Denmark, helped me put some newer theories and technologies into a practical context of Danish seed handling. Finally, I am grateful to Melita Jørgensen for linguistic proofreading of the script.

The illustrations for this book are partly from my own archive, partly from external sources, which, as far as I have been able to trace them, are acknowledged with each picture. Markus Robbins deserves special mention for his excellent drawings, which have been used before in several DFSC publications. Some drawings by Poul Andersen made for the DFSC book have been reused in this publication. I am grateful to authors and publishers who have granted me permission to use their illustrations.

July 2006 Lars Schmidt

Acknowledgments

Contents

Introduction 1

Tropical tree seed handling continuously develops. Scientific research and less advanced, yet persistent practical progress bring about new knowledge and experience on tropical species. Development of new information technology, together with the more traditional writing of textbooks and technical material, bring the new information to a broader user group. Access to better technology and material has characterised many tropical countries over the past 10–15 years. Although many tropical countries are still lagging far behind in economic development, the former habit of making an uncritical parallel between tropical countries and developing/poor countries is not always valid. Progressive and resource-rich tropical countries have shown that it is possible to make well-functioning forest seed supply systems also under tropical climates. Seed research has species by species and topic by topic shown the way towards a more efficient seed handling procedure for individual species, for example in relation to storage behaviour and dormancy (Sacande et al. 2004). Technical facilities are becoming increasingly widely available, and quality is improving. Climbing equipment, storage containers, processing machines and refrigerators are examples of some equipment which can be found in most markets or specialised shops in larger towns throughout the world. Computerised seed documentation systems have revolutionised all documentation and data distribution systems. The technical facilities are thus to a large extent available to provide an efficient seed supply system.

Cheap and simple methods are still a reality in many countries and for particular user groups, and information on how to provide good quality by simple methods still has a place in the extension service. However, on central seed supply level, better equipment, better documentation systems and better distribution systems are often more subject to economic priorities rather than being beyond access, even in the so-called tropical developing countries.

The general advantage of using good quality seed has been well documented (Foster et al. 1995). Where there is a direct economic link between planting material and tree tenure, there should thus be a good incentive to use the best seed available. The incentive would normally justify a good investment in seed technology and improvement. When we can observe that the seed sector is often resource-poor and underdeveloped, and that quality of seed is far from optimal, the main reason should be sought in the lack of a link between planting material and tree harvest. Some frequently encountered constraints are:

1. The relative poverty of seed users. A large and increasing part of tree planting is done on farms and by smallholders (Simons 1997). Many smallholders are unable or hesitant to pay the extra cost for tree seed, which has been claimed, but not necessarily proved, to be of better quality.
2. Lack of a proper distribution system. For lesser-used species there may not be a source and supply at all. For more commonly planted species with improved seed supply, the bottleneck is to get seed distributed to remote areas and particularly to small end users in small quantities. In practice, most seed suppliers distribute seed within a radius of less than 50 km (Nathan 2001).
3. Poor-quality documentation. Seed quality contains a number of components and their relative importance is not always clear. Lack of research trials for most species makes documentation of genetic quality, for example for growth habit, unreliable. Documentation on origin, seed source and mother trees does contain indirect genetic information, but often a blurred concept of the 'best available', which is rather nontransparent for seed users. Since really good, documented quality is obviously expensive, the poor definition of quality obviously invites deceit. Documentation of physiological quality frequently suffers from lack of standards and outdated analyses.
4. Time span from planting to tree harvest. This is the general and ubiquitous problem of forest establishment. In terms of quality seed supply it has implications ranging from corruption and deception to insufficient means of investments in improvement means. Lack of confidence and trust in alleged improved material can almost always be referred back to the lengthy time span required from the purchase of seed or planting material until the trees have reached a reasonable size to be able to judge their growth potential. If there is no real legal procedure to get compensation if cheated, customers cannot be expected to pay for an alleged improved quality. And if customers are unwilling to pay, suppliers are unwilling to provide a better quality; this is the ubiquitous vicious cycle of tree seed supply.

The political-economic trend during the last 10–15 years in most tropical developing countries has been to reduce the public sector and strengthen market mechanisms. This has affected the forest seed sector, since this sector has traditionally been part of the public sector. Market economy necessarily implies generation of profit within a reasonable time span. Short-rotation industrial species in relatively large closed units suit market mechanisms. Long-rotation species, reforestation or supply for resource-poor people and environmental elements of forestry, such as biodiversity or watershed management, do not fit well into a purely private environment. Without an economic incentive or strong public control, the importance of species diversity and genetic quality tends to be neglected. The consequences of neglecting quality control tend to increase, as the quality of random supply gets poorer – the latter due to a general degrading of natural seed sources. The need for regulations and implementation of control systems has thus become increasingly important.

The last 10–15 years has seen a rapid development in the techniques of vegetative propagation. Although mass vegetative propagation requires a fairly high investment in propagation facilities, once it is there it has proved highly competitive with seed propagation for a number of species. Many improved varieties of trees are propagated almost exclusively by cuttings or tissue culture. Vegetative propagation does imply some risk factors compared with seed propagation in terms of genetic diversity. However, provided appropriate control can be maintained, vegetative propagated plants are a good alternative to seed, in particular for species with seed problems and where a uniform performance of a high-bred species is desired. However, although increasingly applied, vegetative propagation has not and will not replace seed propagation as the principal method of plant propagation. The genetic variation contained in seedling plants compared with vegetative propagules is a strong argument to maintain seed propagation in environmental plantings. Seed propagation will almost always be used by small and less equipped nurseries.

Further improvement of seed technology and extension of skills and experiences of seed handling is thus still relevant. It is also necessary to avoid constraints in seed technology becoming a hindrance for diversity of plantations. Far too many planting programmes stick to the 'easy ones' when selecting species (Fig. 1.1). Developing good seed procurement and handling techniques is a method for making potential plantation species 'available' for planting. Experiences have shown that overcoming seed problems can sometimes boost the use of otherwise 'impossible' species.

The need for diversity in planting programmes is becoming more urgent as tree resources in most tropical countries are under pressure. Conservation of gene resources, both species and variation within species, is not done alone in protected areas. Conservation by use implies that conservation becomes integrated in the reforestation programme. Seed handling is one among several approaches to promote diversity.

Rehabilitation of vast areas of deforested land is one of the major challenges of environment rehabilitation and management now and for the many years in

Fig. 1.1. Hard native wood is popular for traditional furniture manufacturing. Natural resources are heavily exploited but the species are rarely planted because they are difficult to establish from seed and are slow-growing

the future. For far too long we have observed the destruction without creating efficient countermeasures. Hillsides turned into unproductive grassland and bushland, siltation of rivers and streams, destroyed coral reefs and thousands of endangered plants and animals not only on a local scale but also on a global scale is what deforestation in sensitive areas has brought. Repairing the damage is what faces our and future generations. Seed handling is one link in the chain to help restore the environment. Though seemingly small, the link is crucial. Seed is the genetic connection between the parent generation and the offspring, and the vehicle that brings progress or recession in terms of genetic quality (Fig. 1.2). The difference between good and poor is very large. For example a poorly managed and degraded shrub may yield less than 1 m^3 of fuelwood per hectare per year – about the consumption of a household. A well-managed forest in the same place may yield 20 m^3 – or from utilising 1 ha just for fuel, the family may, with better genetic material and management, utilise only 500 m^2 (Fig. 1.3).

The supply of quality forest seed has always been subject to a well-known demand–supply problem: customers who demand quality seed but allegedly cannot get it; and suppliers who produce quality seed but claim that there are no customers. Unfortunately both parties could be right. In practice it has appeared quite difficult to make good seed supply operational on a national level containing a broad range of species and containing the best documented genetic quality. Mostly it is a price problem. Genetically improved material is expensive; and any

Fig. 1.2. The seed is the apparatus of regeneration and the vehicle of genes. The physiological quality is influenced by maturity, age and deterioration, and it is manifested by the ability to germinate. The genetic quality is influenced by the parents and crossing, and it is manifested by the growth habit

Fig. 1.3. Fuelwood is one of the most important extracts from forests. Millions of rural people rely on fuelwood as their only or principal source of household energy. As the sources are being depleted, the pressure on the remaining forests is increasing and often results in poor productivity

reasonable selected and documented material is far more expensive than average, randomly collected seed.

This book focuses on seed handling rather than genetic quality. However, the implicit statement is that seed handling is handling of good (genetic) quality seed. The seed is the vehicle of genetic quality whose base camp is the seed source and whose destination is the planting site. Seed handling thus starts from collection from the selected trees in the selected seed source, and it continues to planting and germination in the nursery or the field. Each link in the chain contains risk factors and pitfalls, which can reduce seed quality and thus waste all previous work.

Seed Collection 2

2.1 Introduction

A good tree starts from a good seed. Whatever the succeeding procedures of seed handling, they can only maintain the quality, never improve it. It is thus well justified to pay attention to what is actually collected in the first place. Seed can sometimes be collected from the ground after natural fall. When this is possible without jeopardising quality it is always preferred, as it is by far the easiest and cheapest way. However, all seed collectors have come to realise that good seed must often be collected from the mother tree before it falls or is dispersed. Sometimes it is necessary to ensure that there are seeds (healthy seeds) to collect at all, i.e. before they are dispersed, or have been attacked by predators or even started to germinate – and sometimes to be sure of the identity of the mother tree. Collection from the crown by using long-handled tools and/or short ladders applies to many smaller intermediate-size trees. However, there are a number of species which grow very high and where seeds need to be taken from the crown. How to get up to the top and out to the very thin branches where seeds are usually borne, with minimum effort and risk, has given rise to much invention in tree climbing. Climbing has thus become an integrated part of seed procurement (Yeatman and Nieman 1978; Blair 1995; Barner and Olesen 1983a, b, 1984a, b). How to get to the top without climbing has appealed to even more inventiveness, e.g. balloons, raised platforms or even helicopters (Vozzo et al. 1988). The direct cost and the cost of operation of some of these inventions are so high that they are rarely used unless there are no other suitable alternatives, or where costs are not calculated, e.g. if they are hidden in an institution's core budget or are part of another exercise.

Climbing remains the most suitable way of getting access to the crown, if necessary, but it is both risky and expensive. Genetic considerations suggest collecting from at least 25–50 unrelated good-looking mother trees (Sedgley and Griffin 1989). For large timber trees good-looking trees are large, straight individuals with no lower branches and a small crown with thin branches.

This type of tree is not 'climber friendly' and often has the additional drawback of producing few seeds. Five to six such trees in a day would be a very good achievement for a climber, i.e. a 'genetically safe' collection would take at least a week for a full collection team consisting of two climbers, ground staff and driver. No wonder that alternatives will be considered.

For the most commonly used species, seed collection can be rationalised in established seed sources, where trees are relatively small, managed for large seed production, and are accessible for the various technical accessories designed to ease a collection. In some cases yet another step is taken to reduce seed collection: plants are raised from vegetative material (cuttings or tissue culture) collected in low hedge gardens. In addition to reducing collection cost, established seed sources as well as hedged gardens are normally a part of a tree improvement programme, i.e. using genetically superior material. Labour cost, safety concern and rationalisation tend to reduce routine seed collection by climbing for commonly used forest species. However, species diversity and genetic diversity within species tend to become issues of increasing concern. Local seed collection will still to a large extent rely on seed. Therefore, seed collections that include climbing will remain a necessary element of a broad range of seed procurement programmes.

The choice of collection method thus depends on the biological basis, on the purpose and types of collection, which methods are applicable and available and the economic possibility.

The term 'seed collection' may be somewhat misleading, because in practice we are for the greater part collecting the whole fruit. However, it is the seed with its genetic trait and ability to germinate we want – therefore the term has become common use. Seeds are for most species extracted from the fruit during seed processing (Chap. 3).

2.2 Biological Factors Influencing Collection

Seed is biological material exhibiting a wide range of biological variation in morphology and physiology. Seed is the plant's reproductive material, containing the inherited trait of the parent, evolved and adapted to optimise regeneration in a multitude of niches appearing in forest ecosystems. Seeds are produced and dispersed in such a way as to optimise their survival from predators and in competition with other species. Some species produce a regular bulk crop of orthodox, wind-dispersed seed. Such species offer few problems. Others produce seed crops at long intervals or over long seasons. Animal-dispersed seeds impose particular problems, firstly because animals may eat them or carry them away and secondly because structures attracting animals, e.g. fleshy pulp, are likely to make both collection and extraction more difficult (Chap. 3).

2.2.1 Type of Fruit and Seed

Fruit type reflects an adaptation to dispersal. Most tree seeds are either dispersed by wind or by relatively large animals (birds or mammals). Some mangrove species and coastal palms, e.g. coconut and *Pandanus*, are principally dispersed by seawater. Species with specialised occurrence along rivers (riverine species such as *Acacia nilotica*) show morphological adaptation to water dispersal. However, as rivers necessarily float and thus deposit seed only downstream, water dispersal for such species is generally a secondary adaptation.

2.2.1.1 Wind-Dispersed (Anemochorous) Species

Small size and high air resistance help reduce falling speed and thus increase the time for horizontal displacement by wind. Very small and light seed may be more or less suspended in air (van der Pijl 1982). Tiny seeded species are, for example, *Anthocephalus chinenesis*, *Octomeles sumatrana* and most eucalypts and melaleuca species. Most winged diaspores[1] have wings designed for spiralling when falling, which reduces falling speed significantly. One-winged (mahogany, *Tarrietia*, *Pinus*), two-winged (*Acer*, dipterocarps) and three-, four- or five-winged (*Vatica*, *Shorea*) diaspores possess this feature (Fig. 2.1).

Although dry seeds are necessarily lighter than moist ones, wind dispersal is not entirely linked to orthodoxy, i.e. low moisture content at dispersal. The entire dipterocarp family is an example of a large group of species with mainly desiccation-sensitive seed but with apparent adaptation for wind dispersal. Recalcitrance[2] also occurs in wind-dispersed species in other families, e.g. Sterculiaceae, Meliaceae and Combretaceae. Very small wind dispersed seeds (e.g. Myrtaceae) are always orthodox.

Collection of wind-dispersed species is often easy since fruits and seed are often dry and easily break off the tree. The major observation to be made regarding wind-dispersed seed is time of collection – especially small seed: too

[1] Diaspore is the dispersed unit, which may be a seed, a fruit, part of a fruit with seed or an aggregate of several fruits.
[2] Recalcitrant seeds are seeds that do not tolerate drying to low moisture content (Chap. 4).

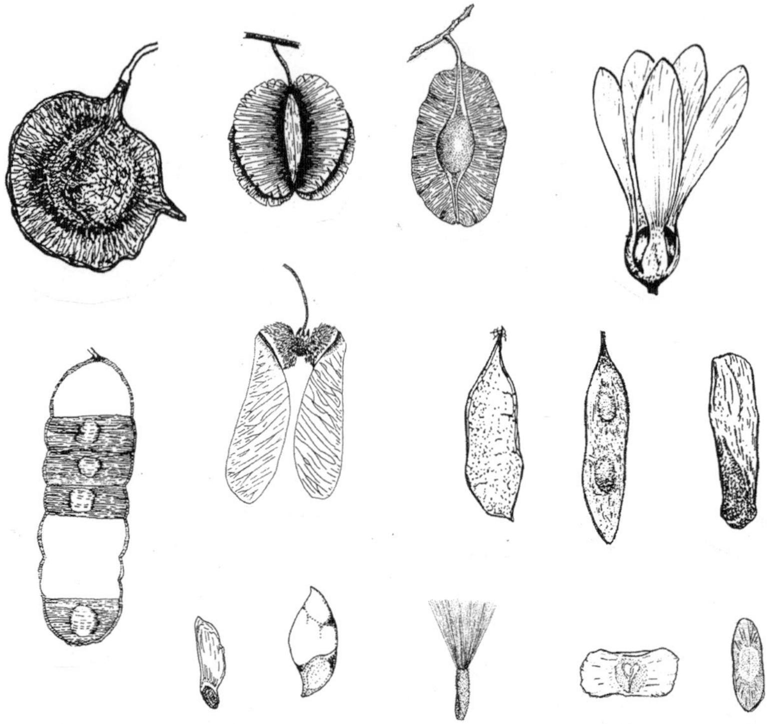

Fig. 2.1. Examples of wind-dispersed diaspores. Wings can be part of the fruits (samaras) or seeds. From *upper left*: *Pterocarpus*, *Combretum*, *Terminalia*, *Shorea*, *Entada*, *Triplochiton*, *Acacia*, *Dalbergia*, *Swietenia*, *Pinus*, *Chukrassia*, *Brachylaena*, *Spathodea*, *Dyera*

late and the seeds are gone! The period for seed dispersal is sometimes very short, in particular, in dry weather. Trying to shake down dehiscent fruits with light seeds may cause most of the crop to blow away.

2.2.1.2
Animal-Dispersed (Zoochorous) Seed

The majority of animal-dispersed seeds have fleshy fruit types, e.g. drupes, berries and various types of aggregate and multiple fruits. In dry areas some zoochorous diaspores, e.g. acacia and prosopis pods and ziziphus drupes, are rather dry. Nutritious appendices (arils) may be dry or moist, but they are usually very conspicuous. Animal dispersal occurs in both angiosperms and gymnosperms, but the morphological adaptations to animal dispersal are much wider in angiosperms. Animal-dispersed fruits and seeds are often quite large

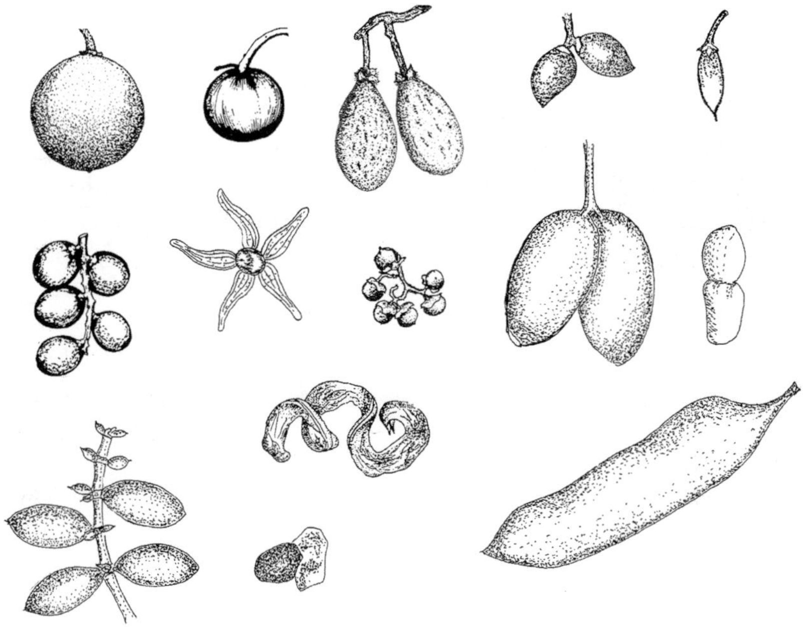

Fig. 2.2. Animal-dispersed seeds. Seeds may be ingested and pass the through the whole digestive track and be deposited with the faeces. In other cases seeds are regurgitated and sometimes they are just sucked free for pulp. From *upper left*: *Diospyros*, *Sandoricum*, *Maranthus*, *Olea*, *Peyena*, *Aglaia*, *Swintonia*, *Cordia*, *Syzygium*, *Dacrycarpus* (arillate seed), *Gnetum*, *Acacia*, *Sindora* (arillate seed), *Tamarindus*

and conspicuous (red or yellow fruits), and often contain protective structures around the seed, e.g. endocarp in drupes or seed coats in many other fruit types (Fig. 2.2).

Animal-dispersed seed may be orthodox, recalcitrant or intermediate. Dormancy is prevalent. Inhibitors in fleshy fruits have the role of impeding germination until after dispersal. Hard structures protect against damage by ingestion, which is the most common mode of animal dispersal.

Animal-dispersed species often have long fruiting seasons, especially those adapted to dispersal by few specialised dispersal agents (McKey 1975; van der Pijl 1981, Janzen 1972). This has two implications for collection: (1) that it is difficult to harvest enough seed in one or two collections; (2) that a fruit crop may be continuously removed by animals, which also has an enhancing effect on the first implication. Harvest of animal-dispersed fruits is often easy because of their large size. Processing, particularly extraction, can, on the other hand, be quite arduous.

2.2.2 Maturity and Seasonality

Physiological maturity of seed, for most species, coincides with readiness for dispersal. As seeds are normally germinable when they are about to be dispersed, natural dispersal time, or indication of dispersal, can be used as a physiological maturity criterion. There are situations where seeds must be collected prematurely and then after-ripened, e.g. if they are easily lost by early dispersal or predation, but even here dispersal maturity criteria are used to determine the best time of collection.

Maturity contains two practical aspects in relation to seed collection:

1. What are the visible or measurable maturity criteria for fruits and seed?
2. How long before potential natural dispersal can seed be collected and, by after-ripening, achieve the same quality in terms of germinability and storability as seed collected at full maturity?

2.2.2.1 Maturity Criteria

Seeds can be released from the tree in two ways:

1. Dehiscence, in which the fruit opens on the tree and the seeds fall out. The fruit here remains attached to the tree until after dispersal. Release of the seed happens via breaking of the seed connection to the fruit, the pedicel. This occurs in many dry fruits, e.g. dehiscent pods and capsules.
2. Indehiscence, in which the fruit is dispersed as a unit. The fruit is here released from the tree by softening or breaking of the fruit's connection to the branchlet, the peduncle. This occurs in both dry and fleshy fruits, e.g. nuts, pods and drupes.

The breaking off of seeds and fruits, and sometimes splitting up of the fruit occur in special layers of cells, the abscission zones, a phenomenon also known in leaf shedding and self-pruning of branches (Osborne 1989; Kitajima et al. 2003). Checking the strength of the abscission mechanism (e.g. breaking off fruits) is a practical way to check the maturity.

Development of dispersal devices as summarised in Figs. 2.1 and 2.2 and Table 2.1 is a reliable maturity criterion. If seed trees are nearby and can be followed currently, seed collection may be arranged when the first seeds can be found under the tree or animal-dispersal agents start to feed on the fruits. If seed trees are remote, waiting until the 'last minute' is risky. In particular, change of weather from moist, cool and cloudy to hot and dry may cause an amazingly

Table 2.1. Practical maturity indices for forest tree fruits

Maturity event	Method of examination
Colour change: dry fruits, green to yellow, brown or black; fleshy fruits, green to conspicuous red, yellow, blue, etc.	Visual 
Dehydration (dry fruits)	Visual, touching or 'weighing' in the hand Measurement of specific gravity
Dehiscence and abscission 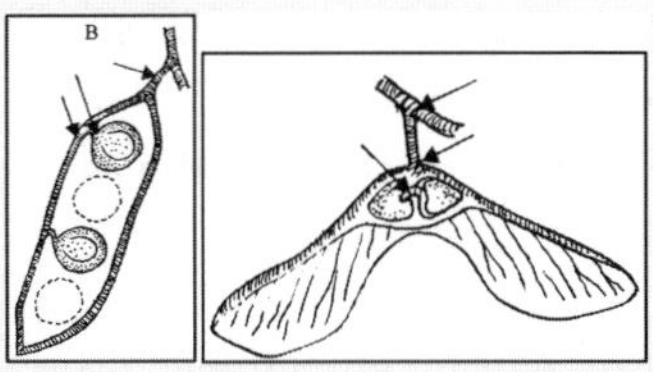	Observation of fruit fall or opening of dehiscent fruits Shaking or beating fruit-bearing branches Beating or manual splitting of dehiscent fruits Examination of opening structures in dehiscent fruits, e.g. valves, scales and margin Breaking off fruit stalks
Hardening of fruit/seed coat	Cutting, pricking, breaking of seed or fruit coat
Hydration (fleshy fruits). Softening of fruit flesh	Squeezing
Loosening of fruit pulp (fleshy fruits)	Squeezing, rubbing or other separation of fleshy part from seed or endocarp
Accumulation of sugar substances (fleshy)	Taste (careful as some fleshy fruits are poisonous to humans) Observation of visiting frugivores
Endosperm and embryo development of seed	Cutting of seed. Squeezing the embryo – the embryo should be firm and hard (Boshier and Lamb 1997)

rapid drying and dehiscence of dry fruit types. Trees which tend to synchronise their crop production may disperse it all in 1 day. At the other extreme are some eucalypt and pine species which retain their seeds in the fruit for a very long time after maturation, sometimes 1 year or even more (Gray 2004).

2.2.2.2 Premature Collection

The latter part of fruit and seed maturation consists of an internal restructuring and denaturation of components, e.g. proteins and hormones, and loss of water (Bewley and Black 1994). The flow of water and nutrients from the branchlet through the peduncle and pedicel to the seed gradually ceases.

The sequence and duration of events leading up to full maturity differ between species and premature collection is always a matter of experience. However, for most orthodox seeds, fruits can be picked and after-ripened when the fruit or seed changes from green to a mature colour (i.e. loses the ability to photosynthesise), which is usually 2–3 weeks before natural dispersal (Boshier and Lamb 1997). Recalcitrant seeds are also in this connection a problem because they continue to accumulate dry matter (i.e. increase in size and weight) up to full maturity (Berjak and Pammenter 1996; Phartayal et al. 2002). These types of seeds must be collected practically at the time of their normal fall or dispersal, as there is only a limited option for after-ripening of nearly mature fruits.

2.2.2.3 Seasonality

Most species have distinct fruiting seasons and seed collection is usually aimed at seasons where most fruits and seeds are available. However, in practice, collection teams often arrive too early or too late, i.e. the crop was either not mature or very little was left. Often collection teams will take whatever little is available. This is not always advisable. Very early, very late or out-of-phase fruiting may be preceded by concurrent early, late or out-of-phase flowering, i.e. isolated in time and thus implying a relatively high risk of inbreeding (Boshier and Lamb 1997). Inbred seeds are often morphologically or physiologically abnormal. Sometimes they are aborted early, sometimes they remain on the tree for a long time after other seeds have been dispersed. The phenomenon of inbreeding differs between species, from species with strong inert inbreeding barriers to species with full compatibility. Most forest tree species are facultatively outcrossing, meaning that foreign pollen has an advantage over their own pollen (Griffin 1990; Sedgley and Griffin 1989). Where inbreeding occurs it is found mainly where flowers are isolated in time and space. The proportion of inbred seed is smaller during the peak season, because peak flowering is the time with the highest chances of outcrossing (Griffin 1990; Sedgley and Griffin 1989). There are other aspects of seed quality affecting especially early and late crops, e.g. maturation (early crops) and insect infestation (mainly late crops). Whatever the cause it is thus generally recommended to collect seeds during the peak season and to avoid very early and very late crops.

2.2.3 Damage to Trees and Future Seed Crop

The method of collection may in some cases directly or indirectly affect the future seed crop. The impact is usually negative, e.g. damage to fruit-bearing branches or damage to the tree, leaving it in a condition of stress that may affect future seed production.

Severe pruning of fruit-bearing branches in connection with seed collection may reduce the number of potential fruit-bearing branches for the next crop. In practice normal seed collection in broad-leaved species, even where it implies pruning of branchlets, has little effect on future crops. The situation is different in conifer species because they often possess cones in early stages of development simultaneously with mature cones. If young undeveloped cones are removed together with mature ones, e.g. by branch pruning, it will affect the next year's production (Fig. 2.3).

On the other hand, moderate pruning can also have a beneficial effect on future seed production as it promotes light exposure to the remaining branches and can increase the average seed weight as more resources are now allocated in fewer seeds. In fact, pruning is often applied in seed orchards to promote flowering and fruiting (Faulkner 1975).

There are species with long fruiting seasons (usually animal-dispersed species) that may require several succeeding collections, where seeds are collected directly from the trees. Here individual harvests must avoid damage to the remaining crop, i.e. avoid collecting still immature fruits or damage to flowers. Moderate shaking will, for most species, make mature fruits fall and leave the immature fruits on the tree.

Fig. 2.3. Branch of *Pinus kesiya*. At the time of seed collection, conelets for next year's seed crop have already developed and can be damaged by some types of collection. Old cones often remain a long time on the tree in this species, so tree stages of fruits occur on the same branches

Scars in the bark left by climbing spurs, and the open end of pruned branches will normally be sealed with resin exuded from the bark and will gradually be recovered completely. There are thus rarely any long-term effects of, for example, spur climbing (de Castilho et al. 2006). However, in some instances insects or diseases may use scars as entry points for attacks. Both damage and the ability to close wounds depends on the species: species with thick bark and excessive resin (e.g. *Pinus merkusii*) are little prone to damage, while species with thin bark and less resin can be easily damaged (Mori 1995). Susceptibility or resistance to damage of an individual species should be considered in connection with the choice of collection method (Fig. 2.4).

Damage to trees can be reduced by appropriate methods of climbing and pruning, e.g. appropriate cutting of branches rather than breaking to reduce the exposed surface and the risk of bark being stripped off (Fig. 2.5).

National parks and other conservation areas often have severe restrictions on operations causing any damage to trees In these cases, less damaging methods must be used, e.g. collecting individual fruits rather than cutting branches and using ladders or advanced lines for climbing rather than spurs.

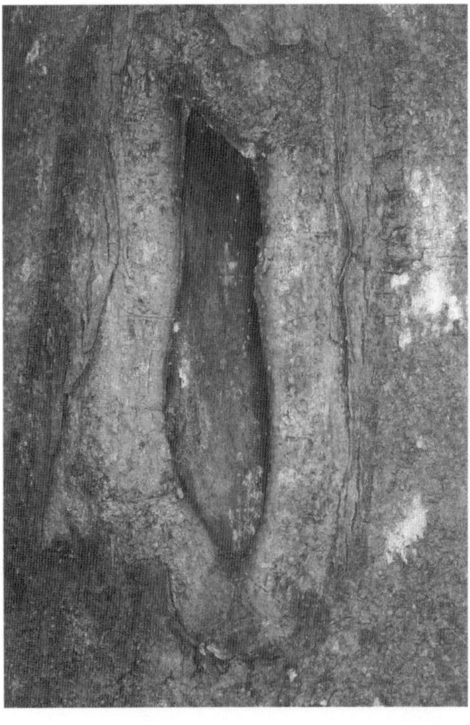

Fig. 2.4. Some species, here *Erythrophloeum fordii* in Vietnam, are very prone to stem damage, which can occur in connection with seed collection

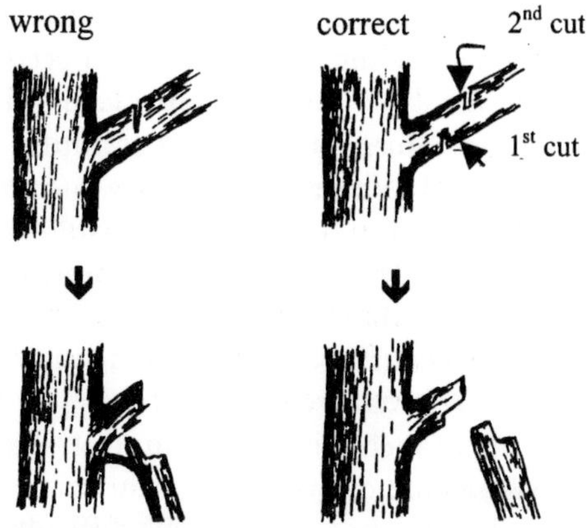

Fig. 2.5. Damage to trees can be reduced by correct cutting of branches during collection: to avoid bark being stripped off and leaving the sapwood exposed, the branch should be cut from below before being cut from above

2.3 External Factors Influencing the Choice of Collection Method

In practice, the choice of seed collection method is often restricted by factors relating to, for example, location of seed source, condition of the stand, individual trees and type of fruits or seeds and their maturity.

2.3.1 Identity of Mother Tree

The exact identity of a seed tree is interesting if it significantly differs from neighbouring trees with which it could be confused. This could occur if it is part of a tree breeding programme, and there could be a desire to come back to the particular tree sometime in the future depending on the progeny raised from the seed. The importance of collecting from particularly good looking phenotypes in stands is, however, often exaggerated. In natural forests, neighbouring trees are often related (Eldridge et al. 1993; Fig. 2.12) and phenotypic selection of seed trees makes little sense where environment and age differences may account for most of the visible difference (Danusevicius and Lindgren 2002). In seed production areas, inferior seed trees are rouged before seed collection, and since seed production areas are thinned to promote seed production they may not exhibit any good attractive timber traits at the time of seed collection. There may be some benefit in selecting seed trees in unthinned plantation sources, where phenotype and genotype have high correlation. Since plantation trees are grown with short spacings and sometimes with interweaving branches, fallen fruits or seeds could be from any mother tree. If one wants to be sure, seed must be collected directly from the crown. Collection from felled trees during timber harvest will allow some phenotypic selection of seed trees. Remember, however, that the seed mother tree only represents half of the genes, the other half come from a usually unknown male.

In tree breeding and other scientific studies, the identity must obviously be established. Where the distance to neighbouring trees is far, seed-rain collected under a mother tree, e.g. on tarpaulins or nets, will quite surely belong to that tree. Where there can be any confusion, seeds should be picked directly from the crown. Manual collection or cutting off fruit-bearing branches from the ground or during climbing is the most common method. For very large trees with very small seeds, e.g. eucalypts, the practice of shooting down fruit-bearing branches is applied in Australia (ATSC 1995; Gunn 2001).

Where the crowns of neighbouring trees have some entangled or crossing branches, it can sometimes be difficult to see from a distance which branches belong to which tree.

2.3.2 Shape and Height of Seed Trees

Many commonly used agroforestry trees and most dry zone trees are short (less than 10 m), and the crown can be reached from the ground by use of long-handled tools with or without use of low elevated platforms, vehicle rooftops or ladders. Many of these can also be climbed easily. There are also trees with a height and a shape that make any effort of climbing questionable; some trees cannot be climbed by conventional methods.

Crowns of shorter trees may be reached by extended pruners or saws, or flexible saws or other equipment operated from the ground or the top of vehicles. Trees with long, relatively straight, clear boles of small diameter can be climbed up to the crown with the help of spurs, ladders or a tree bicycle, after which the climber continues with free hand climbing. Trees with large diameters and trees with large buttresses and overgrown by vines, climbers, stranglers or other large epiphytes are very difficult to climb with spurs and a tree bicycle cannot be used. Here ladders, advanced lines or shooting are usually the only solutions. Large spreading umbrella-shaped crowns typical of many *Acacia* and *Albizia* species make the use of safety equipment very difficult and climbing of these trees may be excluded altogether. Methods of seed collection in relation to height and shape of trees are illustrated in Fig. 2.6.

Tree height and shape should not restrict selection of seed trees. High, straight stems and small branches with high self-pruning are desirable characters in timber species. Avoidance of mother trees with these characters could easily lead to neglect of desirable genetic characters. Poor characters in many exotic species around the world are believed to be due to ignorance or unintentional selection of a poor genotype mother tree during the first introduction (Hughes and Styles 1984).

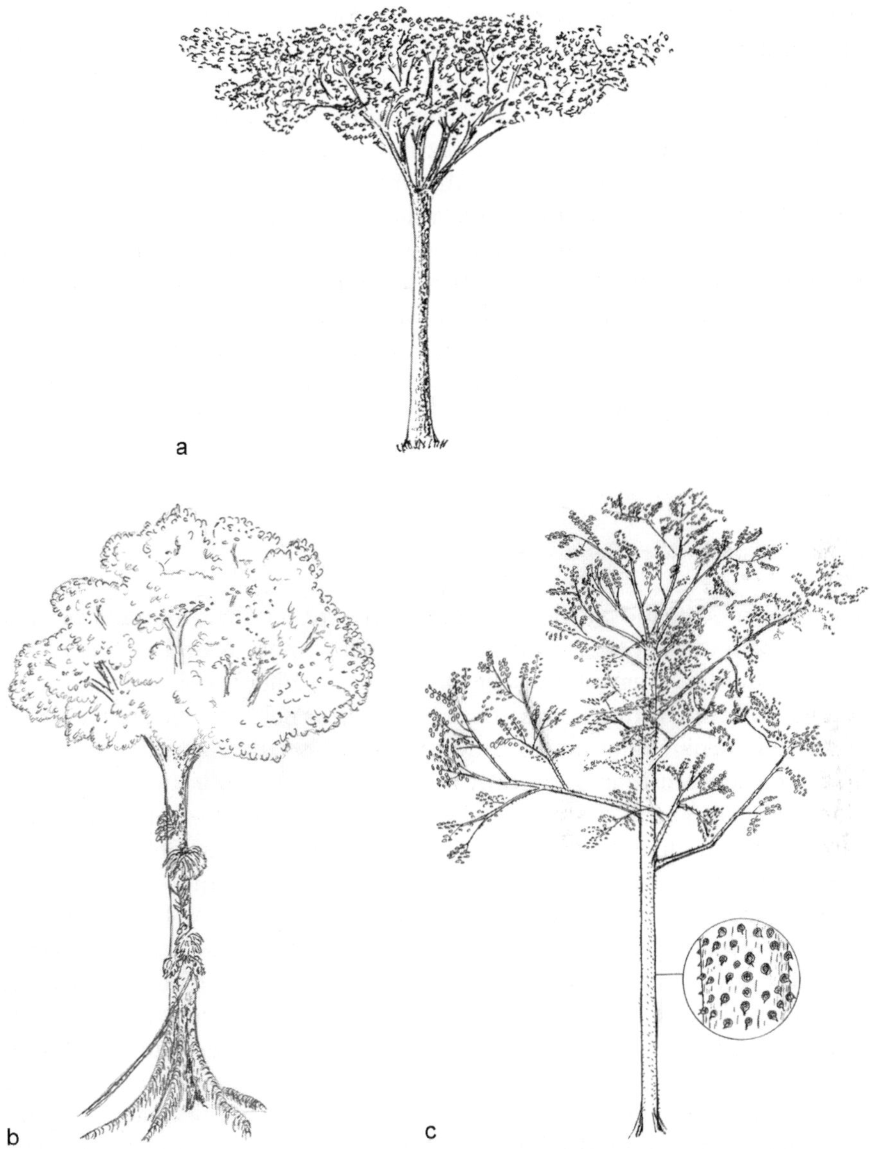

Fig. 2.6. Some phenotypic characters interfering with seed collection methods. **a** The flat crown of many *Acacia* and *Albizia* species makes climbing very dangerous as normal safety lines cannot be anchored high enough. **b** High buttresses and overgrowing epiphytes makes climbing via the bole very difficult. **c** Spiky or thorny stems or branches make any climbing attempt both difficult and very unpleasant

Many dry zone species especially from Africa are extremely thorny and spiky and on, for example, *Acacia polyacantha*, *Zanthoxylum*, *Bombax* and many *Erythrina* spp. large thorns occur on the stems and main limbs. *Albizia*, *Paraserianthes*, *Pterocarpus* and many other Leguminoseae, plus several *Cordia* spp. are often inhabited by extremely aggressive ants, which readily attack climbers. In several African *Acacia* species ants inhabit the thorns – a double protection to the plants but a double nuisance to the seed collector. Thorns and ants can make climbing a rather bloody and painful affair, respectively. Protective clothes, gloves and various insect repellents may at least reduce the nuisance. Tree crowns inhabited by wasps or bees may present a real danger. It is advisable to examine the crowns with a pair of binoculars before climbing: if there are wasps' nests or bees' nests, it is better to stay down (Fig. 2.30).

2.3.3 Climate and Weather Conditions

Dry weather is the most ideal weather for seed collection; movement to as well as within the seed sources is easier. Fortunately most seasonal fruiting takes place during the later part of the dry season. Collection from the ground, whether directly or from spread-out tarpaulins or nets, may be ameliorated by controlled burning of grass vegetation under the seed trees.

Moist or wet weather is generally not suited for seed collection. Accessibility and movement may be hampered, camping difficult or unpleasant. Collection from the ground may be complicated or impossible because of mud and vegetation. Climbing in wet weather is both more difficult and more risky; bark gets slippery when wet, often exacerbated by the growth of epiphytes, mosses or lichens. However, humid weather can sometimes reduce the loss of seeds from dry fruits collected from the crown by shaking or climbing because humidity mitigates fruit dehiscence.

Windy conditions are not suitable for collection. The danger of falling branches or heavy fruits during strong wind makes any stay in the forest risky. Even moderate wind can interfere with some operational procedures. For example, handling of extended pruners, advanced lines, etc. is very difficult; branchlets and wind-dispersed fruits released by, for example, shaking or cutting, may be blown far away from tarpaulins placed beneath the tree.

Fieldwork in the tropics is mostly scheduled in the early morning – this practicality also pertains to seed collection, particularly during the rainy season. Usually there is less wind, lower temperature and brighter light during morning hours.

2.3.4
Accessibility and Terrain

In established and managed seed sources, e.g. seed production areas and seed orchards, accessibility during seed collection is part of the establishment and management scheme and thus rarely produces restrictions on any collection methods. Remote, natural sources with difficult terrain and with no or restricted access by vehicles can only be reached on foot, carrying the equipment. Most collection equipment is heavy and bulky even though development of lightweight material has reduced the weight.

Sloping terrain can sometimes ease access to the crown from an uphill position (Fig. 2.7); however, seed collection can be a nuisance since cut-off

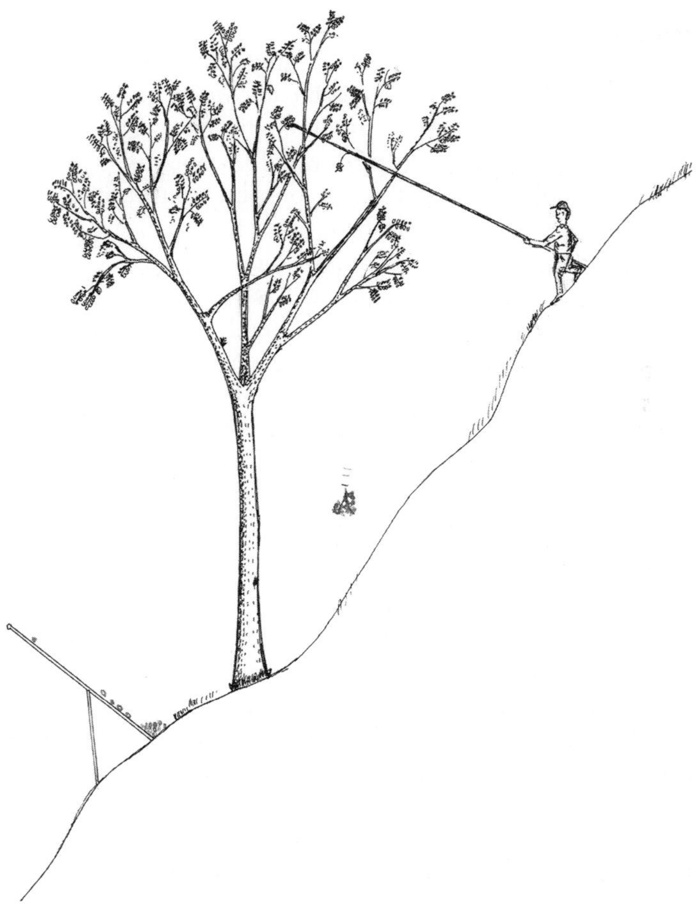

Fig. 2.7. Collecting seeds on sloping terrain

branches and fruits may fall far downhill. Collection nets or tarpaulins must sometimes be strung up or climbers must use collection bags.

Large vegetation may also hinder some collection methods. *Cordia alliodora* is a common shade tree in coffee plantations. The fruits are quite small and when shed naturally or by shaking, the individual fruits are difficult to collect under the coffee bushes. It is thus preferred in this case to cut down the whole infructescense (Boshier and Lamb 1997).

2.3.5 Efficiency, Labour Costs and Safety

Collection is often the most labour-intensive and expensive operation in seed procurement. The basic costs for any collection in remote seed sources are transport and living expenses for a collection team. Tree climbing is likely to increase the cost considerably. Accessory and safety equipment is heavy and using it correctly takes time (Box 2.1). It is not advisable to compromise on either safety or seed quality. The tree climber sets the limit of efficiency – ground staff can usually cope with picking up what drops down. To economise time and effort in climbing, let the climber collect as much as possible from each tree, and let him, as far as possible, finish the job once he is up in the tree. Establish, if necessary, an up/down hoisting system for necessities – tools, bags, drinking water, etc.

Use of local staff has many advantages. It saves expense for transport and accommodation, and it gives some, usually highly appreciated, income to the local community. Involvement in collection is on-the-job training, which may help in raising the awareness of the importance of using good-quality seed and may simultaneously raise the incentive to protect seed sources. There may be tricky balances to deal with. Only trained staff can be expected to use safety equipment correctly and efficiently. But some farmers are amazingly good at climbing without safety equipment and may not be at any higher risk than those with equipment (Sect. 2.6).

2.3.6 Availability and Cost of Equipment

Seed collection equipment, in particular branded equipment from authorised 'overseas' dealers, is considered quite costly compared with seed prices in most tropical countries. Sheets, nets, tarpaulins and funnels are quite expensive, as they must cover a large ground area during the period of seed fall and there is always a risk that unguarded laid-out equipment will be stolen, in particular if it is expensive and can be used for something else. Much equipment available in tropical countries has been provided by development projects. Unfortunately,

Box 2.1

Who climbs trees?

Seed collection is one of several activities involving climbing. What are the other reasons to climb to the top of trees? In rural livelihood, trees are a resource for food, medicine, wood and fodder. Many products are collected in the trees, e.g. honey, fruits, leaves, epiphytes and birds' eggs. Modern forest production tends to concentrate on a few special products. Fruit trees are bred and managed to bear fruit at low height in order to save time, reduce collection cost and reduce the risk of climbing. In tree breeding, climbing is used in connection with seed collection, scion collection and controlled pollination.

Tree climbing is often used in urban forestry. Street trees must be pruned to prevent them interfering with power lines or other constructions and for safety reasons. In crowded cities anything that falls down is a potential hazard for pedestrians, cyclists or motor cyclists as it is likely to hit something or somebody. In many large cities, pruning of park and street trees is done from elevated platforms. Where this is not feasible, trees are climbed.

Tree climbing has, as many other former physical necessities, become a sport and entertainment. Tree climber clubs exist in many developed countries. Some climbers take tree climbing as pure exercise, some as part of an interesting hobby of studying canopy organisms.

Whatever the purpose of climbing, the more people who enjoy or perform the exercise the greater the market for equipment. And with a bigger market follow development and improvement. Hence, tree climbing equipment, including improved rope type, lightweight spurs and protective gear, is continuously being developed (Blair 1995; Arboricultural Association 2004).

While mountaineering has become entertainment for both sexes, women apparently rarely climb trees. This is presumably mostly a cultural phenomenon, since tree climbing is not more physically exhausting than many other jobs performed by women. However, in terms of equipment, female climbers should pay attention particularly to saddle and harness types as they are mostly designed for men.

a lot of donor-provided equipment has rarely found its way from the storeroom to the field. Ill-adapted equipment can be worse than none.

However, a material boom has occurred in many tropical countries and much well-adapted and good-quality material is becoming available. The building industry has created a huge market for equipment that reaches high, from telescope poles, to ladders to elevated platforms. As small manufacturers are also becoming better educated and have access to better material, copies or specially designed products can often be made by order. A 'canvas man' in Vietnam modified his traditional safety belts for the building industry with a type excellent for climbing. A tool manufacturer in Indonesia produced tool heads and climbing spurs on order.

2.3.7 Ease of Prestorage and Processing

Clean seed stores best, and one of the main purposes in processing is to make seed storable (Chap. 3). The ease or difficulty of cleaning seed may influence the collection method, in particular ground collection. Methods in which much debris, such as soil particles, stalks or leaflets, is mixed with the seed may make cleaning or other processing difficult. Ground collection always implies a risk of pathogen contamination (Gray 1990, 2004; Turnbull and Martens 1983).

Collection by raking or vacuum (e.g. in some seed orchards) will inevitably imply contamination with other seeds (herbs and/or trees) and debris (Fig. 2.8). This may be a problem if the debris causes immediate damage, e.g. moisture or fungi during prestorage, and the debris cannot easily be removed

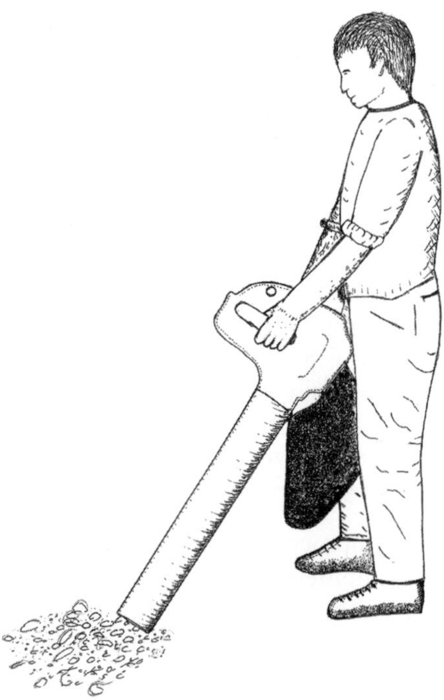

Fig. 2.8. Mobile vacuum cleaner used for seed collection. Powerful vacuum leaf collectors are becoming common in temperate Europe, America and Asia for collecting leaves and debris in gardens and parks. Vacuum collection inevitably implies accidental collection of a large amount of debris, which must be removed later. If cleaning can be done efficiently, vacuum collection can be very efficient especially for small seeds

during cleaning. On the other hand, if cleaning is easy it may turn out to be a very efficient collection method.

2.4 Some Genetic Considerations in Connection with Seed Collection

The objective of any seed collection should be to obtain seed of the best physiological and genetic quality. The former pertains to maturity, health, seed type, etc. The latter pertains to the inherited growth potential and other desired characters, which in turn depends on the genetic quality of the parent trees. Genetic quality is calculated or assessed both on population (seed source) and individual (mother/seed tree) level. In genetic improvement programmes, field trials contain analysis and documentation of genetic quality. Established seed orchards are based on this kind of documented trial. Where such information is not available, good quality often becomes a pragmatic 'best available'. Collections from natural stands, plantations and small woodlots thus adopt routines to 'avoid inferior material'. Though this may seem a modest ambition for procuring good-quality seed, the reality is that the loss by selecting poor/random material is often of the same magnitude compared with the 'average' gain from one or more selecting cycles in a breeding programme (Hansen and Kjaer 1999; Fig. 2.9). Seed collection should pay due consideration to adopting routines which can ensure the best possible genetic quality of the offspring.

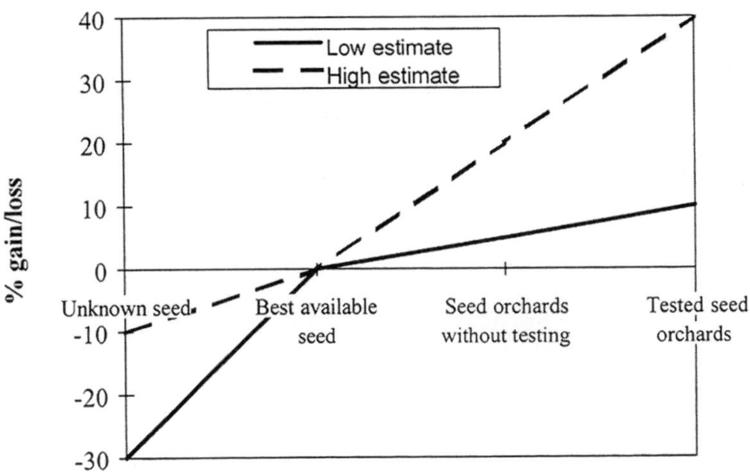

Fig. 2.9. The importance of selecting the best basic seed source is illustrated in the progressive improvement graph. The effect of poor or random seed source selection is detrimental to the performance of the offspring. (From Hansen and Kjaer 1999)

1. *Quality of mother trees.* So-called 'plus tree' selection is the selection of particular attractive phenotypes in a population to be used as seed trees (Fig. 2.10). Whether there are gains by selecting good phenotypes in a population depends on the heritability of the traits, i.e. whether observed traits are mainly genetic, or whether they are mostly caused by differences in age and exposure to different environments, e.g. different soil, exposure or competition (Anderson et al. 1998; Danusevicius and Lindgren 2002). Experience has shown that selection of individual mother trees in heterogeneous natural forest rarely produces much gain, since both age difference and environment tend to overshadow genetic difference (Cornelius 1994). Plantations are different because trees are even-aged and grown at uniform spacing (Boshier 2000; Palmberg 1985 Zobel and Talbert 1984). Selection of

Fig. 2.10. 'Plus trees' of *Dipterocarpus turbinatus*. A 'plus tree' is mainly associated with timber production, where 'plus' genes primarily refer to straightness, small branches, self-pruning ability and rapid volume production. In other connections and for other species, desirable characters can be fruit or foliage production, crown shape or soil-conserving characters

grafted trees in clonal seed orchards rarely makes sense because the phenotype of grafted plants is different from that of plants raised by seed or cuttings. In clonal seed orchards seed trees should be selected on the basis of their genetic records from progeny trials.

2. *Genetic variation.* Genetic variation is not always necessary and mandatory – clonal plantations can perform excellently. Seed collected from a single tree can also perform well – any individual seed is a product of two (and only two) parents, so successful outbreeding will give sibs. When genetic variation is desired in most seed collections, there are three key reasons:
   (a) Genetic variation means genetic adaptation, i.e. increased likelihood for survival in a variable environment, and an assurance against the risk of a poor genotype.
   (b) Trees may deliberately or unintentionally become seed sources in the future and a narrow genetic material is likely to cause inbreeding in such sources. Plantations or scattered plantings in farmland based on narrow material may cause problems in the future if renewed by natural regeneration, or if seed is collected from them.
   (c) Genetic variation is believed to be an assurance against pests, in particular insects. Insects tend to multiply rapidly and a pest adapted to a particular genotype will multiply fast and may cause destruction of a large population. Variation in genotypes is likely to protect the population (Hughes 1998).
3. *Site–source matching.* Trees adapt to their environment like any other organism. Over several generations natural selection will tend to favour individuals best adapted to a given set of ecological conditions. A similar environment tends to select for the same adaptations. Most species contain separate populations which have evolved into separate ecotypes. Selecting a genotype or a provenance matching the conditions of a planting site is thus likely to give better chance of survival and performance than a very different one. While it may be relatively easy to select a 'best match' for a given type of environment to a given set of seed sources, it has proven very difficult to predict over how wide an ecological amplitude a particular seed source can be used. Many seed sources have been established from plant material covering a variation in ecoadaptation, and some seed sources are established in different areas in order to promote flowering. Rare climatic events may be detrimental to performance in a particular environment.

4. *The risk of inbreeding.* Most forest trees are facultatively outcrossing, but selfing and inbreeding also occur in most species (Boshier 2000). Flowering isolation in time and space increases the risk of inbreeding. Single, isolated trees (whether surrounded by farmlands or dense woody vegetation of other species) bear a high risk of producing inbred seed (Fig. 2.11). Trees flowering out of season are functionally isolated. Different rates of inbreeding can also occur within the same tree, e.g. a higher rate of inbreeding at the lower crown compared with at the top (Patterson et al. 2001).

Seed collection for gene conservation will normally aim at collecting as much as possible of the total genetic variation in a population. The population may be defined as a delineated stand or may consist of several separate stands over a larger area. Increasing the distance between mother trees will generally reduce the risk of kinship; the number will increase the likelihood of catching all genes. To catch most genes in a reasonably sized population may require 50–100 unrelated seed trees. For ordinary seed collection for afforestation, the number could be less than half. In practice, if populations are fragmented and scattered it is difficult to comply with the separation and the minimum number of mother trees, simply because there are too few mature trees left. If the

Fig. 2.11. Isolated trees in open farmland have a high risk of inbreeding if the distance between them and the open vegetation restrict cross-pollination. The behaviour of the pollinator in open areas is crucial

population is very small, it would be appropriate to increase the collection area, even if it may then cover more provenances.

Genetic variation in natural forests covers variation between populations, and may be expressed both as growth adaptation (ecotypes) and inherited appearance, in forestry known collectively as provenance variation. This variation can be quite significant and is often so in species with a large distribution range and where populations have been isolated from each other for enough time to cause genetic drift (evolution in different direction). Both time and selection pressure (in natural populations primarily differences in growth conditions) determine provenance variation.

Advances in genetic technology have provided tools for analysing genetic variation and relations on a molecular level. Direct field application is still in its infancy, but research has helped provide documentation for speculated population structures (Williams et al. 2004). However, in practice the genetic history and the genetic structure of most natural seed sources in the tropics are largely unknown. Small populations can thus be reminiscent of larger fragmented populations or they can be small 'satellite'/'island' populations at advanced outposts from the main populations of the species (Ge et al. 2005). The population density can be scattered naturally, or it can be scattered because of genetic erosion, e.g. selective cutting. These considerations collectively determine whether a given stand should be accepted as a seed source and, if so, for which area (site–source matching).

Another factor influencing the genetic structure in natural forests is dispersal and regeneration ecology. Many wind-dispersed species tend to form family groups with neighbouring related trees (Eldridge et al. 1993; Boshier 2000; Fig. 2.12). Animal-dispersed species are often dispersed further and with more random deposit sites and may thus have a less 'patchy' genetic structure. In order to increase genetic variation and avoid family relations there should be a certain minimum distance between parent trees. In wind-dispersed species, the distance should be 50–100 m (Gray 2004; Palmberg 1985); in animal-dispersed species, it could be less. It should be reiterated that this only holds for natural populations.

Stand density or distance between individual trees in an 'open' seed source, e.g. in farmland, becomes critical if the trees are very scattered and where distance between the trees can interfere with pollination efficiency. Isolated trees are believed to contain a high rate of inbred seed and should generally be avoided. Isolation is obvious in farmland, where the distance to neighbouring trees can easily be seen. Although less evident in a forest of mixed species, individual species can exhibit a high degree of isolation, for natural reasons, e.g. scattered distribution at geographical or ecological boundaries – all species are 'rare' at their ecological boundaries. Man's selective logging may 'dilute' stand density to a critical level for cross-pollination and thus increase the risk of

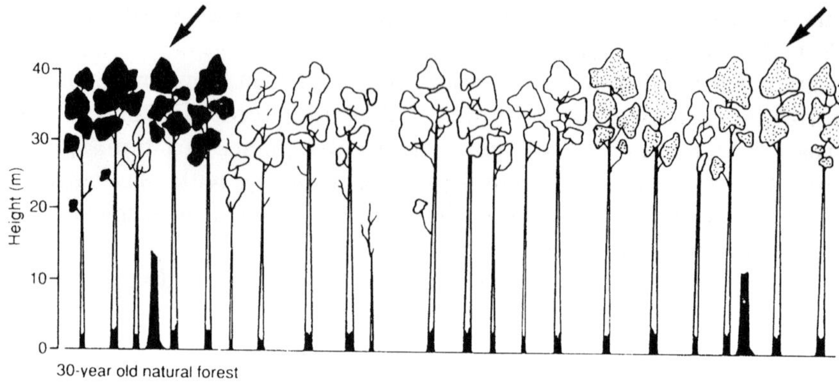

Fig. 2.12. Some forest types and species tend to create groups of related individuals occurring when siblings replace the mother tree. Seed lots of such species may contain a high percentage of inbred seed. (From Eldridge et al. 1993, based on data from Ashton 1975 and 1976; reprinted with permission from CSIRO Publishing and Oxford Science Publishers)

inbreeding (Finkeldey and Ziehe 2004). Eventually, a shortage of pollinators can indirectly affect outcrossing efficiency.

The basic genetic resource of many plantation species is unknown, because they originate from collection in natural forest with no documentation (Williams et al. 2004). Therefore, there has been much focus on recollecting new material for commercial production of such species from their original source with appropriate consideration of genetic width and quality of mother trees. Most of the important plantation species should be established in seed orchards as it is usually far too expensive to launch seed collection expeditions to the original sources for regular bulk collections.

2.5 Collection Methods

The simple rule in seed collection is that the simplest and cheapest method applies, as long as it does not compromise seed quality. Seed is the vehicle of gene dispersal. Yet, in most trees a large part of the seed production will fall under the trees, i.e. the seeds fail to be dispersed. When seeds are about to be dispersed, fruit or seed attachment to the tree becomes weak and detachment can be enhanced by shaking the tree or fruit-bearing branches. All methods of ground collection are based on the philosophy that it is better to wait for the seeds to fall from the crown by themselves than to bother to pick them. Collection from the ground, compared with climbing, is simple and safe and thus untrained casual workers and schoolchildren can collect seeds. There is no damage to the trees and the collection is relatively independent of weather conditions. Where ground collection is not applicable, e.g. because seeds are too small or are dispersed before they can be collected, seeds are harvested from the crowns by various accessories.

2.5.1 Collection from the Ground

Ground collection applies to situations where seeds or fruits are collected either after natural fall or where fall is accelerated by shaking. Bulk collection, in any case, requires that seeds are large enough to be found and numerous enough to make collection rational. Species with large indehiscent fruits or large seeds which fall during a short fruiting season where there is no serious risk of rapid deterioration or germination are suited for ground collection after natural fall. This applies to, for instance, *Afzelia*, *Pterocarpus* and teak (Wasuwanich 1984). If natural fall is relied on, collection may be postponed until most fruits and seeds have matured and fallen to the ground or may be done over several collections. Many seeds may fall during strong winds or heavy rainstorms. Several collections are necessary if seeds easily deteriorate or are destroyed on the ground, e.g. recalcitrant seed, seeds rapidly removed by seed herbivores (Coe and Coe 1987; Lamprey et al. 1974) or seeds attacked by insects (Janzen 1972; Seeber and Agpaoa 1976; Howe 1990). In Ho Chi Minh City in Vietnam the main source of *Dipterocarpus alatus* is park and street trees. During the fruiting season, seeds are collected early every morning before they are run over by the traffic.

2.5.1.1 Accelerating Fruit Fall by Shaking

Shaking trees will improve the efficiency of ground collection by increasing the amount of seed that can be collected at a given time. Shaking fruit-bearing branches accelerates natural fall. Branches can be shaken from the ground by the aid of a hook mounted on a long thin pole (Fig. 2.13) or a rope thrown over the branches, e.g. by the advance line technique (Sect. 2.6). Lines or hooks must be placed relatively distant from the stem where the branches are more flexible.

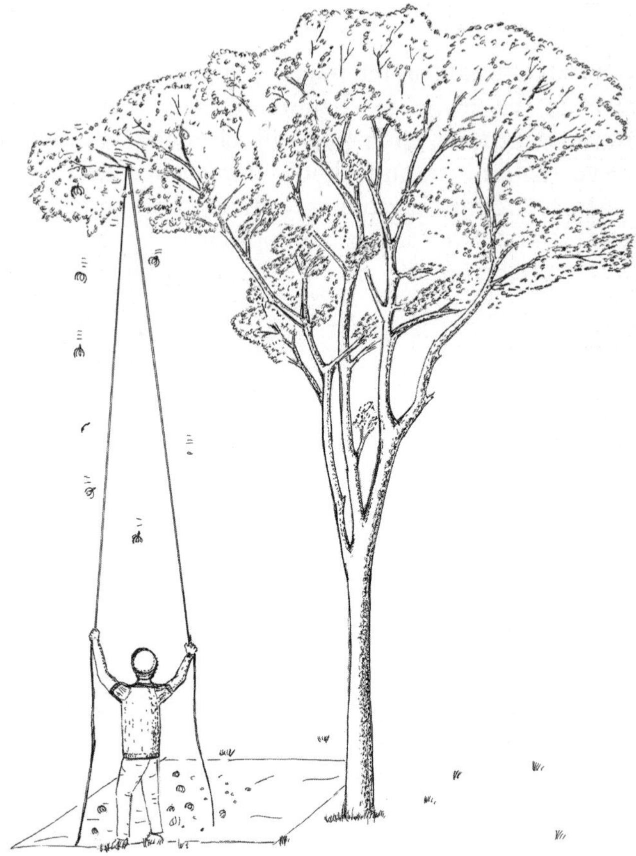

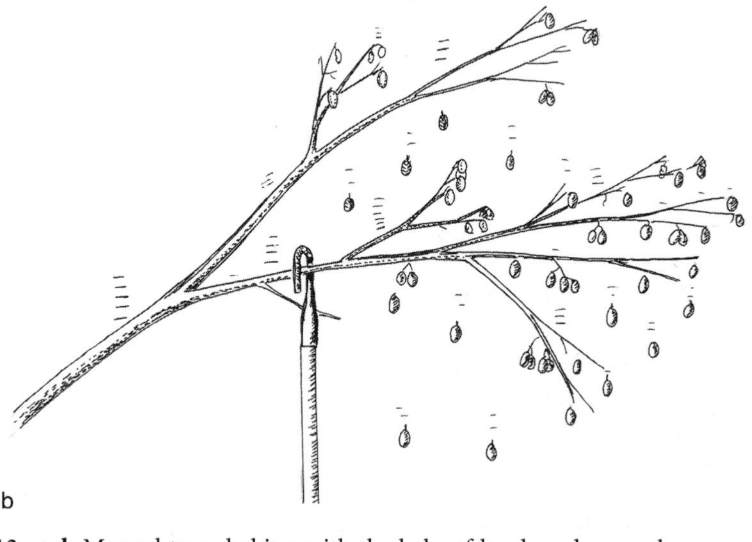

Fig. 2.13. a, b Manual tree shaking with the help of hook and rope; the rope may be placed by throwing or the advanced line technique.

c

Fig. 2.13. (Continued) **c** Mechanical tree shaking by special shaker vehicle (http://www.bragg.army.mil)

In smaller trees, the method can be used for shaking the stem. Manual shaking of branches does not damage trees.

Powerful mechanical tree shakers are used in seed orchards and other high-producing seed sources (and some fruit orchards, e.g. olives) primarily in the USA and southern Europe (Tombesi et al. 1998). Tree shakers are nowadays mostly special vehicles designed to operate in seed orchards. The vehicles are designed to minimise the shaking impact on the vehicle itself. During operation, the shaker's 'arm' is clamped onto the tree trunk and shaking is imposed via the automatic transmission. (Stein et al. 1974; Kmecza 1979). Powerful tree shakers can practically empty a tree of fruit or seed in a few seconds. The impact is highest for relatively large fruits like cones and most drupes where the vibration is easily transferred into the fruit stalk. It is less efficient for small dry fruited species like acacias. In theory, shaking impact can be adjusted to species and conditions, e.g. force only sufficient to release fully mature fruits. In practice, shaking is a 'one-go' process and many immature fruits are likely to be contained in the lot. To minimise seeds blowing away, shaking must be done in calm weather, especially if the seeds are small and the trees high.

A major drawback in mechanical shaking is that damage to the trees can be considerable. Damage occurs as tearing of the bark at the place of clamp attachment. The impact depends on the bark type, the type of clamp and the force of shaking. The clamp should clasp the stem with an even pressure, which will not overpress the cambium, and the operation should be adjusted to the minimum force and time necessary to achieve the effect. With careful adjustment, damage can be significantly reduced (Buker et al. 2004). Since the tree shaker is necessarily close to the tree and thus within the rain of falling fruits, the operator must be protected under a roof cover.

Mechanical shaking is restricted to relatively flat and accessible areas where the specialised vehicles can operate, and to relatively slim trees, which can be shaken. In practice it is only used in seed orchards, where the method is highly efficient.

2.5.1.2
Picking from the Ground

The efficiency of picking up fallen fruits or seeds depends on the size, the ground cover and the ease of later cleaning. Where ground vegetation is short and seeds or fruits are large, seeds can be picked up individually by hand or by using large tweezers, or they can be raked together with a leaf rake.

On flat terrain large seeds and fruits may be collected by mechanical rotating brushes (Hallman 1981). A safer collection method for small and light seed is vacuuming (Riley et al. 2004). Raking, brushing and vacuuming will inevitably imply collection of a lot of debris such as soil, immature and deteriorated seed and other seed. If these impurities are easily removed afterwards by cleaning, the extra bulk and debris may be dealt with later and the ease of collection may outweigh the more time-consuming processing procedure. However, soil collected together with seed from the ground always contains soil-borne pathogens (Gray 1990). The risk of such contamination depends on the type and the ease with which they can be eliminated during processing.

In seed sources with undergrowth, vegetation is preferably removed a couple of weeks before collection takes place. For smaller seed and to avoid contamination with soil and debris, nets, tarpaulins or plastic sheets can be spread under the trees either directly on the forest floor or hung up under the trees (Hallman 1993; Boshier and Lamb 1997). These may be used either to catch natural fall or in combination with shaking. If used to catch natural fall, seeds should be removed regularly to avoid possible deterioration and germination. Plastic sheets have, for example, the drawback that water can easily collect with the seed. For low and relatively narrow crowned species, Doran et al. (1983) suggested sheet funnels should be hung up under the trees (Fig. 2.14).

Collecting from sheets or nets spread out under trees is done by lifting or folding them. Contamination with debris is low. Contamination with soil

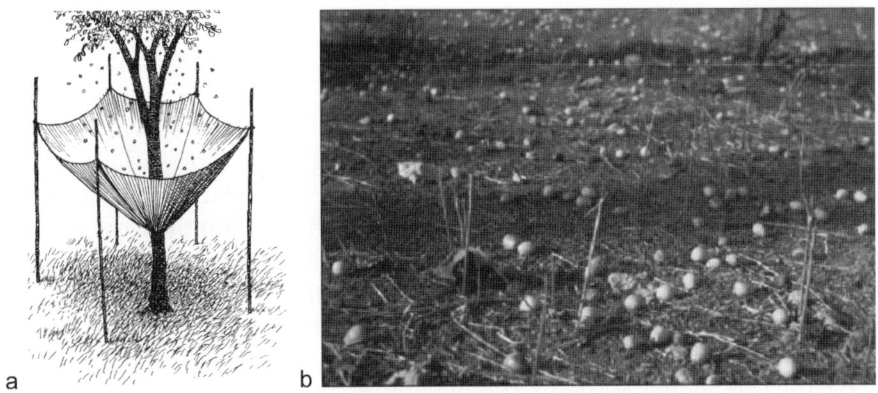

Fig. 2.14. Collection after natural seed fall. **a** Funnel mounted on a small tree for collecting acacia seeds during natural seed fall. **b** Mature fruits of *Gmelina arborea* on the ground. (**a** From Doran et al. 1983; **b** Courtesy of H. Keiding)

particles is a possibility when using nets, but there are no firm reports of this and it may be low and speculative.

2.5.2 Collection from the Crown

Collection from the crown includes any collection where fruits are removed from their attachment to the tree. Fruits are collected by picking or cutting individual fruits or fruit-bearing branches where the collector either stands on the ground using long-handled tools or ascends into the tree. Collection from the crown is applied either where alternative ground collection has some serious disadvantages or where collection from the tree is easy, e.g. in low and easily climbed trees.

Collection of seed from low branches of free-standing trees is generally discouraged as the low branching habit, at least for timber trees, is considered an undesirable character, and lower branches may have a poorer chance of pollination (Hilton and Packham 1997; Patterson et al. 2001). However, where these factors are not relevant, e.g. in low bushy agroforestry trees, seeds may as well be harvested where they are easily accessible.

There are basically two methods of collection from the crown: (1) where the collector stands on the ground or some elevated platforms, vehicle or ladder; (2) collection in connection with tree climbing.

2.5.2.1
Low Trees with Access from the Ground or Low-Elevation Platforms/Vehicles

A person using long-handled tools can reach a height of 5–6 m (Box 2.2). Vehicles and small ladders may add 3–4 m (Fig. 2.16). Longer ladders exist, but their instability means they are not always a good platform for using very long tools. Mobile platforms can reach very high trees but their use is limited by access and operation cost (Box 2.3). Flexible saws are placed via advanced line technique. Their operation height is theoretically limited by the length of the rope ends pulling them – in practice, however, they are very difficult to place in the right position and operate at a height greater than 8–10 m. So in practice this height is about the maximum that can be reached from the ground.

Low-hanging branches can be pulled down, sometimes several metres, if they are long and flexible. Pole-mounted hooks or occasionally ropes thrown over the branches are suitable for pulling down branches. Fruits that can be

Fig. 2.16. Using vehicle roof tops as elevated platforms permits collection from most lower trees

Box 2.2

Reaching an extra inch up and out – using long-handled tools

A person can reach 2–2.5 m above the ground by hand, horizontally only about 1 m. Using a long-handled tool can extend the reach, but unfortunately this is not unlimited. How long can an extension be? What material is the best and strongest? The practical length depends on whether the tool is used in an upright position to reach a certain height or in a horizontal position to reach out, e.g. to a branch (Fig. 2.15). The length is significantly smaller in the latter case as the tool begins to feel very heavy when held in a horizontal position. A maximum length of 2–3 m is what can practically be handled when using extended tools to reach outwards horizontally.

The weight of the device does not impose big problems when carried vertically. However, the two problems limiting the height indirectly pertain to weight, viz. difficulty in raising the device and difficulty manoeuvring the device between branches and putting it in place on what to cut or pull.

One of the strongest, lightest and cheapest natural materials for extended tools is bamboo. Giant bamboos can grow to more than 40 m, but such size is obviously not practically manageable. A dry 5 m high bamboo pole with a base diameter of 5 cm weighs about 2.5 kg. The tool head adds a few hundred grams. A drawback of bamboo is that it is relatively easily damaged, e.g. if stepped on. Cutting the material into shorter sections, which can be fitted together when in use, inevitably weakens the material.

Pipes of various metals and synthetic materials can be used for extended tool handles. Telescopic poles are becoming widely used for many types of tools and are readily available in many hardware shops. Sectional or telescopic poles for window cleaners are up to 10 m in length. For use in seed collection, 6–7 m is about the maximum practical, depending on which tool is mounted and precision requirement. Hooks used for shaking branches do not need to be placed very precisely and long tools can be used. Pruners are hard to place precisely and poles more than 4–5 m long are not practical.

Long-handled tools are easiest to handle from the ground. The extension is also appropriate to use both in connection with climbing, to reach where branches are thin and cannot carry the weight of a person, and from various elevated platforms such as vehicle roofs or ladders. The length and weight of the tools that can be practically and safely handled from such more unstable positions is less than those for ground-operated poles.

The type of tool head depends on the purpose, the tree and the fruit type. Some standards are secateurs, saws and hooks. Various types of cutting devices have been invented for the use in pine cone collection (Fig. 2.27).

(*Continued*)

Box 2.2

Reaching an extra inch up and out – using long-handled tools—Cont'd.

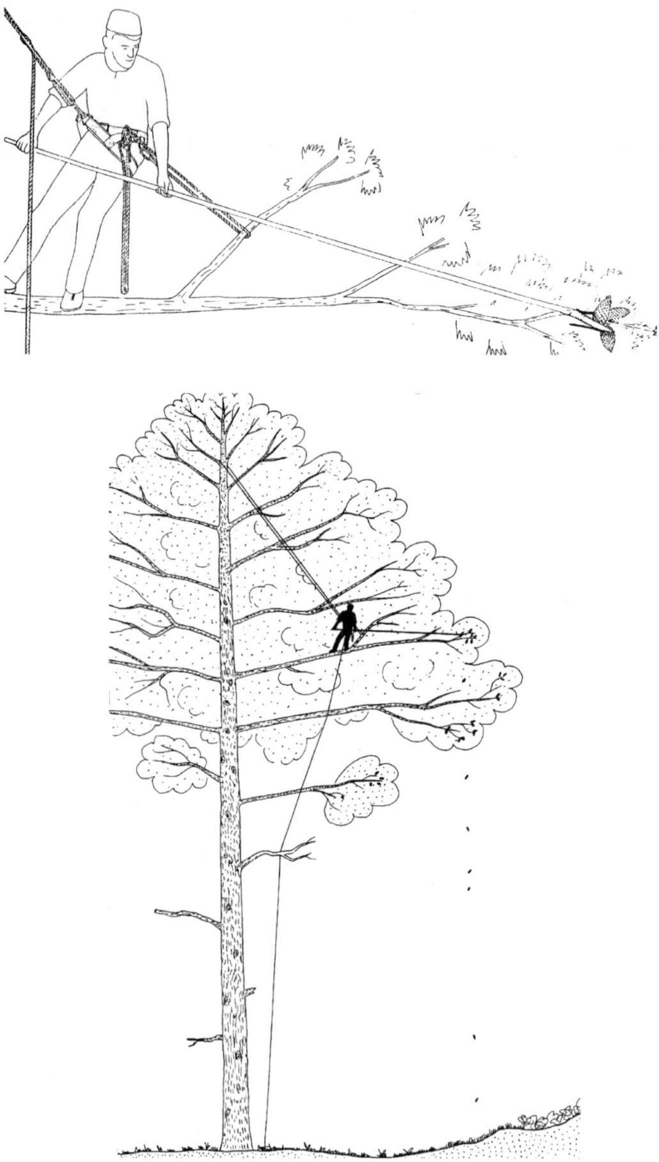

Fig. 2.15. Horizontal use of extended tools. (M. Robbins)

Box 2.3

Elevated platforms

Vehicle rooftops are amazingly versatile platforms for seed collection in small trees. High vehicle roofs are about 2-m high, and used, for example, in connection with long-handled tools, they allow seeds to be collected from most small and medium-sized trees. Mobile hydraulic platforms are operated from trucks. They have capabilities of up to 100 m, but most of those used in construction industries and forestry have capabilities of less than 40 m. The platform consists of a small 'cage' mounted on a hydraulic arm, which is again mounted on the vehicle. Height and direction can be steered from the ground as well as from the 'cage'. The vehicle is equipped with supporting legs, which fix the ground position. Mobile platforms have inbuilt safety devices which prevent the platform from being diverted too much sideways, which would cause the vehicle to turn over. Moving the platform to a new perimeter requires that the vehicle be moved. Terrain presents a limitation to where mobile platforms can be used - they are generally not suited for sloping terrain, soft soil and other factors that restrict access and stability. For operation in tree tops, 'terrain-hardy' types which can be mounted on the rear of tractors or trailers are convenient, and the easy dismantling/disconnection mechanics make it easy to use the vehicles for other purposes when needed (Jasumback 1994). Trailer- and tractor-mounted platforms have, however, limited reach (Fig. 2.17).

The main drawback of mobile platforms used for seed collection is the purchase and operation price. With relatively low labour cost and seed prices, mobile platforms are not competitive in normal routine seed collection. However, the time saving and collection efficiency compared with those of climbing make the equipment highly relevant for high-quality seed collection (Jones 2003).

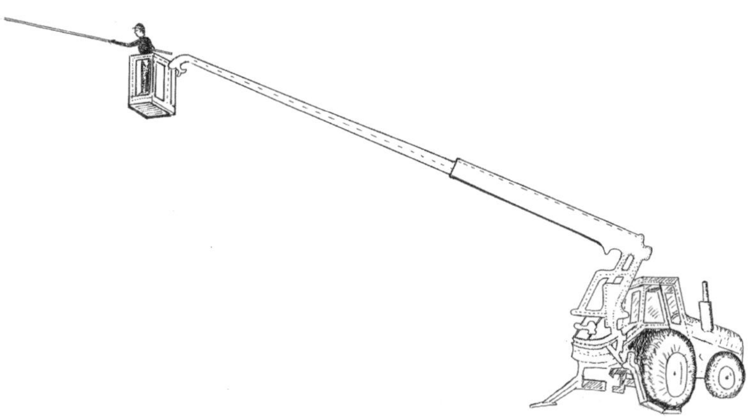

Fig. 2.17. Elevated hydraulic platforms used for operation in tree crowns

(Continued)

Box 2.3

Elevated platforms—Cont'd.

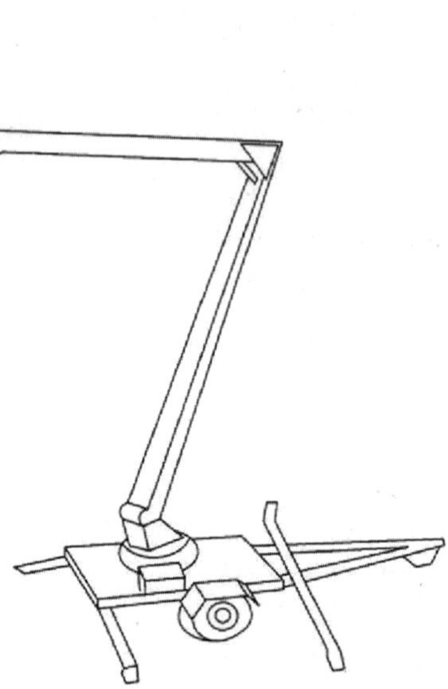

Fig. 2.17. (Continued)

reached from pulled-down branches can be collected by hand. Extended pruners are used for those still out of reach.

Pole-mounted tools are mostly designed to cut off fruit-bearing branches or fruit stalks. Extended secateurs can cut smaller branches up to about 3-cm thick (or 5-cm thick if the wood is relatively soft). Most extended pruners use a wire-pulling device for the cutting operation: once the cutting edge is put in place, cutting is performed with a strong 'japping' pull on the rope attached to the movable cutting edge. In some extended pruner types, the cutting mechanism is constructed with a double pipe so that the cutting is performed with a simple pull on the pole.

Pole-mounted saws consist of a slightly concave saw with teeth pointing downwards – cutting is thus done when pulling downwards (Fig. 2.27). The saws can cut up to 10–15-cm-thick branches, but this operation is very exhausting if the branch is high. During operation the saw is 'resting' on the branch. The sawing position must not be vertical. A problem is sometimes encountered when the branch starts to bend: the saw easily gets stuck. To avoid this, some saws have a lower cutting device that will cut the lower bark. There are telescopic chain saws with a length of up to 4 m which can be operated both from the ground and during climbing.

High thick branches are easier to cut with flexible saws (Fig. 2.18). These saws are similar to the cutting chain of chain saws, but unlike ordinary chain saw chains they cut in both directions and are easier to put in the cutting position because they are designed to orient themselves with the cutting edge down. Thin wire saws are available but they are not used much because they tend to get clogged with resin if used in resin-rich species such as pines (and since they can neither cut nor be pulled back, they are lost!). Flexible saws are operated by alternately pulling down the two rope ends so the saw will cut through the branch. The flexible saw may be operated by one person only standing under the branch and using alternating hands for pulling or by two people standing at some distance.

Cut-off branches and twigs often get entangled on the way down. A hook mounted on a light pole is the universal tool to drag down caught branches and branchlets.

a

Fig. 2.18. A flexible saw is a specially designed chain saw blade which can be placed over a branch via the advanced line technique. **a** The flexible saw in operation.

Fig. 2.18. (Continued) **b** Closeup of the flexible saw

b

2.5.2.2
Reaching the Top of Large Trees by the Way of the Bole

Large trees in this connection means anything which cannot be reached from the ground or from platforms by the methods described in the previous section, i.e. anything higher than 8–10 m. Trees with a relatively small bole (say less than 60 cm) are normally climbed via the bole; very large diameter trees or trees with large buttresses, overgrown by epiphytes or other stem barriers, are climbed via the advanced line technique.

Climbing via the bole is usually much easier than advanced line climbing, because the bole provides support for ladders, spurs, tree bicycle or whatever other accessory may be used. Many rural people can climb straight boles with amazing speed and skill, sometimes using a short string around their feet, sometimes just with their hands and bare feet. Free hand climbing without safety devices is, however, generally discouraged because it does imply an unnecessary risk.

Tree climbing equipment and accessories have two functions: (1) safety, i.e. preventing fatal injuries in the case of accidents; (2) ascent, i.e. easing climbing and making up a 'working platform' from where collection work takes place (Box 2.4).

Ascent accessories are used for climbing the boles up to the height where there are sufficient branches to provide support for further free hand climbing. Ladders provide the 'missing steps' on the bole; spurs and tree bicycle provide grip on the bole where steps are missing.

Ladders exist in numerous forms and designs and lengths (Fig. 2.20). Locally manufactured ladders are usually made of bamboo. Industrial ladders are made of light metal, usually aluminium. All-round extendable ladders usually have a maximum length of 8–12 m. Such ladders are of the leaning type, which use the tree as a support. Specially designed tree ladders consist of sections each about 4 m in length. The ladder sections are placed on top of each other in a fitting and are tied to the bole as the climber ascends (Barner and Olesen 1983b). The climber hoists sections up when he reaches the end section. Three to four sections reach about 10–15 m. On the way down the climber takes apart the ladder sections and passes them down.

Climbing spurs exist in a variety of designs and qualities (Fig. 2.21). The spurs should fit the climber; especially large (European) sizes are useless for

Box 2.4

A working platform

A mountaineer uses safety lines and belts primarily to prevent a possible fall. His lines are designed to stretch about 50% in the case of a fall, and belts are designed to be tight and not to interfere with smooth movement. In tree climbing, ropes are also working ropes and saddles and harnesses are used to sit in or to hang in during work (Fig. 2.19). This creates some different requirements for the equipment. A number of saddle harness designs are available. The three main types are:

1. Double hip saddle type without leg belts
2. Double hip saddle with tight applications
3. Full harness where belts are designed for both hip and upper-body part

The critical point in design is how convenient various types are to hang or sit in. It depends both on design and on the person using them. A double hip saddle type without leg belts is easier to climb in but tends to press legs together, put much pressure on the outer side of the hip and block blood circulation there. Skinny people find them painful to hang in. The second type tends to distribute the force more evenly between hip and thighs, but the leg strops appears to create some discomfort for some users. A full harness is basically a saddle with supporting strops for back and shoulders. This gives a better distribution of weight and pressure in some situations. Harnesses also have the possibility of moving the safety lines to a breast attachment, which can be convenient in some situations.

Some extra rings and attachment applications are convenient when working at height. The collector often needs different tools and as he exchanges, for example, saws and pruners or needs free hands; tools are conveniently hooked to the belt.

Box 2.4

A working platform—Cont'd

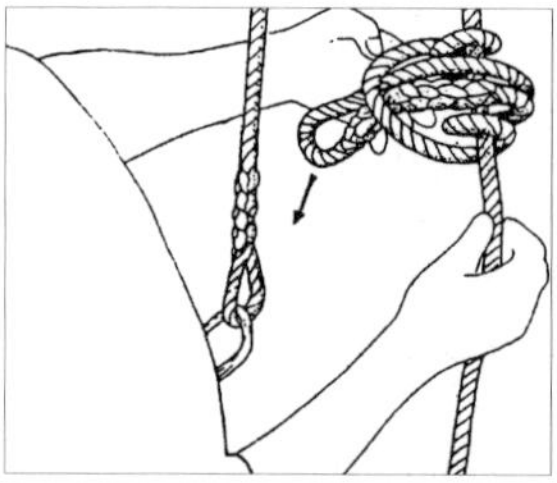

Fig. 2.19. The rope end to which the climber is tied by the prussic loop hangs freely. The prussic loop is easily loosened but will grip firmly in the case of a fall (Whitehead 1981; Ochsner 1984)

small legs (e.g. Asian). The wrong size is both inconvenient and potentially dangerous. When climbing, spurs should be tied tightly to the feet of the climber (right spur to right foot!) by the strops. During ascent the climber walks his way up by kicking the spurs into the bark (Robbins 1983b). He supports himself by holding and moving upwards his safety strap, which is attached to his safety belt or harness (Barner and Olesen 1984b).

The tree bicycle is a Swiss-designed device, named so for its origin and its pedals (Fig. 2.22). It consists of two parts, a short one for the left leg and a long one for the right. Each consists of a pedal, a pedal arm and a ring, which encircles the trunk when in use. The climber works the tree bicycle in a similar way

Fig. 2.20. Different types of ladders. (M. Robbins)

as the spurs, moving the legs alternately up and down, i.e. moving one ring up while pushing down the other leg. The tree bicycle is an easy-to-use and safe device (Barner and Olesen 1983a; Mori 1984). The purchase price is high, but it is very durable if well maintained. It requires a fairly clean and regular bole, and passing side branches is very difficult (in practice they are cut down until the climber reaches the crown).

A safety belt and safety strop is always used in connection with ascent by vertical ladders, spurs and tree bicycles. The safety strop goes around the tree trunk and the two ends are connected to the climber's harness/safety belt (Fig. 2.23) by carabiners, which can easily be disconnected when the climber passes a branch. For safety reasons the climber has two strops, so he can place the second one safely above the branch before loosening the first one when a branch blocks the way and has to be passed. When reaching the crown, where

Fig. 2.21. Climbing with spurs

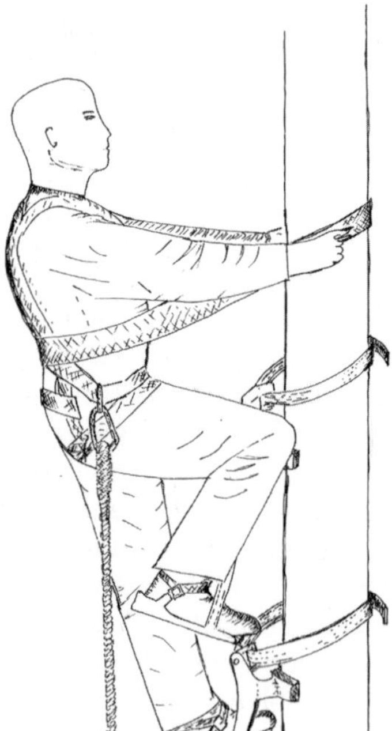

Fig. 2.22. Swiss tree bicycle

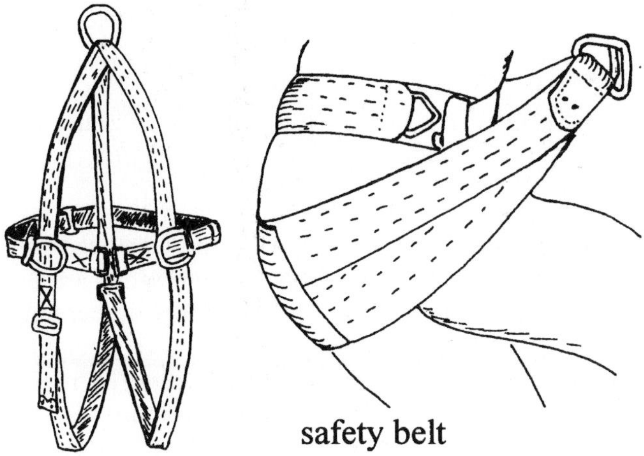

Fig. 2.23. Safety belts. (M. Robbins)

there are regular side branches, the climber continues free hand climbing to the top, where he secures himself for further work by a safety line.

Spur climbing requires some training to use without them slipping off, tying them without blocking any blood circulation, and removing them with one hand when reaching the crown. All three types of equipment are easiest to use in relatively small diameter trees. The maximum diameter for tree bicycle is about 80 cm (diameter of the ring in its widest position); using spurs at large diameters makes positioning the feet very strenuous; tying of strops for ladders is difficult if you cannot reach around the stem. An additional problem in large trees is moving up the safety strop. This applies for all equipment. A relatively stiff strop or strap is easier to move than a flexible rope.

2.5.2.3
Reaching the Top of Large Trees by Advanced Lines

The advanced line technique consists of placing a rope or other ascent device in the tree and then using this device to get access to the tree without using the tree trunk support. The method is more difficult, more physically exhausting and more time-consuming than bole climbing and thus is generally used only when there is no other way, e.g.:

1. Large-diameter trees
2. Trees with large buttresses or overgrown by epiphytes
3. Trees where the crown spreads into several slant stems or branches
4. In protected areas where there are often prohibitions on any activities that can be potentially damaging to the trees, e.g. use of spurs, and where, for example, ladders cannot easily be carried.

Most open-crowned trees can be climbed by the advanced line technique. Trees with dense crowns, e.g. cypresses and some pines are less suitable – arrows and lines are difficult to place and easily become entangled. Placing the line and avoiding it becoming entangled in the crown requires reasonably free sight and space. Advanced lines are placed either by manual throwing or, for greater height, with the aid of a certain shooting or ballistic device, e.g. a bow, a crossbow, an airgun or a catapult (Box 2.5). A thin line, e.g. a fishing line, is tied to the sandbag, arrow, lead bullet, rod or whatever is shot or thrown. The object is thrown over a strong branch as high as possible. When it falls down it will pull up the thin line. Via an intermediate line, this line is then used to pull up the working line/safety line/lifeline or rope/wire ladder (Blair 1995; Stubsgaard 1997; Arboricultural Association 2005; Fig. 2.25).

A precise throw or shot is difficult to make as it pertains both to aiming precision and shooting/throwing distance. The object should just pass the target branch and fall on the other side. Too powerful shots imply a risk of the weight and line going too far and becoming entangled in the branches. Friction of the line during firing or throwing may affect both throwing distance and precision. Therefore the line should unwind freely, e.g. from a fishing reel or, for thicker lines, from an open bag. With some exercise and using, for example, centrifugal throwing technique, most people would be able to throw a 50–100-g sandbag over a 10–15 m-high branch. This technique can also be used when working in the crown to reach higher positions. Powerful shooting devices are required to reach the top of the high trees (Fig. 2.26). The object to be shot over the branch must be quite heavy as it should be able to pull up the string, overcome the friction and pass possible small hindrances freely on the way down.

Traditional climbing techniques use the prussic loop technique (Fig. 2.19): one end of the line is attached to the climbers safety belt by a carabiner, the other is hanging free but attached to the climber via the prussic loop, which is attached to the harness. Ascent consists of stepwise pulling oneself up via the free rope end and securing oneself by the prussic loop, which is gradually pushed upwards. (Blair 1995; Ochsner 1984; Whitehead 1981). This technique requires good strength and stamina.

Box 2.5

Ballistic or shooting devices

The types of equipment used for placing advanced lines are mostly 'old fashioned weapons' such as bow and arrow, crossbow and slingshot/catapult (Fig. 2.24). Small rifles with blank cartridges or air guns are sometimes used to fire arrows or steel rods.

Crossbows and guns have very good sighting mechanisms, but the distance and height are difficult to adjust. A perfect shot just passes the target branch and falls down on the other side. A shot sent too far results in the arrow or rod passing far beyond the target branch and it may get entangled in the branches above. Using arrows or rods with different weight allows some adjustment.

Ordinary bows require a bit more aiming skill and experience than the crossbow, but by adjusting the pull on the bowstring the distance can be adjusted to the force necessary for the arrow just to pass the target branch. A problem with bows is that they need relatively long arrows, which get entangled more easily than the short crossbow arrows. Modern 'sport' bows are very powerful and allow the use of heavy arrows.

Catapults are the most versatile shooting devices as they use spherical 'bullets', which do not get stuck so easily. The power of the one-hand sling shot (Stubsgaard 1987) is, however, too low for high branches. A very powerful catapult, the 'big shot' is used in, for example, Australia and Papua New Guinea (Gunn 2001; Gunn et al. 2004), and is specially designed for placing advanced lines. 'Big-shot' catapults have a very strong pull and can thus shoot far and with a relatively heavy (100–200 g) sandbag 'projectile'. Shooting devices are potential weapons and in some countries are illegal or subject to special licenses. However, the 'big-shot' catapult can hardly be considered a weapon and should thus not be subject to such restrictions.

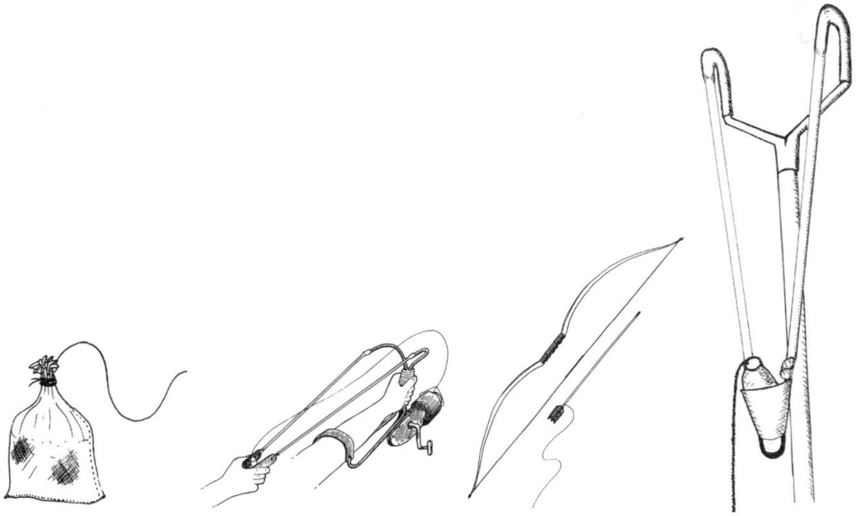

Fig. 2.24. Different types of shooting devices used for advanced line techniques. (M. Robbins)

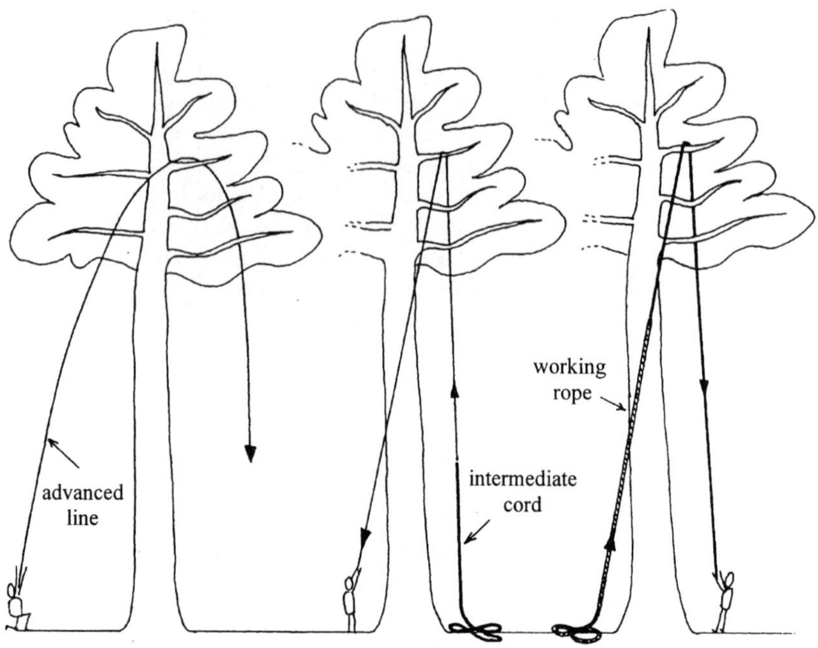

Fig. 2.25. Practical use of advanced lines. (M. Robbins)

Rope ladders are easier to climb but are heavy to carry and raise. Rapid development of thin steel wires may provide some options for use in tree climbing. A rolled steel wire ladder of 15 m weighs less than 5 kg. It is easy to pull up and climb.

The advanced line system has been developed further by using special rope and rope-lock systems. There are two rope locks, one connected to the safety belt/harness, the other attached to a loop in which the climber places his feet. The climber raises his body resting on the foot loop, and pushes the upper lock upwards. He then rests on the safety harness and the upper lock, while the legs push up the lower lock. The climber ascends by alternating the upper and lower locks. The ground end of the rope must be held down so that the leg lock can be pulled up smoothly (Blair 1995; Perry 1978; Perry and Williams 1981, 1985; Arboricultural Association 2005).

All the methods are quite arduous and time-consuming because of the positioning of the advanced line system. The rope-lock system is the most convenient, safest and quickest way of ascending but the locks normally require a special type of rope, which is quite expensive and easily gets worn by the locks and hence needs regular replacement.

Fig. 2.26. Advanced line technique using a 'big shot' catapult. (Courtesy of B. Gunn, Australian Tree Seed Centre)

2.5.2.4
Climbing Within and Harvesting Seeds from the Crown

Fruits are normally borne on the outermost branches often out of reach of the stem, and it is sometimes necessary to climb or walk on horizontal branches. Vertical climbing is much easier than horizontal movement on branches since in the latter case there is no natural support for the hands.

While working at great height it is most important to be secured by a safety line, which also serves as a support line while moving around. The safety line, which is a strong, usually three-stranded nylon climbing rope, is placed at the highest possible safe position, either manually during vertical climbing or via the advanced line technique. The line will arrest a possible fall via the attachment to the prussic loop, which is attached to the safety belt via a carabiner (Fig. 2.19). The safety line is at least double the length of the height above the ground of the branch over which it is placed, so it can be used for descent (Yeatman and Nieman 1978; Stubsgaard 1987, 1997; Ochsner 1984; Blair 1995).

Several methods and accessories are available for harvesting fruits from the crown, some of them are the same as those used for harvesting with access from the ground (Fig. 2.27). The technique applied for cutting down the fruits or branchlets depends on tree species, fruit type and maturity stage. The more that can be done on the ground the better and if possible the climbers should cut fruits or fruit-bearing branches, or shake branches and let fruits and fruit-bearing branches fall down to be picked up and fruits removed by the ground staff. Some fruit types can be pulled off branches by pulling the branches through a stationary 'rake' (Fig. 2.28). Secateurs and long-handled tools are applicable for fruit harvest in the crown. It should be noted, however, that handles must be much shorter (less than 2 m, depending on weight) because they are much more difficult to manoeuvre when working in the crown (Box 2.2). Tarpaulins or sheets placed under the trees facilitate picking from the ground.

Fruits that are likely to dehisce and lose their seed content if they fall to the ground must be picked manually and put into bags carried by the climber. The method is also applicable where ground collection for other reasons is difficult, e.g. rough terrain or vegetation. For example, dry dehiscent fruits like cones of *Araucaria*, *Agathis* and temperate *Abies* spp. may be collected directly in bags because they easily disintegrate upon falling. The collector carries the bag attached to his belt. Filled bags are lowered down by a tool line and are emptied by the ground personnel.

2.5.3 Some Special Collection Methods

2.5.3.1 Collection from the Crown of Felled Trees

Although trees should not be felled for the sole purpose of seed collection, there are cases where seed collection can be done in connection with operational logging or tree felling. Where applicable, such collection is both time- and resource-efficient. The method combines the advantages of collecting from the trees directly (mother tree identity, avoiding loss of seed) with the low cost of ground collection. In practice the method is used only for plantation species.

Good-looking (phenotype) seed trees are selected and marked before logging operation. The trees are sometimes left during the normal cutting operation and

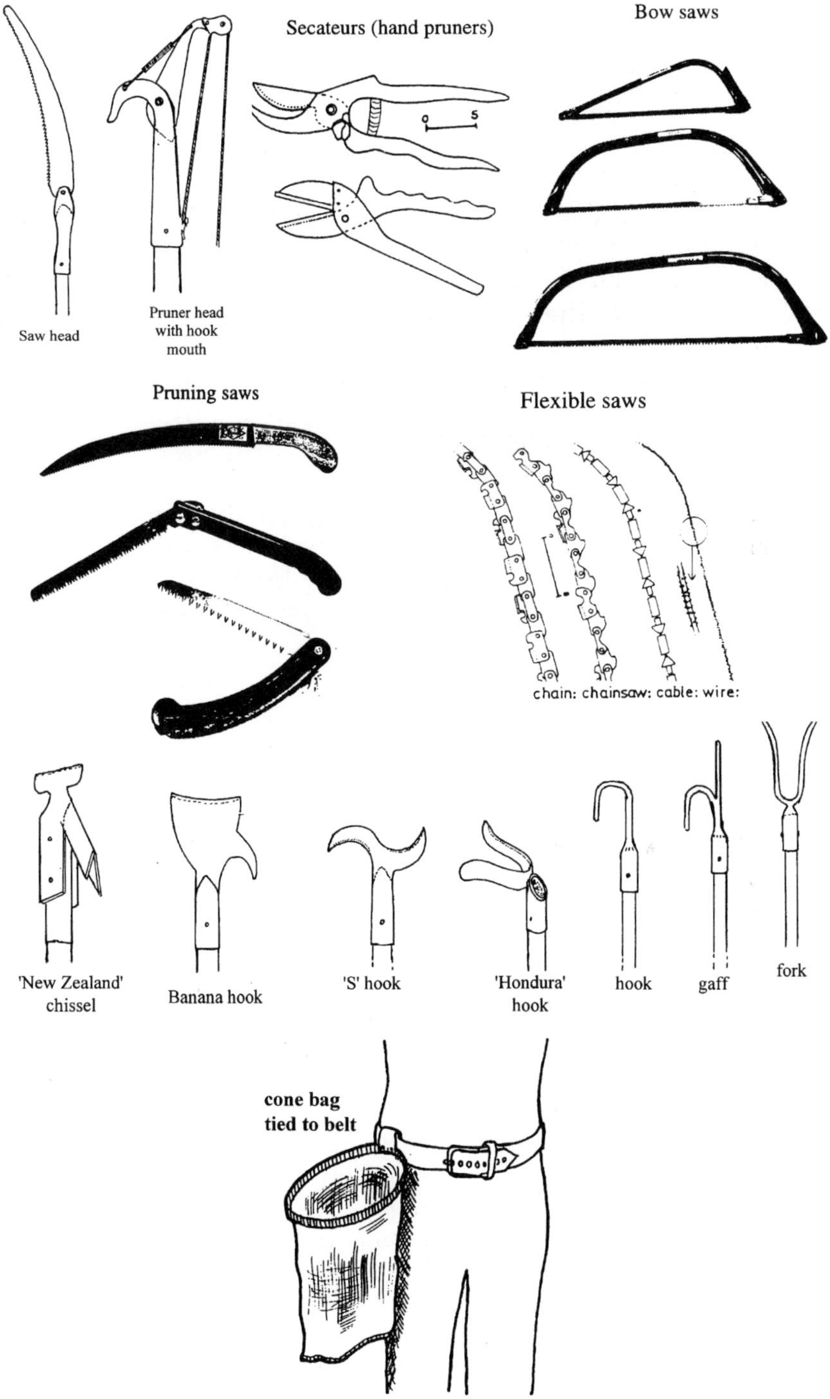

Fig. 2.27. A selection of tools for harvesting tree seed. (From Robbins 1984)

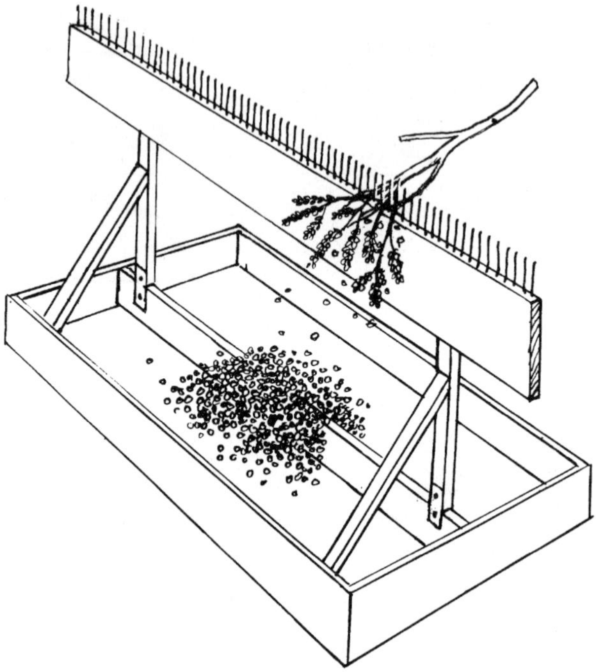

Fig. 2.28. A stationary rake device used for pulling off small fruits from branches. The branches are pulled through the teeth of the 'rake'

cut separately at a time that suits seed collection (optimal maturity). Timing is often critical. Small mature fruits may easily get lost when the trees fall. Felling in the early morning hours when the crowns are relatively moist may reduce loss in some species.

In New South Wales, Australia, most *Callitris* seed is acquired that way, and the method is widely used in West Africa, e.g. for *Triplochiton scleroxylon*, *Terminalia ivorensis* and *Khaya ivorensis* (Brookman-Amissah 1973). In species with regular abundant fruiting, seed availability can be assured by scheduling logging at the best time for seed collection.

2.5.3.2
Shooting Down Branches

Shooting down fruit-bearing branches for seed is hardly used much outside Australia and Papua New Guinea where it, however, is commonly used for sample collection of, for example, very tall eucalypts (Fig. 2.29). As eucalypts have

Fig. 2.29. Shooting down fruit-bearing branches is primarily used in Australia and Papua New Guinea for collecting seed of small-seeded eucalypts from tall mature trees. (Courtesy of B. Gunn, Australian Tree Seed Centre)

very small fruits and seeds, a considerable amount of seed can be brought down with one branch. The method is quick and safe compared with climbing. Guns and cartridges are lightweight compared with much climbing equipment and collection can be done by one person only (Gunn 2001; Gunn et al. 2004).

To maximise impact, shooting is done by large-calibre rifles (0.222, 0.243 or 0.308), 'pointed soft point' type ammunition (hunting type – not military ammunition!) and a shooting angle of as close as possible to 90°. The first shot is made at the down part of the branch to cut the lower bark and hence avoid hanging of the branch (like an undercut of a pruned branch). The second shot is made at the upper part and the remaining shots are made at intervals across the branch. Branches up to 15–20 cm diameter can be brought down from the highest trees by five to 15 well-placed shots (Boland et al. 1980; Gunn 2001; Green and Williams 1969; Kleinschmidt 1989). Branches of some species with stringy bark (e.g. *Eucalyptus globoidea*) can be difficult to dislodge by the method (Gunn 2001).

Except for the technical constraints and the price of ammunition, shot-cutting of branches is subject to various safety concerns. The general rules and restrictions on the use of firearms apply, e.g. safety measures, licence requirement and prohibited use near populated areas or in wildlife conservation areas. The noise produced may be annoying to people and disturbing to wildlife (the shooter and other members of the collection team should use earmuffs for protection).

2.6 Safety and Routines

In dangerous operations like tree climbing, safety must not be sacrificed to save time or money or for other short-term convenience. In fact, appropriate safety measures are likely to make collection more efficient. Risk of injuries cannot be eliminated, but can be greatly reduced by exercising the appropriate safety measures. As in most other work functions, most accidents happen to beginners and very experienced staff. The former because of insufficient practice, the latter because of overconfidence and less alertness.

Safety precautions should be observed particularly in relation to the following points:

1. *Use of ground operated equipment.* Although minor cutting injuries happen to users of, for example, secateurs and saws, there is probably a higher risk of being injured by others (and of injuring others) than of injuring oneself. Long-handled tools like extended pruners and hooks to pull down branches can be perilous as the user often miscalculates the weight when the tools are held in a nonvertical position. Bows, crossbows, catapults, etc. used in advanced line techniques are potential weapons and should be treated with the safety measures weapons deserve, i.e. assembled and available only when in use, with fellow workers safely behind the person shooting, and safe aiming (not at people – not even for fun!). For firearms, special rules and safety measures apply for use and storage of both gun and ammunition; these should be known and complied with by the licensed holder.
   In a group of more than three people it is difficult to keep track of the position of all the others during operation; therefore, large teams should be avoided.
2. *Collection near transmission lines.* Climbing or collecting seed from trees growing close to transmission lines is probably one of the highest-risk factors and should generally be avoided unless an agreement can be made with power authorities to temporally shut off the power. If collection has to be done from such trees, metal equipment such as ladders and long-handled pruners should not be used. Tree climbers should keep a safe distance from such lines.
3. *Danger of falling objects.* The danger of being hit by accidentally or miscalculated falling objects applies both to ground staff and to climbers. Climbers may be more in control but cannot escape an accidentally falling branch. The direction of falling branches is often unpredictable as they may be diverted by wind or other branches. Climbers may easily

break off branches and heavy fruits and accidentally drop equipment like secateurs or hand pruners. Hand tools should be attached to the climber by a string, which both protects ground staff and avoids the annoyance of losing the equipment. Both climber and ground staff should observe safety precautions. The climber should be aware of the position of the ground staff, who in turn should never place themselves under a climber. Ground staff should be notified when, for example, bags of fruits are thrown down. In addition it is advisable that ground staff use safety helmets/hard hats.

4. *Maintenance and careful use of climbing equipment.* Climbing implies the obvious risk of falling down. A fall is always painful, often injurious and in worst cases fatal. Safety precautions for climbers imply correct use of safety equipment such as safety belt, harness and safety line. Safety devices should be manufactured from high-quality materials, preferably they should be acknowledged brands. Special attention should be paid to locally manufactured ropes, which do not always reach the standard required for tree climbing. Safety equipment should be inspected regularly, i.e. before each climb, and possible damage should be repaired before use. If necessary, the equipment must be replaced. Special attention should be paid to the sewing of safety belts, and fibre damage of belts and ropes. Because nylon ropes do not rot, their durability is often overestimated. Nylon rope is easily damaged by heat created by friction, and long exposure to sunlight weakens the material.

   Leather is prone to rotting, especially under humid conditions. Tearing and rotting often start near the holes of rivets, which must therefore be examined particularly carefully. Longevity of leather is improved if it is kept dry and possibly preserved by leather grease when not in use.

   Carabiners are used to fasten safety strops and safety lines in D-rings of the harness. The carabiners should be easy to open and close, and should have a safe lock system.

Use of safety equipment during climbing comprises a number of special techniques which should be known and carefully observed. For details reference is made to Blair (1995), Arboricultural Association (2005) and technical notes from Danida Forest Seed Centre (e.g. Stubsgaard 1997). A few points are emphasised here:

(a) Safety belts or harnesses are always used during tree climbing. The safety belt consists of a broad belt fastened around the waist, and a saddle. A harness is provided with shoulder strops. D-rings of the safety belt serve to connect safety strops, safety lines, working rope, tool line and equipment.

(b) The safety strop (or strap if it is a belt) is used during vertical ascent. It is connected to the safety harness with carabiners and goes around the bole. When a branch or a fork is passed, the strop has to be disconnected; therefore, a second strop is connected above the branch to be passed before the lower one is disconnected.

(c) Safety lines are used during ascending, horizontal climbing and descending. A safety line is connected to the climber's harness/safety belt, then looped above the climber at an anchor point, and attached to the climber again by a prussic loop (Fig. 2.26). The connection to the safety line is adjusted so that it does not impede the free movement of the climber, but is simultaneously tight enough to suspend a possible fall within 2–3 m. Care must be observed that the appropriate safe knots are used.

5. *Personal fitness.* Climbers should be physically fit, have a good sense of balance, fast reaction and not suffer from height dizziness (acrophobia). A climber who does not feel well, or who is recovering from debilitating illness, e.g. malaria, should not be allowed to climb. Obviously drugs, alcohol or hangovers should be banned in connection with climbing.
6. *Personal clothing.* Clothing should be strong and fit well without being tight, and be without loose straps, belts, pockets or other appendices that might get entangled in branches. Climbing in tropical temperatures can be physically exhausting and very hot, and the clothing must allow adequate ventilation. Climbing trees with myriads of aggressive ants is a nuisance; well-fitting clothing with zippers rather than buttons and elastics around wrist and ankles yields some protection. Footwear should be strong and well fitting with high-friction soles. If climbing spurs are used, the best footwear is boots which protect the shins. They should be provided with a marked heel to avoid the spur slipping off. Gloves may be advisable to protect the hands and increase the friction when ropes are used.
7. *Tree defects/weaknesses.* Trees differ in the physical strength of their branches; some species have very brittle branches. The tree climber should be familiar with the strength of the species he is climbing. Lower branches are normally self-pruned by abscission from the tree; therefore, there are often several dead branches below the crown which are not safe and cannot carry a person. The climber should also observe any disease in branches which could make them weak. When climbing in branches the climber should have three points of support at all time, i.e. one hand and two feet or two hands and one foot, moving one limb at a time.
8. *Insects.* Although ants may be an extreme nuisance during climbing in some trees, they rarely pose a real danger other than that of paying a certain

amount of attention. However, bees and wasps can be really dangerous. It is advisable to examine crowns with binoculars before ascending: trees with bees' nests or wasps' nests should never be climbed (Fig. 2.30).

9. *Equipment when climbing.* Any excess weight or projection is an obstacle to safe and smooth climbing as it may get tangled in the branches and be dangerous in the case of a fall. While climbing up and down, the climber should be free of any excess equipment. Especially dangerous is anything tied around the neck. A pair of secateurs and a folded pruning saw are practical for removing obstacles when climbing. All other equipment is preferably left on the ground and hoisted up via a thin tool line once the climber is in place, connected to the anchor point and ready to start the collection. The equipment is lowered before the climber descends.

   Rope and sharp equipment is a potential dangerous combination. Great care should be exerted when using secateurs, saws or the like near cuttable material. Also climbing spurs can cause damage to ropes.

It is mandatory that climbers have received appropriate training in techniques and safety precautions before they start climbing, and that very difficult tasks are carried out by experienced climbers only (Box 2.6). Safety training for climbers should include rescue operation from the crowns in case a climber is not able to get down by himself. Accordingly, two sets of climbing equipment

a

Fig. 2.30. Sign of dangerous insects in trees. **a** Leaf ants make a nest of connected leaves. The nests are not always conspicuous and are easily overlooked, but the ants can be extremely aggressive. (By courtesy of M. Schotz).

Fig. 2.30. (Continued) **b** Wasps' nests can vary from very small hosting a few individuals to rather big and home to thousands of wasps. **c** Termites can build huge nests in trees, though annoying they are less aggressive than ants.

Box 2.6

Learning to climb – licence to climb!

Practice makes perfect, in tree climbing as in any other skill. There is no impediment in learning by doing and many problems associated with techniques can be solved as and when faced. However, it is generally advised to learn basic skills during training sessions by professional and experienced instructors. Training sessions have the purpose of teaching special routines and techniques. Examples of what is best learned during formal training sessions are equipment inspection routines, knot applications and rescue operations. Tree climbers should pass a test in these skills before they are allowed to do professional tree climbing. Tests may be upgraded to legal certificates, i.e. only climbers with formal certified qualifications can obtain a licence for tree climbing. Seed collection is a special application of tree climbing skill, and training in and licence to climb trees is not necessarily linked to seed collection.

should always be available at a climbing site, and at least two trained climbers should be in a collection team. Further, the staff should be trained in basic first aid and the team should be provided with a first-aid kit.

2.7 After Collection

In connection with collection and subsequent handling procedures it is necessary to ensure that all necessary data for the collection are noted (time, location, etc.) and that seeds are stored or preprocessed so that they will remain viable until processing. The latter is especially relevant for remote collections and easily deteriorating seed.

2.7.1 Field Records and Sampling

The seed documentation system is described in Chap. 8. Documentation starts in the field and follows the seed lot during processing, storage and testing, to distribution. The following information should be recorded during collection:

1. *Species (and subspecies or variety).* If there is any doubt in the species identification, herbarium material should be collected together with the seeds for later botanical identification. The material should preferably include flowers (if any are left at the time of seed collection), leaves and intact fruits. It is important that the material is preserved appropriately in a plant press, and labelled so that it can be related to the seed lot (individual tree).
2. *Location.* Geographical coordinates should include, as precisely as possible, the site of the actual collection. Geographical coordinates are found on large-scale maps, e.g. 1:25,000 or 1:50,000. A more advanced system, Global Positioning System (GPS), uses the position of satellites for geographical coordinate finding (Chap. 8). Altitude is indicated in metres above sea level, which can be found on topographical maps (note that some older maps and some countries indicate feet), or with the aid of an altimeter. The altimeter works on the principle of decreasing atmospheric pressure with increasing altitude. Since the pressure varies with weather conditions, there is always a certain (but rarely important) error in altimeter indications. Adjusting the altimeter at a point of known altitude reduces the error. In hilly areas an appropriate altitude range (e.g. 400–600 m above sea level) is recorded.

3. *Soil type.* Soil samples may be taken for analysis and documentation of soil type. Alternatively, simple analysis such as pH and structure may be undertaken on site.
4. *Number of parent trees.* The number of parent tress from which seed is collected should always be recorded for documentation of genetic diversity.

Appropriate labelling goes hand in hand with seed documentation. Maintaining the identity of the seed lot is especially important for single-tree collections where the identity of the mother tree must be known. Each seed lot is labelled with a number that serves as a preliminary identification number. The number on the label corresponds to the number appearing on the seed collection form (Chap. 8). The label must be written with non-water-soluble ink on weatherproof labels, and safeguarded from blowing away or otherwise getting lost during handling (Fig. 2.31). While the seed collection form is normally kept safely with the seed collector, the seed labels are prone to be lost during seed handling, after which the identity is lost. It is advisable to put duplicate labels inside as well as outside each fruit container.

2.7.2 Preprocessing, Field Storage and Transport

If fruits are to be collected from the crown directly, the climber normally carries a small cone bag attached to his belt (Fig. 2.27). When the bag is full, it is lowered via the tool line and emptied into a large bag. Normally only dry fruits are collected that way. Fruits or seeds collected from the ground, directly or from pruned branches, are preferably collected in open containers such as baskets, buckets, tarpaulins or canvas sheets (Fig. 2.32). Baskets are used for relatively large fruits and seeds. Baskets can be made of wire, rattan, bamboo or the like.

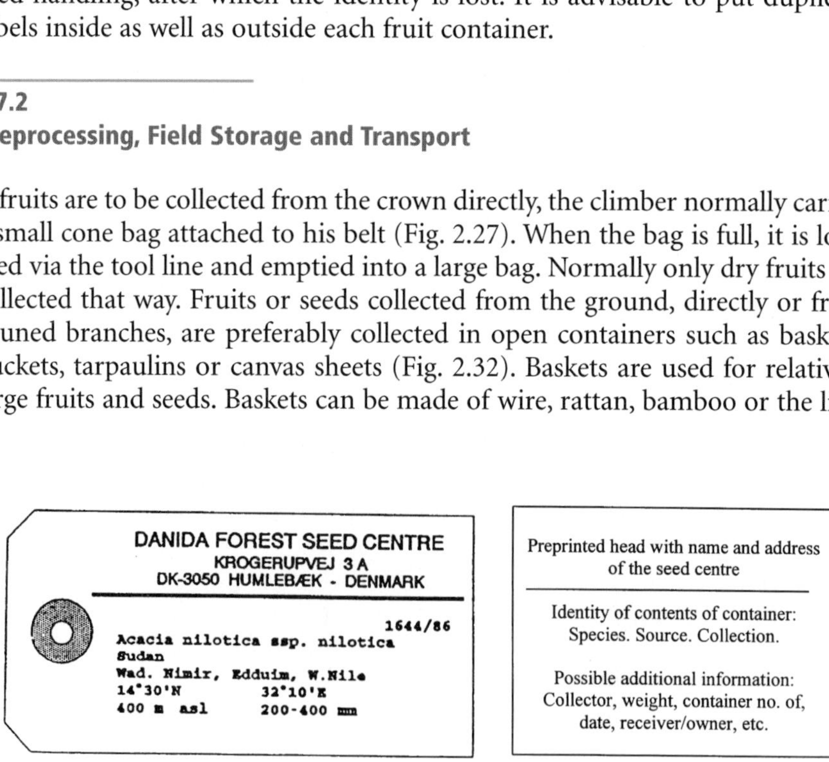

Fig. 2.31. A practical seed label with essential information about the seed lot

Fig. 2.32. Temporary storage and field processing equipment. (M. Robbins, P. Andersen)

Fig. 2.32. (Continued)

Wire baskets are usually preferred since they are easier to clean and stack, but they are only usable for large fruits. Buckets are used for fleshy fruits. Metal buckets are preferred because they are stronger than plastic ones and can be repaired.

Tarpaulins can be used for most fruit types. After collection the fruits are raked or swept together and put into the storage container. Care must be taken not to perforate the tarpaulin by stepping on it as twigs or other sharp material will readily perforate it.

Canvas sheets are convenient for quantities of small and dry fruits and seeds like eucalypts and casuarinas. The sheets (approximately 2 m × 2 m) are spread on the ground and fruits and fruit-bearing branchlets are collected on the sheet. When the work is finished and the sample has been labelled with a number corresponding to the seed collection form, the sheet can be folded together and tied carefully corner to corner. The sheets can easily be opened to expose the fruits to drying when required (Fig. 2.32).

If large quantities of fruits are collected, collection containers may preferably be emptied into sacks, barrels or other larger containers.

Collection containers or sheets as well as those used for bulk storage should be tight and cleared of all seeds and debris from previous collections. This is especially important for small seeds like eucalypts and casuarinas whose seeds easily slip out of small holes, and leftover seeds may be stuck in corners or canvas with risk of contamination. Fruits are preferably removed from the twigs before being put into the sacks as twigs may easily perforate the material. Storage containers should not be too large, both for convenience of carrying and because aeration may be insufficient in large containers.

Seed Processing 3

3.1 Introduction

Seed processing[1] refers to the handling procedures between collection and storage (or sowing), which aims at achieving clean, pure seeds of high physiological quality (germinability) which can be stored and easily handled during succeeding processes, such as pretreatment, transport and sowing. Processing includes a number of handling procedures, whose applicabilities differ, e.g. according to fruit and seed type, condition of the fruits or seeds at collection and potential storage period. Processing can be grouped into the following seven procedures:

1. *Precleaning.* A rough cleaning to remove larger debris such as leaves, twigs and empty fruit parts.
2. *Precuring.* A prolonged gradual drying procedure applied to complete fruit and seed maturation and to ease extraction.
3. *Extraction.* The physical separation of fruit and seed (or other storage unit).
4. *Dewinging.* The removal of wings, hairs and other appendices from seeds or other stored units.
5. *Cleaning.* The removal of impurities, i.e. all non-seed material, e.g. foreign seed, fruit parts, floral parts and 'dust'.
6. *Grading.* Separation of the seed fraction into several parts based on, for example, size.
7. *Adjustment of moisture content.* Usually drying as preparation for storage. For desiccation-sensitive seed, remoistening may occasionally be applicable if seeds have been dried to below a safe level.

Seed processing normally follows the above order, but certain steps may be irrelevant and hence omitted for particular species or seed lots. Processing

[1] The term 'conditioning' is often used in American literature.

often starts in the field as a continuation of seed collection, e.g. to reduce bulk and prevent deterioration. Any delay in processing may hamper both the ease of processing and the quality of the seed in species with moist and easily deteriorating fruits. Processing is not necessarily one continuous process. Partial processing and intermediate processing, like precleaning, drying or removal of fleshy fruit parts, are typically undertaken either in the field or immediately upon arrival at the processing depot in order to impede deterioration during transit. Final processing may then be postponed until later.

Any step in the processing procedure should be adjusted to the particular fruit or seed type. Processing implies a risk of losing seeds both by undertreatment and by overtreatment. Undertreatment means that a procedure is insufficient to achieve the desired result (e.g. insufficient extraction of seeds); overtreatment means that a procedure is 'overdone' with consequent damage to seeds, e.g. loss of viability or reduced storability. Since a seed lot[2] typically contains variation, e.g. in degree of maturity, size and shape, it is in practice rarely possible to avoid damage by undertreatment or overtreatment of a mixed seed lot. If physical and physiological variation is systematically linked to different parent trees, e.g. seed trees with a different maturity stage or different seed size, processing may affect the genetic composition via overtreatment or undertreatment (Hellum 1976; Silen and Osterhaus 1979).

Hygiene and documentation are essential elements of seed processing: hygiene, because many seed lots are handled together and the risk of spread of contaminants (foreign seed or pathogens) is high; documentation, because the risk of losing seed lot identity is high.

3.2 Use of Technology in Seed Processing

The principle of forest seed processing is similar to that of agricultural crops, viz. to get clean storable seed. Since crop seeds are often handled in large quantities, there has been a high incentive to develop mechanical equipment that can process bulk quantities rapidly and efficiently. With simple adjustment, some of this equipment can be used for forest seed. Technology is in other words often available. However, the fascination of equipment and what can be done with it, and the prestige of having a collection of machines have often led to overprocurement of forest seed processing machinery. 'Museums' of idle forest seed equipment all too often document inconsiderate procure-

[2] A seed lot is a consignment of fruits and seed collected together, for example at the same time and from the same seed source.

ment policy by, for example, donor projects. Before procuring (expensive) mechanical equipment, some general consideration on applicability should be undertaken:

1. The quantity of forest seed is usually small compared with that of agriculture seed. Even where large quantities are to be processed, variation in seed lot characters often implies that seed is processed in separate seed lots. Several small-capacity pieces of equipment rather than one large-capacity piece of equipment are usually preferable.
2. Agricultural species are herbs with small seed, and most equipment is built for that type. Many forest species have large seeds which are less suitable for any mechanical handling.
3. The diversity of species is often large. Adjusting equipment to fit a species where only a small amount of seed is to be processed may be more time-consuming than simple manual processing. The time needed for cleaning equipment after use is often underestimated. In addition, it can be difficult to use large machines for small quantities.
4. Labour cost and staff requirements differ between different countries and seed suppliers. However, current rationalisation generally tends to reduce manual labour because of both actual direct cost (salary) and derived cost, e.g. requirement for working space, administration, and compulsory social commitments.
5. Maintenance of equipment may require both skill and spare parts. Where these are not available, equipment may soon stand idle with all investment wasted.
6. Thrashers and other extraction equipment for agricultural seed often cause great damage. Most crop seed is used for consumption, and in this case mechanical damage is not necessarily important.
7. The critical stages or need of processing is often highest in the field right after collection. Mobile (small) equipment may make the need for central processing facilities relatively small.
8. The implications of not reaching the maximum result, e.g. lower than 100% purity, should be considered. Forest seeds often have a much simpler distribution system than agricultural seeds. Seed transported more or less directly from collection to nursery does not require much processing to reduce bulk or increase purity.

Procurement of equipment should thus be according to what is needed (i.e. the process to be carried out, and in what capacity) rather than what is available (the catalogue!). That being said, it should also be added that better and more efficient equipment has been made available owing to progress in technology and production of machines. 'Copy' machines like depulpers and dehuskers previously only available from international suppliers can now be found in most countries.

3.3 Precleaning

Precleaning is the removal of larger matter such as leaves, twigs and empty fruits before extraction. If collection is done by relatively 'clean' methods such as picking up individual fruits or seeds from the ground, ground cover or trees, the seed lot may contain little foreign material – larger material can just be removed by hand. Seeds collected by vacuuming, sweeping or raking the forest floor, on the other hand, contain large amounts of debris. Such matter is redundant bulk and may moreover hamper both processing and seed quality. Precleaning has the following rationales:

1. Mechanical extraction may be hampered by any foreign material, but in particular larger dry material such as branch pieces and leaves are a nuisance in, for example, depulpers.
2. Small, seedlike material like fruit stalks, leaf pieces and broken twigs are easier to separate from fruits than from seeds and thus are preferably removed before seed extraction. An example is the scaly leaves (cladodes) in casuarinas, which are quite similar to seed. Breaking off fruits from the branchlets and removing the dry branchlets during a precleaning process makes subsequent cleaning much easier (Gunn 2001; Turnbull and Martens 1983).
3. Threshing breaks material into smaller pieces and is used for extraction in many indehiscent fruits, e.g. indehiscent legume pods. However, if the fruit lot contains branchlets and twigs, threshing is likely to break this material into fragments of similar size and weight as seed. The fragments are then difficult to remove by any cleaning method after extraction. (Gunn 2001).
4. Pathogens such as bacteria and fungal spores attach themselves to soil and organic material. If pathogens appear in large amounts and conditions are conducive to infection, they can cause serious damage both in seeds and in germinating seedlings. Clean material is preferred. As many pathogens are attached to non-seed material, precleaning can reduce the contamination significantly.

Much precleaning can be done manually by picking branches with large fruit pieces and leaves sometimes in connection with maturity sorting (Bowen and

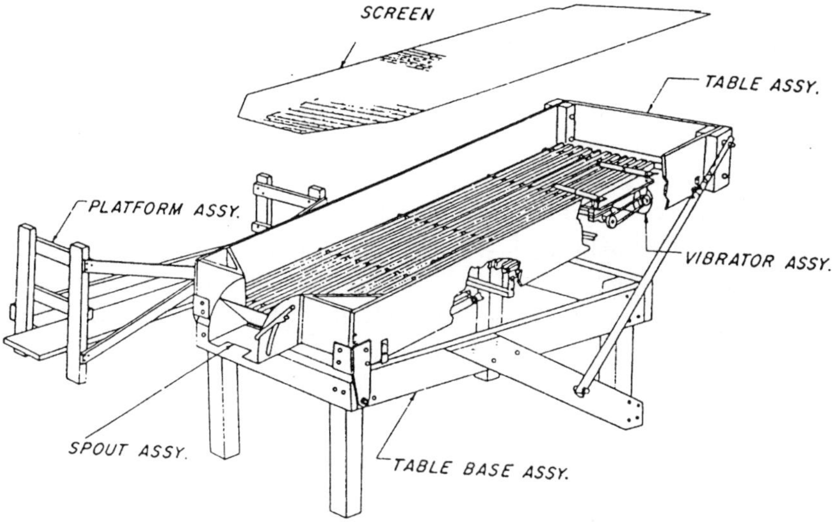

Fig. 3.1. Simple mechanical precleaner with a vibrating screen. Different screens can be used for different fruit sizes. (From FAO 1967)

Eusebio 1982). For larger quantities precleaning is done mechanically, e.g. on vibrating or oscillating screens or in tumblers (Figs. 3.1, 3.8). Precleaning separates the material into three fractions:

1. Material larger than fruits, which is discharged.
2. Fruits, which go to further processing procedures, i.e. extraction.
3. Material smaller than fruits, which typically contains a mixture of seed and small debris. Depending on seed content, this fraction may be discharged or cleaned by the same procedure as extracted seed.

3.4 After-ripening

The two main events of the later maturation development are achievement of germinability and development of dispersal structures. The former has importance for the physiological seed quality, including storability, the latter has importance for extraction. Immature fruits are thus fruits with undeveloped dispersal features where seeds are typically firmly retained inside green fruits. Immature seeds are seeds which have not achieved germinability. The two events are linked: seeds are normally dispersed only when physiologically mature.

Physiological maturation and dispersal mechanisms normally take place on the tree and are a continuation of previous events in fruit and seed setting. Premature collection may be deliberate to avoid seed loss by dispersal or predation, or not deliberate, e.g. if collection was done early or the collected fruits contain a large fraction of not fully mature fruits. Some events of completion of the maturation process after dispersal are not unusual. It is thus often observed that seeds tend to germinate better after a short storage period (Rizzini 1977; Masilimani et al. 2002).

A small group of species tend to disperse their seeds before full maturity, i.e. the aforementioned normal synchronising of seed physiology and dispersal development is skewed. *Ginkgo*, *Fraxinus* and *Taxus* are some of the genera where seeds are often dispersed with immature or underdeveloped embryos. The phenomenon is often categorised as a type of dormancy (Chap. 5), but the pretreatment is in this case a simple after-ripening.

The process in which fruits and seed are exposed to conditions which make them complete the maturation process under controlled conditions after being detached from the tree is called precuring or simply after-ripening. After-ripening is typically a slow drying process, which simulates the type of maturation drying that would normally take place on the tree as part of the natural maturation process. Only the maturation processes connected to drying can take place during precuring; fruits and seed must thus have completed their growth, i.e. be mature in size. Small, underdeveloped fruits cannot be after-ripened. Increase in size (in physiological terms, increase in dry weight) is thus one crucial maturation parameter. Orthodox seeds typically reach peak size long before maturation and dispersal. Recalcitrant seeds on the contrary continue to accumulate dry weight practically until the dispersal time (Berjak and Pammenter 2002). Consequently the chances for successful after-ripening of recalcitrant seeds are limited.

A fruit lot typically contains a large variation of maturity stages, and only fruits that are not fully mature should be after-ripened. Fruits are thus sorted in fractions prior to processing: small underdeveloped fruits are discarded; fully mature fruits go directly to the next step in the processing chain, usually extraction; an intermediate portion may consist of fruits of mature size capable of after-ripening. The latter portion may be subdivided into maturity stages, e.g. according to colour, and exposed to after-ripening for a period necessary to complete maturation. For example, in Malaysia pods of *Acacia mangium* are separated into three classes according to colour, viz. greenish brown, brown and black. Greenish-brown pods are after-ripened for 120 h, brown pods for 72 h, and black pods go directly to extraction by kiln drying (Bowen and Eusebio 1982).

Along with dehydration of storage material and enzymes during maturation drying, the components are packed and structured in the most efficient way for restart after imbibition; therefore, after-ripening is not mere desiccation. With the point of departure that maturation normally takes place on trees, the following environmental factors should be addressed during precuring:

1. *Temperature.* The temperature is kept within physiological range, i.e. extreme high temperatures should be avoided. Under natural maturation, shade by leaves and evaporation regulate temperature. Under precuring conditions, temperature is regulated in the same way, i.e. by shade and regular water spraying.
2. *Humidity.* Aerial humidity regulates drying rate. During the natural maturation process, drying is also regulated by water flow from the mother tree. That source is cut off during precuring and humidity is regulated entirely by air humidity with the consequence that fruits can dry too quickly. Drying rate is reduced by increasing the humidity by shade and regular spraying. The humidity is initially high and is gradually reduced as maturation progresses.
3. *Light.* For dry fruits the mature colour is usually yellow, brown or black (Fig. 3.2). Fleshy fruits can have any bright mature colour. A bright mature colour tends to develop less strongly during after-ripening than during natural conditions. It probably has no effect on seed quality but can sometimes mislead determination of the maturity stage.

Fig. 3.2. Maturity stages in dry fruited *Tarietia javanica.* Maturation drying is responsible for change of colour, abscission and dehiscence. The colour changes from dark green to pale green to light brown to dark brown. Drying is easily recognised in wings – dry wings break rather than bend. In dehiscent fruits drying leads to dehiscence

4. *Atmosphere.* Natural maturing fruits are exposed to a normal atmosphere, Under precuring conditions air exchange is restricted and gases can accumulate. Certain gases, primarily acetylene and carbon dioxide, are known to be important for fruit maturation in fleshy fruits.

In order to ensure the most uniform after-ripening conditions, fruits should be spread in a thin layer (for large fruits a single layer) on concrete floors or in trays. During precuring the fruits are regularly turned and sprayed with water. Fleshy fruits will usually exude liquids, which must be drained off to avoid rot.

The duration of after-ripening depends on the species, maturity stages and conditions. Dry, dehiscent fruits will, once mature, start opening and releasing their seed. Indehiscent fruits must be broken open and the seeds examined, e.g. by a cutting test. Note that tetrazolium is not a good indicator in this context – the chemical will stain all live tissue and thus does not distinguish between mature and immature fruit. After-ripening may take from 2 days to 2 weeks. In Kenya, seeds of *Azadirachta indica, Thevetia peruviana* and *Ximenia americana* are after-ripened for 2–3 days after collection (Ahenda 1991). In Thailand, precuring of *Pinus merkusii* and *Pinus kesiya* is routinely done by storing freshly collected cones in loosely tied gunny bags or bamboo baskets for 7–14 days (depending on maturity) (Granhof 1984; Sirikul 1994). The longer time required for precuring of pines is primarily due to extraction problems: rapidly dried cones tend to develop 'case hardening' where seeds get trapped inside the cone scales. A summary of after-ripening steps is given in Box 3.1.

Once after-ripening has been concluded, fruits go the next step in the processing chain, viz. seed extraction. In dry, dehiscent fruits, extraction is in practice a continuation of after-ripening where desiccation is allowed to proceed, i.e. water spraying is stopped and shade removed.

Box 3.1

Procedure for precuring

- Separate fruits in two or three maturity classes
- Store at ambient temperature at a ventilated place and high humidity; stir regularly to allow ventilation and spray to avoid rapid desiccation
- Reduce moisture (less spraying and reduced shade) as the fruits approach a mature colour
- Conclude the process as the fruits attain a mature colour

3.5 Seed Extraction

Seeds are typically enclosed in a fruit or other structure during collection. Seed extraction is the procedure of physically releasing and separating the seeds from this structure. Seed is here used in the wide sense as the stored and sown unit, which is the morphological seed plus sometimes additional attached or enclosing structures, e.g. endocarp, the entire pericarp or arils.

Extraction has three key objectives:

1. *Reduce bulk.* The seeds sometimes make up only 1–5% of the total fruit volume and 5–10% of the weight. Bulk reduction helps to reduce the cost of storing and transport (Table 3.1).
2. *Ease of handling.* Seeds are normally tested, pretreated and sown individually, which necessitate their separation from the fruit. In many cases fruits contain inhibitory substances, which must be removed for germination to take place (Chap. 5).
3. *Improve storability.* Easily decomposable fruit parts such as the pulp of fleshy fruits or arils must be removed to avoid their decomposition during storage. Moisture contained in dry fruit types and cones may attract fungi and insects, especially if stored under ambient temperature. In addition, drying of seeds to safe moisture content becomes difficult if the seeds are not extracted.

The collective name 'seed' refers to what is stored and sown, the 'rest' is removed during extraction (Box 3.2). However, what is removed and what is left often depends on practical considerations, cf. above objectives: Seeds which are not to be stored, where extraction does not reduce bulk significantly, where fruits contain one or few morphological seeds and where fruit structures do not hamper storage or germination may not need full extraction. Partial extraction is, for instance, removing the wing of *Pterocarpus* pods, but leaving the seed inside the pericarp.

Extraction is usually done as soon as possible after collection in the normal seed processing chain; however, extraction may be accelerated or delayed according to specific conditions. Accelerated extraction is relevant, for example, for:

1. Very bulky fruits like mahoganies, which are inconvenient to transport in large quantities. The more that can be extracted in the field, the lower the transport cost.

Table 3.1. Bulk reduction during extraction of seeds from 1 l of unprocessed fruits

Species	Fruit type	1 l of fruit		Seed content in 1 l of fruit		
		Number	Weight (g)	Number	Volume (ml)	Weight (g)
Pinus merkusii	Cone (closed, moist)	20	200	400	40	12
Pinus merkusii	Cone (open, dry)	13.5	72	270	25	7.8
Acacia nilotica	Indehiscent pod	110	240	1,400	300	186
Dalbergia cochinchinensis	Indehiscent pod	450	210	700	30	22.7
Swietenia macrophylla	Capsule, winged seed	2.4	160	200	312	115
Swietenia macrophylla	Capsule, dewinged seed	2.4	160	200	195	115
Khaya senegalensis	Capsule	16	190	320	147	83
Pterocarpus indicus[a]	Samara/indehiscent pod	18	180	35	Not available	150
Lagerstroemia speciosa	Capsule	1,000	220	18,000	210	142
Canarium album[a]	Drupe	105	1,000	105	65	186
Tectona grandis	Drupe with involucre	145	75	145	30	115
Tectona grandis	Drupe	1.100	210	1.100	175	770

One litre of fruit is used as the volume of loosely packed fruits – in practice 1 l of large fruit is based on a larger volume, e.g. 10 l or more.
[a] Species are incompletely extracted, e.g. by removing fruit wings or exo-mesocarp

Box 3.2

Fruit taxonomy

In a strict botanical definition, a fruit is the mature carpel with enclosing seed. Seed is the mature embryo with surrounding endosperm and the enclosing seed coat. The seed coat originates from the integuments. Arils, which originate from the integuments, are thus part of the seed and not the fruit. In the same way hard endocarps enclosing seeds in drupes are part of the carpels and are thus botanically part of the fruit. In practical seed handling, botanical definitions are often used in a more relaxed way. For example, true fruits only occur in angiosperms, as the fruit botanically is what characterises this group; gymnosperms are 'naked seeded', i.e. without enclosing fruit. However, seed-bearing structures in gymnosperms, e.g. cones, are similar to fruits as far as seed handling is concerned. Arils in gymnosperms such as *Gnetum*, *Taxus* and *Podocarpus* are practically like fleshy berries or drupes. In angiosperms, botanists distinguish between true fruits, which consist of carpels with enclosed seeds, and aggregate, multiple and false fruits. Aggregate and multiple fruits (together called compound fruits) consist of several morphological fruits in a larger fruitlike structure, which have taken over the dispersal function of simple fruits. False fruits contain structures of non-carpel origin, e.g. the enlarged placenta in fruits with inferior ovaries (e.g. pomes). Compound and false fruits are handled as any other multiple-seeded true fruits, e.g. capsules for dry structures and berries for fleshy structures. The traditional fruit classification are according to structure (fleshy or dry), number of seeds (one or many) and opening mechanism (dehiscent or indehiscent). Most fruits can be classified into the main categories berries, drupes, capsules, pods, follicles and nuts, but many intermediate forms exist, e.g. dry drupes (e.g. teak) and fleshy indehiscent capsules (e.g. durian).

2. Soft or fleshy fruits which deteriorate, ferment or rot. This is both highly unpleasant and potentially deteriorating for seed. Where fruits contain high levels of germination inhibitors, delayed extraction can cause development of strong dormancy (Chap. 5). Portable depulpers, e.g. those designed for coffee depulping, can do the first depulping in the field.
3. Cones will open when dried, but if transported in containers or sacks with limited space for expansion, they will dry without opening. This so-called case hardening implies difficulties during later extraction.

Extraction is usually undertaken prior to storage (as one of the rationales is to reduce storage volume), but in some species it may be delayed until just before sowing or omitted altogether, e.g. where extraction implies high risk of seed damage with possible effect on seed storability. This is mainly the case in seeds with fragile, papery seed coats. For example, extracted *Cedrus* seed have reportedly inferior storage compared with that of seeds stored with the cones and extracted just before germination (Stubsgaard and Moestrup 1991). *Vitex* species have berries where soft material is usually removed by depulping and washing; however, as depulping inevitably causes reduced storability, the fruits can be dried and stored dry in a cold place. Fruit flesh is then softened and removed prior to sowing to eliminate germination inhibitors. A very hard endocarp or pericarp is often combined with a fragile seed coat. Seeds from drupe stones and hard indehiscent pods are therefore preferably not extracted before storage despite their bulkiness. Very hard fruit structures may restrict germination, and extraction is sometimes carried out prior to sowing. This holds, for instance, for drupes of *Melia volkensii* and *Pterocarpus* species (Kamondo and Kalanganire 1996). Extraction to promote germination as in these cases is the same as a dormancy pretreatment (Chap. 5).

There are principally two methods of extraction, viz. dry and wet extraction, which correspond to dry and fleshy fruit types, respectively. Dry fruit types are dried and this causes dehiscent fruits to open and dehisce their seed, and it eases mechanical extraction of seed from indehiscent fruits. Fleshy fruits are extracted by various types of washing. Some fruit types contain both dry and fleshy material and extraction is here a combination of the two types. For example, some *Prosopis* spp. are first extracted from the pod by dry extraction, i.e. drying and threshing; the enclosing pulp is then removed by washing. Seeds of *Afzelia* species are extracted dry as normal for dry pods, but the enclosing aril can be removed only by subsequent softening in water. *Magnolia* and *Michelia* spp. are extracted from the compound fruits by dry extraction and the fleshy enclosure of the individual seed is then removed by washing. In many-seeded drupes like *Melia azedarach*, depulping is carried out according to the procedure for fleshy fruits, while an extraction procedure for dry fruit must be used if the seeds subsequently are extracted from the stone.

A summary of extraction procedures according to fruit types is shown in Table 3.2. A breakdown on major species groups is summarised in Appendix 1.

3.5.1 Seed Extraction from Dry Fruits

Dry extraction encompasses two procedures, viz. drying to low moisture content and threshing or disintegration. Drying alone is used for dehiscent fruits, i.e. fruits that naturally open on the tree at the time of dispersal. However, the desiccation rate at which fruits open depends much on species. Many capsules, pods and cones open at maturity at a moisture content of 20–25% and are fully open with a moisture content of 10–15% (Stubsgaard and Moestrup 1991).

Table 3.2. A summary of extraction methods for various fruit types. Compare with the species list in Appendix 1

Fruit type	Characteristics	Seed extraction method and extracted unit
Dry fruits		
Dehiscent pod	Carpel splits into 2 halves at maturity. In legumes the seed remains attached to the carpel by its funicle Examples: Legumes: *Acacia* and *Albizia* species, e.g. *Acacia senegal, Acacia reficiens, Acacia mellifera, Albizia lebbeck, Albizia procera* Non-legumes: Many Bignoniaceae, e.g. *Spathodea* and *Marchamia* spp. Species with very hard, woody fruits are *Afzelia, Baikaea plurijuga, Ormosia, Sindora, Brachystegia* and *Millettia* species 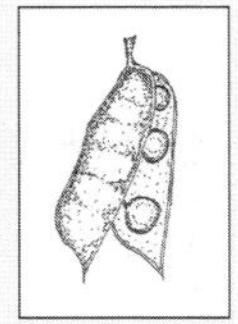 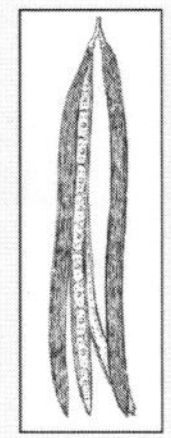	Pods will dehisce when dry and most of the seed will fall off easily by gentle mechanical impact such as shaking or tumbling. Species which split open only when very dry or at high temperatures (e.g. *Acacia mangium, Acacia auriculiformis* and *Afzelia* spp.) may be extracted by threshing or flailing. Woody pods need very high temperature and often mechanical impact ('blow')
Indehiscent pod	Pods remain closed at maturity. Inner layer of pod often consists of fleshy material or the seeds are embedded in a soft nutritious substance Examples: *Tamarindus indica, Acacia nilotica, Prosopis juliflora, Cassia fistula, Pithecellobium dulce, Samanea saman, Inga* spp. and *Dialium* spp. 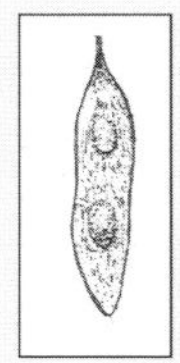 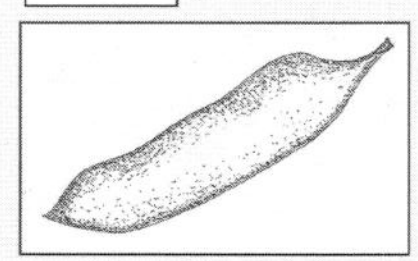	Extracted unit is the seed. The seeds are adapted to animal dispersal and are quite resistant to mechanical impact. Mechanical disintegration of the pods is necessary, e.g. using hammer mills or mortars. High-pressure water helps clean seed free of sticky substances. Ingestion by animals or manual sucking and spitting for small quantities

(*Continued*)

Table 3.2. A summary of extraction methods for various fruit types. Compare with the species list in Appendix 1–Cont'd.

Fruit type	Characteristics		Seed extraction method and extracted unit
Follicle	Consists of 1 carpel that splits open at 1 suture Examples: Single follicles occur, e.g., in *Grevillea* and *Sterculia* spp. Follicles are common in compound fruits, e.g. *Illicium verum, Magnolia* and *Michelia* spp.	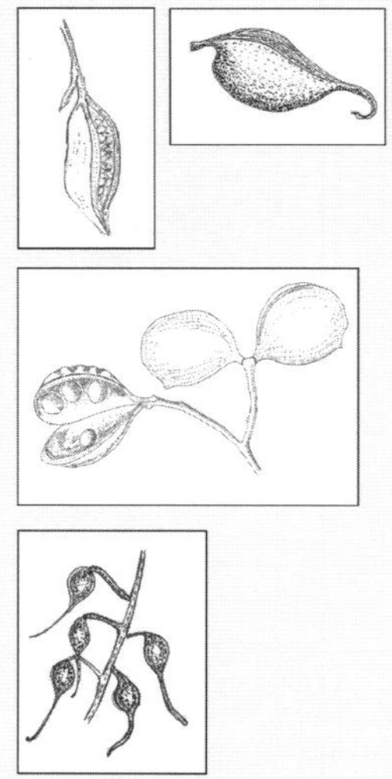	Seed extraction similar to that of pods, i.e. drying until natural dehiscence and mechanical tumbling or flailing to split the fruits open. Note that some seeds of, e.g. *Sterculia* spp. have a very thin and fragile seed coat that is easily damaged by mechanical treatment with consequent damage to the seed
Capsule	Consists of several carpels that split open at maturity or open via special structures Examples: Meliaceae (*Khaya, Swietenia, Cedrela*) and Myrtaceae (eucalypts), *Lagerstroemia*	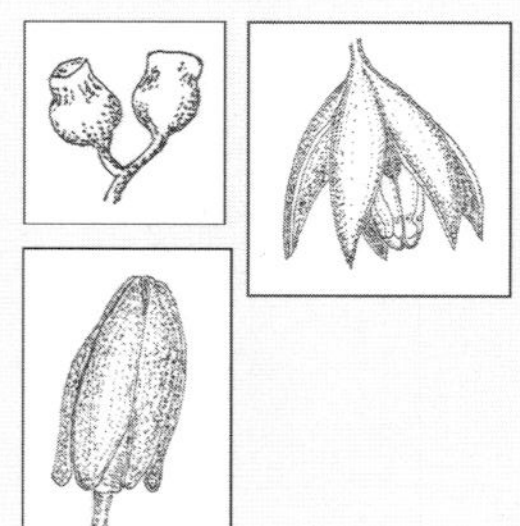	Most capsules open upon drying. Seeds may be retained by strong funicle attachment which requires some mechanical impact for breakage. In species with superior ovaries, e.g. several eucalypts and melaleucas, seeds are retained by an only partly split operculum and floss in the capsule. Tumbling and occasional threshing needed to split up fruits

Table 3.2. A summary of extraction methods for various fruit types. Compare with the species list in Appendix 1

Fruit type	Characteristics	Seed extraction method and extracted unit
Samaras and nuts	Fruits are indehiscent and contain usually only 1 seed. Samaras have wings, while nuts are 'naked' or enclosed in a dehiscent structure, e.g. cupula in Fagaceae Examples: Nuts: *Quercus*, *Lithocarpus*; samaras: *Acer*, *Terminalia*, *Pterocarpus*, *Tarrietia*, dipterocarps 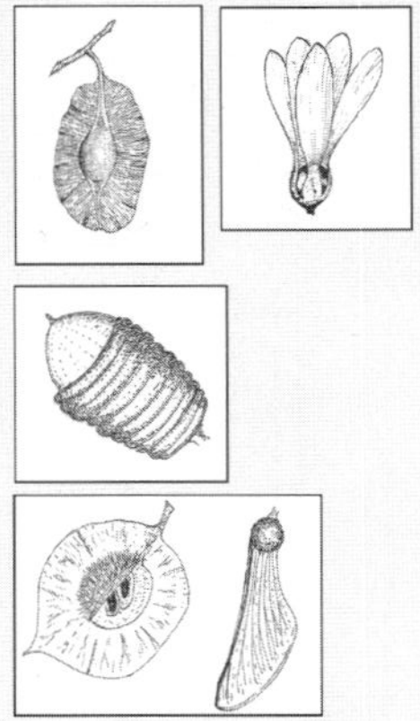	Seeds are not extracted from the pericarp, which technically is the seed. Nuts usually dehisce from the enclosing structure by drying, but in some species they remain firmly attached and are extracted manually. Samaras are dewinged manually or by some mechanical rubbing
Cones	Typical 'seed-bearing structure' in conifers consisting of cone scales. Cones are dehiscent structures that open in one of three ways: (1) splitting, open at maturity (2) cone scales fall off, (3) serotinous – opening requires high temperature Examples: The prevalent type in conifers including *Pinus*, *Cupressus*, *Thuja*, *Abies*, *Keteleeria* and *Foekinia* 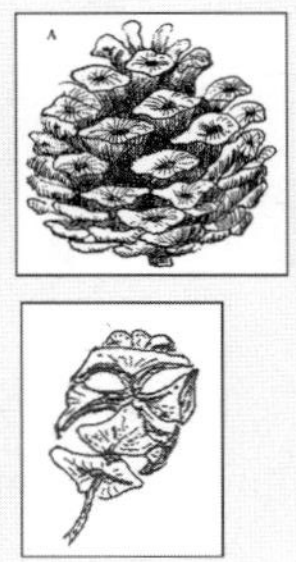	'Normal' cones dehisce or split apart by drying and seeds are released by gentle tumbling or raking. 'Serotinous' cones have their cone scale sealed with resin, which is melted by kiln-drying at high temperature

(*Continued*)

Table 3.2. A summary of extraction methods for various fruit types. Compare with the species list in Appendix 1–Cont'd.

Fruit type	Characteristics		Seed extraction method and extracted unit
Compound dry fruits	Multiple and aggregate fruits consisting of many small fruits together in one structure. Compound fruits open at maturity to release individual fruits, usually nuts or drupes Examples: *Casuarina*, *Banksia*, *Magnolia* and *Manglietia*	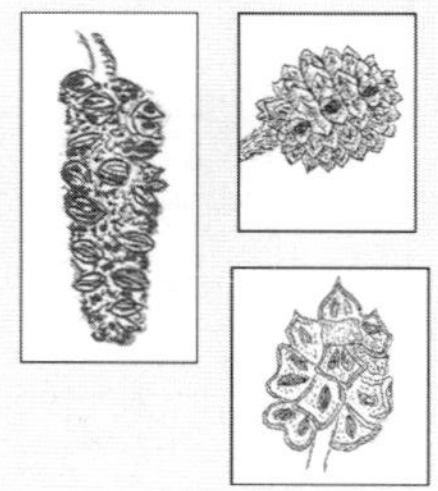	Fruitlets are extracted from conelike infructescenses by drying – in difficult species such as *Banksia* by scorching at high temperature. In many magnolias, seeds are surrounded by a fleshy aril which must be removed by washing, e.g. with high water pressure
Fleshy fruits			
Drupes	Consist of an outer fleshy part and an inner hard layer (stone or pyrene). The stone contains 1 or more seeds A few drupes are dry and are extracted as dry fruits, e.g. teak and coconuts Examples: *Canarium*, *Cinnamomum*, *Maesopsis*, *Prunus*, *Mangifera*	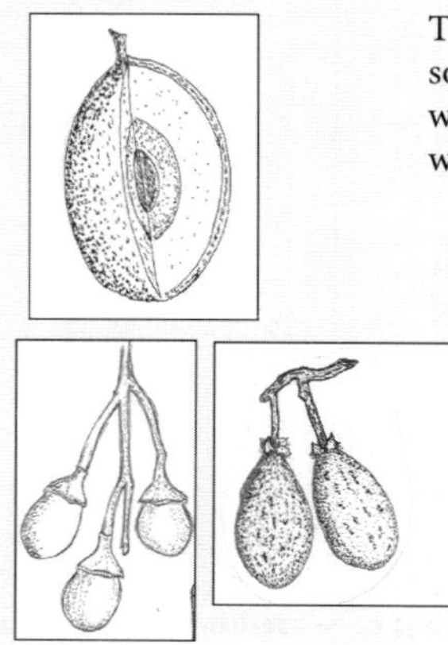	The stone is extracted by wet extraction, which implies softening of the fruit flesh, e.g. by fermentation and washing, e.g. by wet tumbling, stirring or high-pressure water, or in mechanical depulpers

Table 3.2. A summary of extraction methods for various fruit types. Compare with the species list in Appendix 1

Fruit type	Characteristics		Seed extraction method and extracted unit
Berries and arillate seed	Seeds are surrounded by or are embedded in fleshy fruit substance. Often there are several seeds. Seed coats are often thin and easily damaged Examples: Berries: *Rubus*, *Persea*, *Madhuca* and *Diospyros*. Some gymnosperms have arillate seeds, which are extracted (i.e. aril removed) in the same way as fruit flesh in berries, e.g. *Ginkgo*, *Taxus*, *Gnetum* and *Podocarpus*	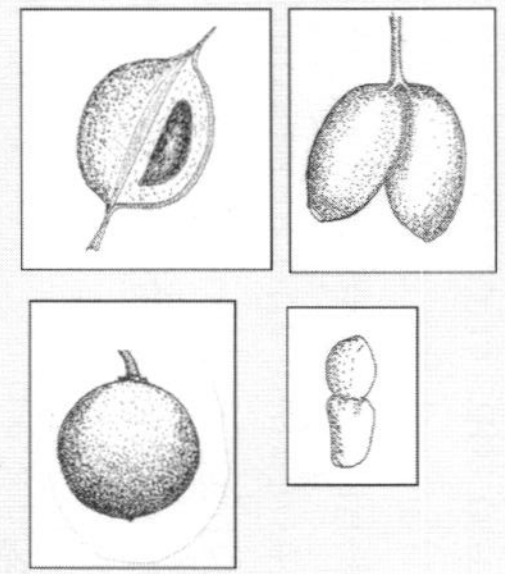	Wet extraction by softening of the fruit flesh, e.g. by fermentation and washing by wet tumbling, stirring or high-pressure water, or in mechanical depulpers. As seed coats are generally much more fragile than pyrenes, mechanical treatment can easily be overdone with consequent damage to the seed coat and hence seed viability
Compound fleshy fruits	Multiple and aggregate fruits embedded in a fleshy substance. Fruitlets are usually drupes Examples: *Arthocarpus* and *Annona*	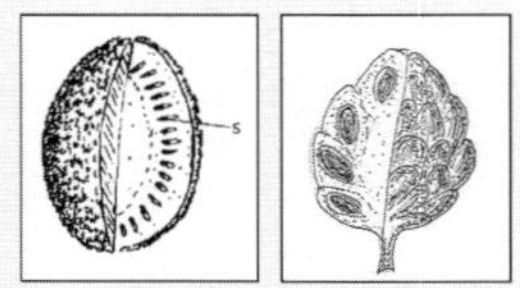	Seed or stone extraction is similar to that for drupes and berries

Others need very high temperature and in some species dehiscence occurs only after a brief exposure to fire or other high temperature that breaks a resin 'seal'. The moisture content of fruits is in equilibrium with air humidity such that when the air is dry, the fruit loses moisture and splits open; when the air is humid the fruit may regain moisture and close again. A moisture content of 15% is equivalent to a relative humidity of around 70%. On sunny days even in humid areas the air humidity is usually below 70%.

The moisture content in organic material is in equilibrium with air humidity: the drier the air, the drier the tissue (see later). Under ambient conditions, air humidity is dependent on the atmospheric conditions and temperature: under moist conditions, the humidity will reach 100%. High atmospheric humidity is experienced in mountainous areas in the so-called mist zone, the elevation where clouds usually form – the humidity is here close to 100%. Desert areas, in contrast, can experience extremely low humidity, approaching almost complete dryness (0% relative humidity) during daytime. Humidity is closely connected to temperature. So the easiest way to reduce humidity and the equivalent fruit moisture content in order to promote natural dehiscence is to raise the temperature. Sun-drying is 'free' and thus energywise the cheapest method, and it may in many cases be fully sufficient for extraction. Examples of some easily extractable seeds are those from many species of *Eucalyptus, Acacia, Albizia, Callitris, Cedrela, Swietenia, Casuarina*, and most conifers.

Where humidity is high and desiccation dehiscence consequently difficult, ambient drying can be made more efficient by some simple adapted methods:

1. The drying site is away from moist places, e.g. a nursery or forest. In some cases it would be appropriate to move the extraction location physically, e.g. if the seed processing station is located at a humid highland site.
2. Drying takes place on a dry platform, e.g. concrete or an elevated platform.
3. Drying platforms are elevated wire meshes which are naturally ventilated.
4. Temperature is increased by the 'greenhouse effect', e.g. drying under a transparent plastic cover (Fig. 3.3a).
5. Fruits are placed in a thin layer and regularly turned to promote even drying of all fruit parts (Fig. 3.3b).
6. Artificial ventilation, e.g. by a fan, is applied in very calm wind conditions.
7. Fruits are covered under conditions of increased humidity, for example, at night or in moist weather. The covering prevents the fruits regaining moisture.

Drying under very hot and dry conditions implies a risk of overheating or too rapid drying, especially for fruits and seeds with relatively high moisture content. Too rapid drying can cause moisture to be trapped inside fruits or seeds,

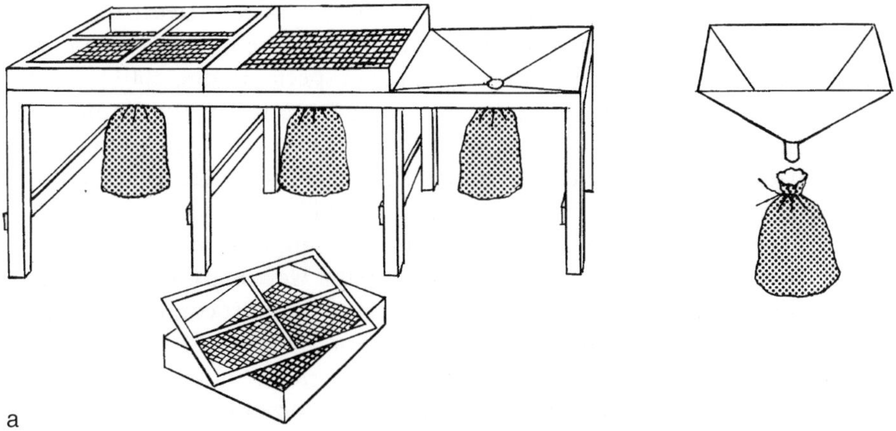

a

b

Fig. 3.3. **a** A low-cost extraction trays for extraction by sun-drying. The trays are covered with a 'non-closing' frame of transparent plastic sheet, which increases the temperature ('greenhouse effect'), yet allows moisture to escape. **b** Dry extraction of *Cedrela odorata* capsules on wire-mesh tray. (**a** Redrawn from Granhof 1984)

so-called case hardening. For fruits it interferes with the dehiscence mechanism, for seed it can cause reduced viability or storability. Overheating damage occurs when immature and moist seeds, which are metabolically active, are exposed to temperatures that can disrupt the physiological mechanism. Once seeds are dry and the metabolism is low or quiescent, temperature sensitivity is small. The simple way to prevent both temperature damage and case hardening is to control the drying process. Relatively moist material is predried for 1–2 days under shade and then extracted by sun drying. The Australian Tree Seed Centre recommends an initial drying for 2 days at 25°C, then increasing to 35°C for immature and moist material of most Australian species (Gunn 2001).

Hot-air devices, kilns, increase drying efficiency. Increasing the temperature of air reduces the relative humidity and thus increases drying capacity. Kilns may be used as a standard drying method or where seeds are retained in the fruits after normal drying (Bowen and Eusebio 1982). Simple kilns consist of stalked drying trays through which a current of dry, heated air is led. The heated current of air is created by an electric heater with a fan. Such devices are becoming almost universally available in regions with short cold seasons and in highlands. The air-current temperature of the hot-air blower should be around 60°C (Fig. 3.4a).

More efficient are rotating-drum kilns used, for example, in temperate regions for extracting pine seeds. The temperature in rotating-drum kilns is regulated up to about 80°C and kiln drying is applied for 1–6 h (Fig. 3.4b).

Higher-temperature drying kilns are used for extraction of so-called serotinous fruits. Serotinous fruits need temperatures above what is normally experienced even under very dry and hot conditions. Species with serotinous fruits

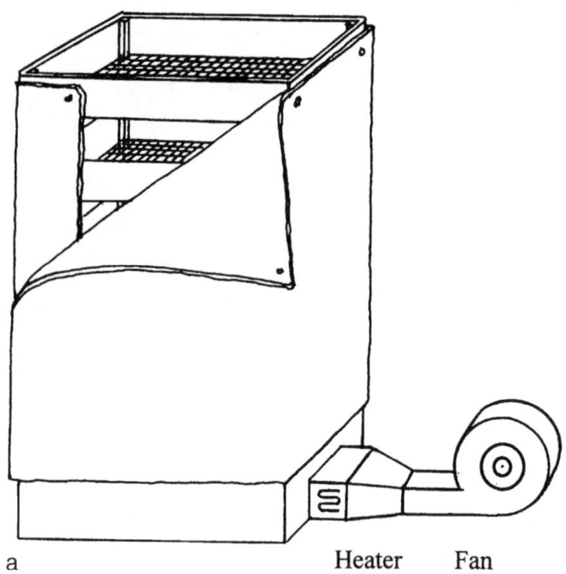

Fig. 3.4. Kiln types used for seed extraction from dry fruits. Most kiln types use electrical heating devices to achieve the high temperatures necessary for drying. A low-cost kiln consists of stacked trays provided with an electrical hot-air blower **a**. The lower trays dry first because the incoming hot air has a high water absorption capacity. As the air absorbs and cools its absorption capacity reduces as it passes the trays upwards.

b

Fig. 3.4. (Continued) **b** In the rotating kiln type the hot air is evenly distributed. The mechanical impact during rotation helps release the seeds if they are retained by any mechanical barriers (fruit/cone apertures and/or funicles). In some kiln types seeds fall through a grid once extracted – this is in order to avoid unnecessary heating of the seed after extraction. (**a** M. Robbins, **b** courtesy of Dorthe Jøker, Danish Tree Improvement Station)

are, by means of opening mechanisms, normal dehiscent fruits, but they require very high temperature. In nature, extreme temperatures are experienced during grass and bush fires. Ultrahigh temperatures are known to promote germination and regeneration of several of these species. Some serotinous fruits are resinous, and the high temperature is necessary to melt the resin. This is the case mainly for the pines like *Pinus taeda*, *Pinus brutia*, *Pinus halepensis* and *Pinus contorta*. Others are hard and woody, e.g. some *Eucalyptus*, *Casuarina*, *Brachystegia*, *Millettia*, *Baikaea*, *Banksia* and *Hakea* species. Species with serotinous fruits are mostly from very dry areas with regular fire.

There are various degrees of serotiny and thus extraction conditions. Many serotinous cones and fruits open after heating to at least 70–80°C for several hours. A shorter time with higher temperatures may be just as efficient. Seeds of *Pinus contorta* var. *latifolia* were briefly exposure to a maximum of 1.5 min of superhigh temperature of 220°C provided by a gas flame. Longer exposure seriously hampered viability (Wang et al. 1992). Fruits of *Banksia* and other extremely hard fruits are opened by placing them on a wire mess over hot coal until they split open. When the fruits have opened, they are immediately immersed in water and then sun-dried (Kabay and Lewis 1987; Gray 1990).

Less serotinous fruits, or just difficult dehiscent ones, can sometimes be opened by one or more cycles of drying and wetting at normal or kiln temperatures. In some species a short dip in boiling water (a few seconds to 10 min) eases normal dehiscence upon drying.

Capsules of, for example, *Swietenia* and *Entandophragma*, cones of, for example, *Abies*, *Araucaria* and *Agathis* and folicles of, for example, *Markhamia*, *Fenandoa* and most other Bignoniaceae disintegrate completely and the carpels fall off during normal dehiscence. Capsule types in species with an inferior ovary (e.g. eucalyptus) have particular aperture mechanisms, which open upon drying. In compound fruits of, for example, casuarinas, seeds fall through an opening or slot and in pines and similar cone types mature seeds fall from the opened cone scales. In these fruit types seeds may be retained within the fruits even after dehiscence, if the opening is insufficient. The retention is accentuated, for example, if fruits are infected by insects. An insect web may physically trap seeds released or obstruct the normal opening mechanisms. Chaff and flower residues may physically restrict the aperture. This phenomenon is frequent in eucalypts. Species with half-superior ovaries like *Eucalyptus camaldulensis* release their seeds more easily than those with inferior ovaries, e.g. *Eucalyptus delegatensis* (Boland et al. 1990).

Seeds of some species maintain a strong attachment to the fruits via the funicle after dehiscence. That applies especially for dehiscent legumes like *Acacia*, *Albizia*, *Acrocarpus* and *Paraserianthes*, which are naturally dispersed attached to half of the dehiscent pod. In mahoganies and related species, the seeds remain attached to the central columella by the funicle sometimes after the carpels have fallen off.

Seed retention within dehiscent fruits or via funicle attachment is overcome by mechanical treatment.

3.5.1.1
Mechanical Extraction of Dry Seed

Mechanical extraction is used for species where drying is not sufficient to release seed; however, thorough drying is usually a precondition for efficient mechanical extraction. Mechanical treatment is necessary for species where seeds remain attached to or enclosed within the fruits after drying. It applies in particular for species with indehiscent dry fruits, i.e. where fruits do not have a natural opening mechanism and are dispersed either as entities, usually by wind, or by ingestion. Examples of wind-dispersed indehiscent legumes are *Dalbergia*, *Pelthophorum*, *Ormosia* and *Pterocarpus*. *Sindora* is a transition group which usually remains closed until after dispersal. Most of these species are functionally samaras, i.e. winged nuts. Samaras where seeds are not extracted but fruits are often dewinged occur in, for example, *Acer*, *Terminalia*, *Combretum*, *Triplochiton* and *Tarrietia*. There is usually an inverse correlation between seed coat hardness and pericarp hardness. Where the pericarp is very hard, e.g. in *Pterocarpus*, the seed coat is thin and fragile. Where the pericarp is thin and papery, for example in *Dalbergia* spp., the seed coat is thick and hard. Ingestively dispersed seeds, e.g. those of *Acacia tortilis*, *Acacia nilotica* and *Cassia* species, always have a very thick seed coat.

Mechanical treatment is used for partial extraction, e.g. dewinging of samaras, removal of involucre and felty exocarps and mesocarps in, for example, teak, and removal of cupula in fruits of *Quercus*, *Lithocarpus* and *Fagus*.

Mechanical extraction or treatment is most efficient for dry material. Initial drying, e.g. 40–45°C for 24 h, or heat treatment promotes brittleness of the pods, which will facilitate subsequent extraction (Gunn 2001).

Flailing or beating is an old and simple method of seed extraction (Fig. 3.5). It is effective, for example, to detach funicle attachments to fruits after

Fig. 3.5. Beating bags of fruits for seed extraction. (M. Robbins)

dehiscence and to split up dry, fragile fruits. Fruits are filled in bags and beaten or flailed with sticks or clubs. The impact is easy to regulate by time and strength.

Fruit types needing harder treatment can be pounded in mortars. This is, however, only applicable if seeds are very resistant to mechanical damage. Large hard fruits, e.g. hard pods of *Afzelia* and *Brachystegia*, may be opened by holding and beating individual pods with a wooden club.

Mechanical threshing works on the principle of a revolving beater mounted on a horizontal cylinder disintegrating fruits. Fruits are fed from one side and torn apart by the beater. Seed and small fruit parts may pass perforations in the bridge (the concave plate beneath the beater) or the sieve behind, while larger material is removed (Fig. 3.6). The seeds are thus precleaned together with the mechanical extraction. Mechanical threshers designed for agricultural crops like rice or grain are designed to release grains from the straws, which is a quite different task from splitting fruits. For most forest seed, the flailing thresher is most efficient because it splits up the material completely. The type is used, for example, for several Australian *Acacia* species (Doran et al. 1983).

Mills are mostly designed to grind material, but can with proper adjustment be used for extraction. When used for extraction, the grinding stones or steel must be adjusted to a distance slightly larger than the seed. Hammer mills work in principle like a flailing thresher, where the beater is replaced by a revolving steel cross pulverising the material against the drum (Fig. 3.7). They can be used for threshing hard fruits if the revolutions of the hammer are reduced to 250–800 per minute and the outlet screen is replaced by holes that will let the seeds out (Stubsgaard and Moestrup 1991). Although most hard seed is quite resistant, some damage is difficult to avoid. A way to minimise mechanical damage is to remove seeds that have already been extracted. Applying two to three rounds of threshing with intermediate removal of extracted seeds rather than one complete threshing at high speed can also reduce damage.

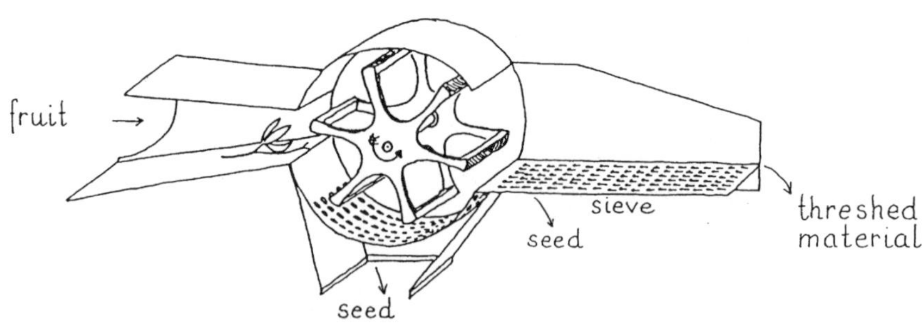

Fig. 3.6. Working principle of the flailing thresher. The beater revolves at high speed and splits up material. (From Stubsgaard and Moestrup 1991)

Fig. 3.7. Working principle of a hammer mill

Seed tumblers are versatile pieces of equipment in seed extraction and cleaning (Fig. 3.8). They are often used as a combination of drying and mechanical extraction. When fruits are dried in slow rotating tumblers, e.g. drum kilns, where seeds drop out continuously, mechanical damage to the seed is minimised. Mechanical impact in tumblers is the effect of fruits falling down when the cylinder revolves. This type is used, for example, to extract seeds from dehiscent cones or capsules.

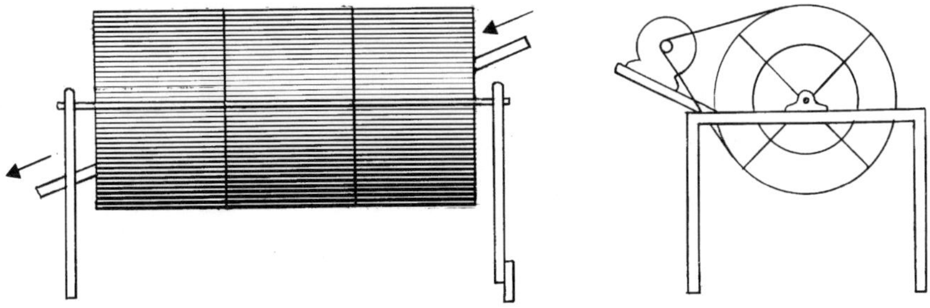

Fig. 3.8. Low-cost seed tumbler consists of a grid made of wire mesh, which can be rotated by a handle. (M. Robbins)

Greater mechanical impact is imposed by mixing seeds with some heavy abrasive material such as wooden blocks or gravel in a cement mixer (Fig. 3.9).

Brushing machines consisting of a fixed cylinder and a revolving brush are specially designed drum types (Fig. 3.18). The brushes split up fruits by grinding between the brushes and the cylinder. The impact can be regulated by using different brush types. Both cement mixers with abrasive material and brushing machines are used also to remove small adhesive material like arils, wings or hairs.

Tumblers are often combined with a precleaning function, e.g. open mesh cylinders which retain most of the fruits while seeds and small material fall out.

Fruits consisting of dry material only are extracted by drying and threshing. Where seeds are embedded in a pulpy material, several subsequent treatments including wet extraction are often necessary.

Fig. 3.9. Cement mixers are universal types of equipment with multiple applications in seed processing. Tumbling together with wooden blocks gives a mechanical impact as a 'mild threshing'; tumbling with sand or other abrasive material gives a 'grinding' effect, e.g. for polishing, removal of residual pulp or dewinging. Tumbling with water and some abrasive material is used for depulping

3.5.1.2
Abrasion

Dry appendices, involucre and felty pulps are often incompletely removed by threshing. Abrasion is a process in which fruits are ruptured by a grinding material. In India, a depulping machine with a drum tightly wrapped with barbed wire was found suitable for teak (Bapat and Phulari 1995; Fig. 3.10). Tumbling with course-grained sand in a concrete mixer abrades the surface of, e.g. teak stones. Brushing machines are versatile types of equipment to remove dry fruit parts and appendices (Fig. 3.18).

3.5.1.3
Removal of Sticky Substance

Several types of multiseeded, indehiscent pods designed for animal dispersal consist of an outer dry brittle cover and an inner sweet, sticky layer. Examples are species of *Tamarindus*, *Pithecelopium*, *Prosopis*, *Samanea*, *Inga*, *Cassia* and *Dialium*. In some species the pods are more uniformly leathery with no distinct layers, e.g. *Acacia nilotica* and *Parkia biglobosa* (Fig. 3.11, Table 3.2). Seeds must be extracted both because of practical handling (they contain many seeds) and

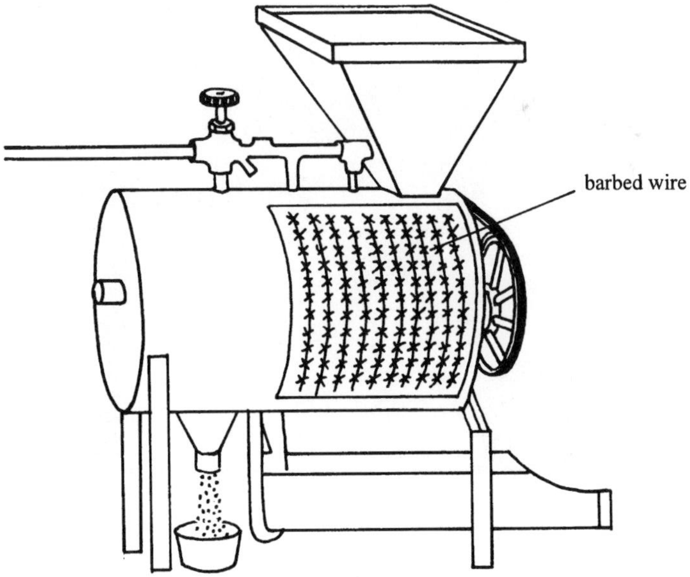

Fig. 3.10. Abrading machine used to remove involucre and dry pulp of teak seed. The rotating drum is lined with barbed wire. (Redrawn from Bapat and Phulari 1995)

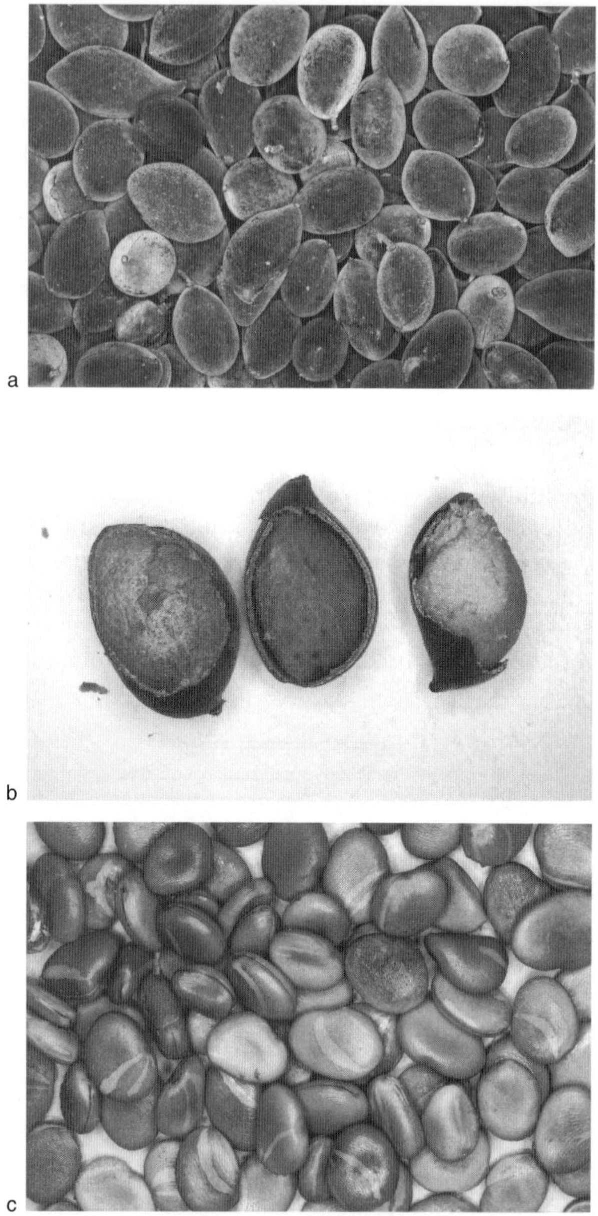

Fig. 3.11. In fruits of *Dialium cochinchinensis*, the seeds are embedded in a sweet and sticky substance, which is difficult to remove mechanically. The pulp tends to stick to both seeds and mechanical equipment. Methods used for extraction of fleshy fruits, e.g. high-pressure water, are most effective after some type of softening or dissolving the substance. Softening can be a mild acid or bleach, or the fruits can be left to partly decompose under humid conditions. **a** The entire fruit, **b** some fruits with the dry pods removed and **c** the clean seed

because the sweet pulp easily decomposes and harbours infections. Pulp may also contain germination inhibitors. Since species with indehiscent pods are usually dispersed by animals, the seeds have developed an extremely hard coating to be able to withstand teeth and passage through the digestive tract. This eases extraction options because the seeds are quite resistant to mechanical impact and chemicals.

For *Prosopis cineraria*, Bonner et al. (1994) suggested several rounds of threshing with intermediate drying, or the pods should be run through a coarse meat grinder to extract the seed. However, sticky material tends to paste to mechanical parts, which can make them both ineffective and hard to clean. A type of wet extraction is thus applicable. The following steps were suggested by Bonner et al. (1994):

1. Breaking the pods
2. Soaking in 0.1 N hypochloric acid[3] for 24 h
3. Washing and drying
4. Pounding the dry material with a hammer or in a mortar

Biological extraction can be a realistic alternative. Pulp of *Inga*, *Tamarindus* and *Dialium* is edible and sweet. It may require a bit of organisation to collect spat-out seed, but for small quantities it is probably a reasonable alternative. Goats will eat most pods and many seeds will be left undamaged in the droppings. However, the animals will crush many seeds between their molars when ruminating, and collecting and extracting seeds from the dropping is an additional operation. Seeds from large fruits like those of *Parkia biglobosa* may be extracted by individually splitting each fruit by hand (Some et al. 1990).

3.5.2 Seed Extraction from Fleshy Fruits (Depulping)

Seed extraction from fleshy fruits means removal of fruit pulp. Fleshy fruits types are, for example, drupes, berries and several compound fruits. Some dry fruits have fleshy parts or appendices, which are sometimes removed dry, sometimes wet (Box 3.3).

Fruits with soft and water-soluble pulp are usually easy to extract in water. In many species the pulp separates readily from the seeds – in other types and if the fruits are not fully mature, the pulp needs to be softened

[3] Sodium hypochlorite is sold as bleach under various trade names.

Box 3.3

Drying without extraction?

Fleshy fruits in many dry-environment species, e.g. *Ziziphus*, *Balanites*, *Diospyros* and *Grevia*, contain little water The species can often be stored as dry fruits without depulping. If the temperature can be kept low and possible pests controlled, there is little gain in extraction in terms of storage. Arils of, e.g. *Sindora* and *Afzelia* are relatively dry and seeds can easily be stored without removing the aril. However, leaving pulp or arils has a drawback as these structures usually contain germination inhibitors, which prevent seeds from germinating before dispersal (Fig. 3.12). The inhibitors impose a type of dormancy which prevents germination as long as they are present and active. Seeds sown unextracted from the fruits will thus usually not germinate or germination may be delayed until water has diluted or precipitated the inhibitors. Extraction to precipitate inhibitors prior to sowing can be done as dormancy pretreatment. However, it is known that seeds stored with fruit pulp often develop strong dormancy because of inhibitors, presumably because the inhibitors tend to move from the fruit flesh into the embryo or inner layer of the seed.

Fig. 3.12. *Dacrycarpus imbricatus* with fleshy sarcotesta. Seeds can be dried and stored without removing the arils, but inhibitors in the arils will prevent or delay germination. The pulp is easiest to remove when the seeds are fresh before drying

before extraction. The fruit 'skin' (exocarp) physically protects the soft parts below from desiccation during maturation. This layer must be broken or removed to soften the pulp below. Depulping is usually a combination of two extraction processes: mechanical depulping squeezes or grinds fruit flesh; the remaining flesh can then be removed by, for example, high-water pressure.

Species with relatively firm pulp, e.g. *Azadirachta indica, Aleurites* spp. and *Santalum* spp., need softening to ease seed extraction. Softening is in practice carried out as a soaking process, where fruits are submerged in water permitting some fermentation or rotting of the flesh.

Fleshy fruits are dispersed by animals and removal of the pulp happens in nature by animal ingestion. Fruit flesh of fruits that fail to be dispersed will gradually decompose and be washed away by rain.

Although soft pulp is usually readily released from seed or stones, a short softening treatment will ease extraction of most species. Natural softening occurs during decomposition of the pulp and can be accelerated by soaking in water for one to several days. Fruits of species with relatively dry pulp, e.g. *Ziziphus mucronata* and some *Diospyros* species, need such treatment. Soaking for several days has been recommended for *Santalum* spp., for example, the pulp of which is generally difficult to remove (Gray 1990). Mechanical rupture of fruit skin prior to soaking often speeds up softening and decomposition (Albrecht 1993). If the fruit pulp has only been partly removed during field handling and then dried, it is necessary to rewet the fruit to remove the remaining pulp at the processing depot.

Decomposition may be aerobic or anaerobic. Anaerobic decomposition will take place in the absence of sufficient oxygen, e.g. in a heap of fruits or in a container of soaking fruits in water. Anaerobic decomposition is fermentation, which produces alcohol, which is toxic to seeds. For example, seeds extracted from fermented *Gmelina arborea* drupes showed a significant reduced germinability compared with seeds extracted mechanically from fresh fruits, an observation suspected to be due to alcohol accumulation (Liang and Yong 1985). In *Syzygium cuminii* optimal results of several quality parameters were obtained after 1 day of fermentation followed by thorough washing – a longer duration of fermentation had a negative effect (Srimathi et al. 2003). Controlled fermentation appears generally to be an effective method to soften fleshy pulp before depulping, but there is a high risk of overtreatment by too long fermentation.

Aerobic decomposition does not produce toxic compounds but is often a slower process. To prevent fermentation, sufficient aeration must be provided during soaking, e.g. (1) via an air pumping system, (2) by regularly changing the water, for example once every 12 h or (3) by providing the container with a continuous water flow. Aerobic soaking for a long time implies a risk that seeds will germinate after inhibitors have been removed from the pulp.

Once the pulp has turned soft, the pulp can be separated from the seeds. Two constraints must be considered when choosing the depulping method:

1. In species with fibrous fruit pulp, e.g. mango and peaches, the pulp or some pulp residues tend to stick firmly to the seed or stone.
2. Species with very fragile seed coats are easily damaged especially by mechanical depulping. This is particularly the case for berries, while drupes usually have hard and resistant stones that are tolerant to mechanical treatment.

The following depulping methods are applicable and cover most applied devices:

1. *Individual manual extraction.* Gentle, manual extraction is preferred for species with very fragile seed coats, e.g. *Syzygium cumini.* Dry-sticky pods of the aforementioned tamarind–dialium type are often also the easiest to remove manually (Fig. 3.13).
2. *Washing in deep bowls or drums.* Fruit pulp that loosens easily tends to come off with little/low impact. Regular stirring or using a strong water stream while the fruits are submerged in water for softening often suffices to remove a large part of the pulp. The impact can be increased by more vigorous stirring or increasing the water pressure. Kitchen utensils such as electric mixers and blenders (low speed) are applicable for small seeds and seed lots.
   Concrete mixers are also used for wet extraction. Fruits are mixed with an abrasive material like gravel plus excess water and are rotated in the drum for various lengths of time, typically from 5 to 20 min. In Kenya, blocks of wood were used as abrasive material to remove pulp of species with sticky pulp, e.g. *Vitex keniensis, Maesopsis eminii* and *Cordia* spp. (Ahenda 1991). In some Euphorbiaceae, e.g. *Antiaris toxicaria* and *Bishofia javanica*, fruit flesh is very sticky and removal is facilitated by adding some detergent or alkali to the water during tumbling, e.g. 1 N hypochlorite.
   Pulp plus skin and seeds are separated by flotation: seeds or stones remain at the bottom, while pulp and skin tend to ascend to the surface where they can be skimmed off (Fig. 3.14). In some species, very clean seeds can be achieved by washing, in others washing may be used as a preextraction procedure.
3. *Washing on wire-mesh screens.* Screens with a mesh size that will retain the seeds while the pulp passes through are used. The pulp is released

Fig. 3.13. Fleshy pulp removed manually by washing

by manually rubbing the fruit against the grid while washing (Fig. 3.15). Fruit skin and firm fibres retained with the seeds or stones after rubbing and washing may be separated by flotation in excess water as described above. If the fruits are rubbed manually on the screen, the mechanical impact can easily be adjusted when extracting fragile seeds. To separate pulp from tiny seeds, fine mesh screens must be used, e.g. for *Anthocephalus cadamba* 1/16-in. mesh (Seeber and Agpaoa 1976).

High-pressure water enhances the effect of washing and can sometimes be used as an alternative to mechanical depulping. Water is generally gentle to seed surfaces and the risk of damage or overtreatment is small.

However, the fruits must be physically fixed, e.g. in a wire-mesh bag with mesh sides smaller than the seed size. The high pressure will otherwise cause fruits and seeds to blow far away.

High water pressure machines are becoming readily available in most countries. There are generally two types, viz. ordinary air compressors

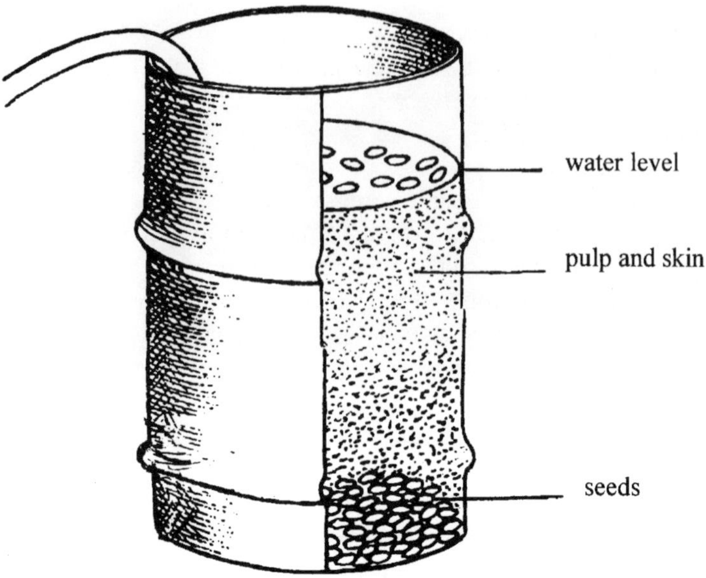

Fig. 3.14. Drum for softening of pulp and separation by floatation. The fruits are left soaking in the water for 1–2 days and are occasionally stirred to loosen the pulp. The floating pulp and skin are skimmed off. (P. Andersen)

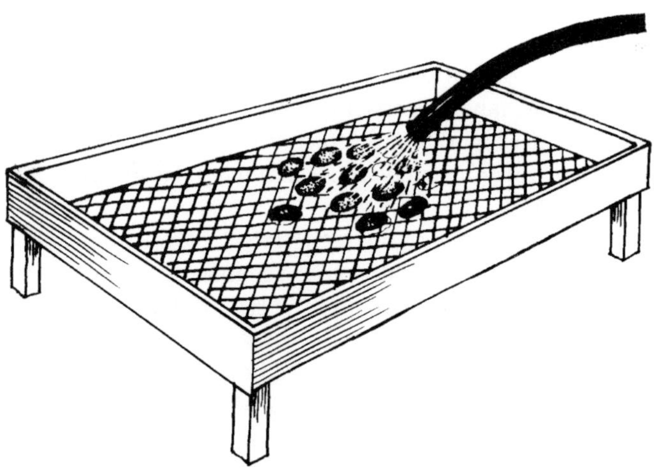

Fig. 3.15. A current of high-pressure water is an effective and mild extraction form for many fleshy fruits. Water pressure is increased by using compressed air, e.g. from ordinary air compressors. (P. Andersen)

with connection to a water hose and high-pressure water devices used for high-pressure washing. High-pressure cleaning is very efficient and is used, for example, for cleaning seeds of *Prunus* and *Vitis* spp. (Bonner et al. 1994).

4. *Mechanical depulping devices.* The coffee depulper is a versatile low-cost machine readily available in all coffee-growing regions (Fig. 3.16a). The device is used either manually or with engine power and can be adjusted or modified to fit a wide range of species (Bowen and Eusebio 1982; Liang and Yong 1985). During operation the fruit pulp is ruptured and squeezed against or between its mechanical parts.

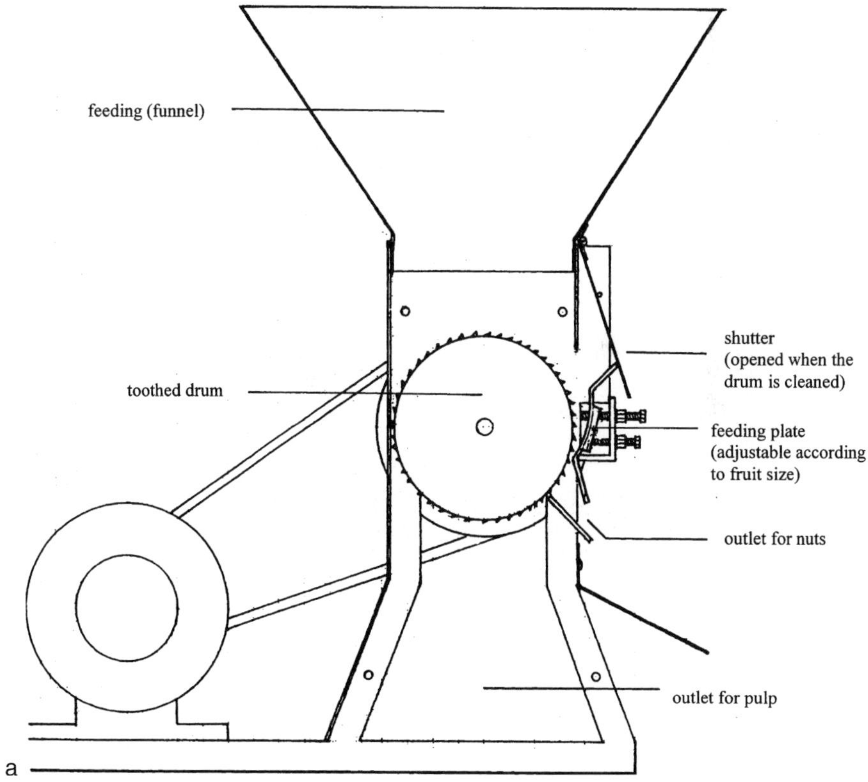

Fig. 3.16. **a** Coffee depulpers are industrial low-technology depulpers with high applicability.

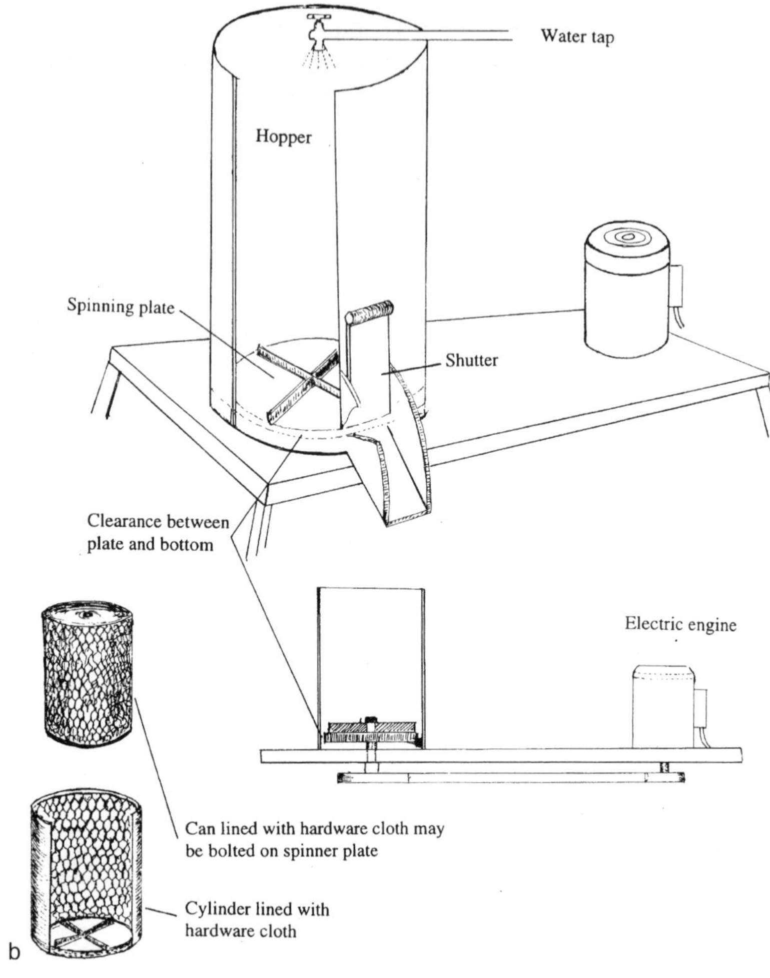

Fig. 3.16. (Continued) **b** Dybvig depulper. Both devices use a dual function of squeezing/rupturing and washing. (Redrawn from Amata-archachai and Wasuwanich 1986)

Because of the risk of mechanical damage, the depulper is particularly applicable to fruits with relatively hard seed-coats or stones, e.g. most drupes.

Another widely used mechanical depulper is the Dybvig macerator (Amata-archachai and Wasuwanich 1986). During operation the fruit pulp is abraded on a flat spinning plate provided with four bars arranged in a 90° cross at the bottom of a cylinder (Fig. 3.16b). Fruit pulp is washed away by a continuous water stream. To increase the rupturing ability, the inside of the cylinder may be lined with a wire net and the spinning plate provided with a bolted can, also lined with wire net. In this way the fruits are squeezed and ruptured between the two rough surfaces of the cylinder and the bolted-on can.

Where seed coats or stone surfaces are smooth, practically all pulp can be removed by washing. Where seeds or stones have a rough surface and fruit flesh is fibrous, some pulp may remain after washing. Residual pulp can have two drawbacks:

1. It sometimes harbours bacteria and fungi, which can cause serious seed damage; this is particularly a problem under ambient storage under humid conditions.
2. Chemical inhibitors may delay germination; the effect depends both on species and the amount of pulp left (Dransfield 2001).

It is thus generally recommended to clean seeds as much as possible before storage. Residual pulp can be removed by abrasion, polishing or brushing. Dry or moist tumbling with sand or other abrasive material in a cement mixer is effective, but may damage sensitive seeds. Seeds of *Gmelina arborea* have been successfully cleaned for residual pulp by polishing them in a coffee dehusker (Liang and Yong 1985; Bowen and Eusebio 1982).

3.5.3
Biological Extraction

A concomitant aspect of animals' ingestive dispersal is that seeds get removed from fruits and fleshy appendices. Animals may regurgitate seeds after digesting the fruits or the seeds may pass through the entire digestive track and are left with the droppings (Fig. 3.17). In some cases part of the seeds are digested in this process. Seeds of some animal-dispersed fruits are sometimes very difficult to extract by mechanical means, e.g. those with sticky fruits, indehiscent pods with seeds imbedded in a sticky pulp (tamarind type) and seeds with firmly affixed aril (e.g. *Afzelia*). Some examples where biological extraction and cleaning have been used in practice, or where deposits from dispersers have been used as 'seed sources' are:

Fig. 3.17. Biological extraction. Goats readily eat pods of most acacia species. Although some seeds are crushed between the molars, a large part of the seeds pass through the digestive tract and are deposited clean and 'pretreated'

1. Accumulated manure in goat enclosures in areas with prolifically fruiting acacias, e.g. *Acacia nilotica* and *Acacia tortilis*, often contains large amounts of seeds from these species.
2. Large amounts of clean seeds (stones) of *Maesopsis eminiii* regurgitated by hornbills are often found under the birds' favourite resting trees in East African forests. The stones are large and conspicuous.
3. Ants and termites are not so much dispersers as they are too small to remove seeds far from the mother tree. However, they can be efficient in cleaning seed from dry or moist appendices. For example, in East Africa, termites attack fruits of *Kigelia* (sausage tree), the seeds of which are very difficult to extract except manually. In Brazil, ants have been reported to efficiently remove pulp from pods of *Hymenaea courbaril* (Oliveira et al. 1995), and in the Philippines it has been observed that pods of *Samanea saman* piled up in a dark place will readily be attacked by termites that will consume the fruit parts only and leave the seeds behind (Seeber and Agpaoa 1976). In Vietnam, ants have been observed actively removing the arils of *Acacia* seed.

4. In temperate regions, rodents and some birds often collect and store seed for survival during food shortage periods, and these so-called caches have been used for collection in, for example, North America (Dobbs et al. 1976; Stein et al 1974). Seed storage also exists in tropical regions, e.g. in highland areas.

In addition to extracting seed, animal deposits have some advantages:

1. Seeds are 'collected for free'. This can be a great advantage in large trees with widely dispersed fruits, where direct collection is difficult.
2. Most animals feed on mature fruits only, so deposits are likely to contain only mature seeds.
3. Seeds are usually free from insect attack (Coe and Coe 1987; Lamprey et al. 1974).
4. Possible dormancy is often broken. In fleshy fruits, the pulp with possible inhibitors is removed (Oliveira et al. 1995). Hard seed are 'pretreated' by digestive juices and the seeds often have a significantly better germination than seeds that have not been ingested (Coe and Coe 1987; Lamprey et al. 1974).

A drawback of using seeds from natural deposits is that the identity of the seed is unknown. The 'provenance' is most likely close to the deposit site as animals do not move very far with their stomachs full. But the seed trees could be any of the trees in the area and deposits sometimes contain seeds of several species, which may be difficult to both distinguish and separate, e.g. acacias.

Managed biological extraction implies that both intake and deposit are controlled, i.e. animals must be 'domesticated' – they are fed with the fruits and droppings with seed must be deposited where they can be collected. This type of management does include some problems:

1. Animals damage part of the seed. Ruminants like goats and camels often digest part of the ingested seed. The fraction varies with species but even in species adapted to ingestive dispersal, a high fraction of seed may be digested. Ants usually consume soft parts only, while termites often attack seed as well.
2. Seeds deposited in animal droppings must be extracted from the manure, which implies an additional workload. Seeds can be extracted from manure by wet or dry extraction. During dry extraction, the manure is initially dried and fractioned, e.g. by gentle pounding in a mortar or the like,

then cleaned by tumbling and sifting. During wet extraction, the manure is soaked and washed in water. The seeds that gather at the bottom of the container are then separated by sifting under running water. Wet extraction gives the cleanest seed, but if scarified by the ingestion, the seeds may readily imbibe, which may make them sensitive to further treatment.

3. Keeping a stock of extracting animals implies some inconvenience as their service is only needed seasonally and they must be 'fed' also out of season. Neither ants nor termites are popular close to wooden premises and nurseries.

3.6 Dewinging

Dewinging, in a broad sense, is removal of any dry seed appendages, including wings, spines, hairs and some aril types. These structures rarely hamper germination but can have a negative effect on storage, e.g. in cases where they tend to collect moisture, which can attract fungi, and thus indirectly influence storability. In any case, both wings and other appendices are redundant structures and are inconvenient in handling. The main purpose of dewinging is thus to reduce bulk, ease handling during storage, pretreatment and sowing, and in some cases to prevent fungus attack.

Wings and hairs occur in wind-dispersed species, arils in animal-dispersed ones. Some winged seeds are illustrated in Fig. 2.1. Wings can be strong and woody as in samara fruits of *Terminalia, Pterocarpus, Triplochiton, Heretiera, Kokoona, Casuarina* and dipterocarps, and seeds of many Meliaceae (e.g. *Cedrela, Chukrasia, Khaya, Swietenia*). Thin membranous wings are found in Bignoniaceae (e.g. *Markhamia, Tabebuia, Tecoma* and *Spathodea*), and prevail in conifers. Hairs (floss) occur mainly in Bombacaceae (*Bombax* and *Ceiba* (Kapok)) and Salicaceae (*Salix* and *Populus* spp.). In some *Pterocarpus* spp. the samara has thin spines. In many Australian acacias, e.g. *Acacia mangium* and *Acacia auriculiformis*, the funicle has enlarged into a soft aril (Table 3.2, Appendix 1).

The strength of attachment and thus the ease of removal differ between species and type of wings. Wings of samaras and seed wings of Meliaceae are integrated structures of fruit and seed coat, respectively, and dewinging implies physically breaking the structures. Wings of pine seed clasp the seed and are normally lost before germination, usually after wetting.

Woody wings can often only be removed by breaking off the wing by hand or cutting it with secateurs. Some seed processing units have successfully used coffee dehuskers or hammer mills.

Wings of conifers are removed by mechanical abrasion during tumbling. Because wings are more hygroscopic than dry seed, a slight wetting by spraying, for example, 1 l of water on 50 l of seed during tumbling often facilitates dewinging (Tanaka 1984). Special mechanical dewingers are available where wings are abraded between brushes. Species with delicate seed coats that are easily damaged are preferably dewinged by tumbling in closed drums where the mechanical impact is reduced by slow revolutions where seeds rub against each other (Edwards 1981). Casuarinas and other species with papery wings may be dewinged by tumbling in cement mixers together with some abrasive material like sand or gravel. The same procedure may be used for removing hairs and spines. If the seeds are mixed with abrasive material, it should be considered if it can easily be removed from the seeds after tumbling and if it does not damage the seed. Abundant hair like that in kapok may be removed by burning.

A very efficient machine for dewinging and detachment of dry appendices such as hairs, floss, arils, floral or fruit parts from the seed is the brushing machine shown in Fig. 3.18. During operation the seeds are rubbed by revolving brushes against the wall of a cylinder consisting of wire mesh. Rotation speed, distance between brushes and cylinder, type of brushes and mesh wall of the cylinder can be adjusted according to seed type and wing or appendix to be removed (Karrfalt 1992; Barbour 2006).

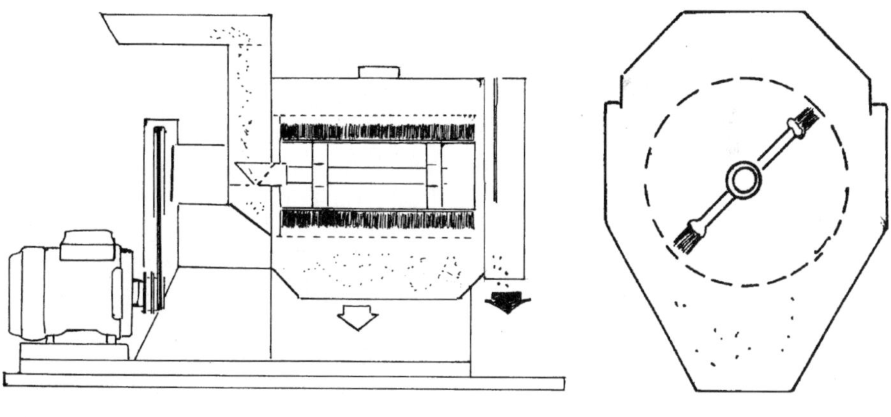

Fig. 3.18. The brush dewinger consists of a cylinder with a rotating brush. The seeds are rubbed against the wire cylinder (shell) by two (in some types four) rotating brushes. The dewinged or deflossed seeds pass through the opening in the wire cylinder together with removed wings, hairs, etc. Various types of brushes provide different treatments for the seeds, and brushing machines have several applications in seed processing, e.g. extraction and dewinging. (Redrawn from Jensen 1987)

3.7
Seed Cleaning

Separation of seed and fruit parts is often part of the extraction process. Most seed lots require additional cleaning to get rid of impurities or inert matter mixed with the seed (but not matter attached to the seed). Impurities are, for example, twigs, leaf, flower and fruit fragments, soil particles, empty and foreign seeds, dust, chaff and seed fragments (please refer to Sect. 7.5 for pure seed definition). The aims of seed cleaning are (1) to eliminate foreign material to reduce bulk, (2) to improve storability and (3) to make seeds easier to handle during subsequent processes. Hence, the ideal cleaned seed lot consists of all viable seeds of the target species, and is free from any other matter.

Cleaning is a separation process. Material can be separated from the seed if it differs in any distinct physical characteristic, e.g. size, form, surface structure or specific gravity. Seed cleaning is thus subject to the trivial precondition that the more the inert matter differs from the seeds in these physical characteristics, the easier it is to separate. And the more similar the impurities are to the seeds, the more difficult they are to eliminate. Objects can be very similar in some aspects and different in others. As cleaning separates according to differences, the method using the largest differences is the most effective (Fig. 3.19). For example, twig fragments and seed that have very similar weight must use another variable factor for separation, e.g. specific gravity. Variation in seed size and morphology of the seed adds another constraint to seed cleaning: the larger the variation in the seed lot, the more difficult it is to clean. For a seed lot containing a large variation, e.g. of seed size, it is difficult to achieve high purity without eliminating viable seeds if cleaning is based on size only. A purity of say 80% may be fairly easy to achieve for most seed lots; further cleaning can be very hard and laborious. When a certain purity has been achieved, the balance must be considered: either to continue cleaning to achieve a higher purity with the implications of higher processing costs, possible damage to the seeds and possible loss of viable seeds, or to accept a certain degree of impurity with the implied disadvantages of handling impurities (storage, pretreatment, sowing, etc.), and a possibly reduced price for the seeds (Box 3.4).

In addition to these individual considerations, official rules may set minimum standards of purity. In seed trade, cleaning may further have a more psychological rationale: customers feel cheated if they pay for seed that contains impurities.

Seed lots in which seed and debris are very different, e.g. dust in seed lots of relatively large seed, may be efficiently cleaned by one cleaning method, e.g. sieving or winnowing. More often seed cleaning consists of a series of processes during which impurities are gradually removed and the seed lot concurrently achieves a progressively higher purity (Figs. 3.20). If more than one method

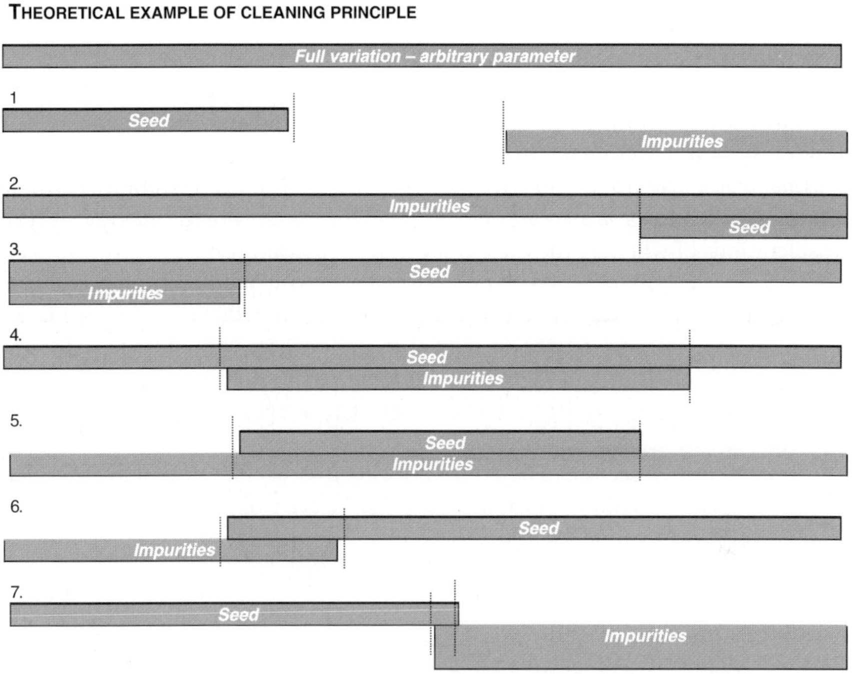

Fig. 3.19. Some impurity–seed differences in relation to cleaning. The full variation of the seed lot is illustrated by the *bar*, which could represent a diameter, specific gravity, friction or other range. Variations may represent various compositions. **1** Impurities and seed have no overlap. Seeds can easily be cleaned. **2** Impurities represent the full variation. Seed cleaning gives two parts: one part of impurities only, which can be discharged, and one part of mixed seed and impurities, which may be cleaned further. **3** Opposite of 2: there is a pure seed fraction and a mixed fraction; the latter is subject to further cleaning. **4** Impurities make up the middle section of the scale. The lower and the higher fractions of seed are clean. **5** Opposite of 4: impurities at the lower and higher ranges are discharged; the middle fraction is subject to further cleaning. **6, 7** Cleaning gives three fractions; a pure seed fraction, a fraction of pure inert matter and a mixed fraction. The examples do not include the relative quantities of seed and inert matter, only the variation related to a specific cleaning parameter

must be applied, the order is chosen so that as much debris as possible is removed by the first method. This is in order to reduce bulk, which will ease subsequent processes. An example of a sequence of cleaning is sieving → winnowing → flotation. Some cleaning procedures separate the seed lot into only two fractions, one containing (mainly) the seed and one containing (mainly) inert matter to be discharged. Other methods may separate the seed lot into several fractions with various purities. Intermediate fractions typically contain both seeds and inert matter and must be cleaned further.

Box 3.4

How clean is clean?

By using several cleaning parameters it is possible to achieve practically 100% purity for a large number of seed lots. On the other hand, handling a certain amount of inert matter is not necessarily a big problem and even a certain inconvenience may make up for the saved cleaning cost. Technical possibility and economic benefit do not necessarily fit together. It is worth making certain considerations of what are the costs and benefits of cleaning:

1. *Relative bulk.* Extra bulk will necessarily have to be handled in all subsequent procedures, i.e. carried around and transported whenever the seeds are moved. Extra cost related to bulk is in connection with storage space and energy (electricity), and postal fee for long-distance transport.
2. *Potential damage to pure seed.* This depends on the character of the matter and storage conditions:
   (a) Other tree seeds can be quite annoying in a seed lot especially if the germinated seedling looks like the target species. Examples are other pine species in a pine lot, hybrid seed in a pure seed lot or visa versa. Infertile seed hardly does any direct damage but may be required to be removed for other reasons.
   (b) Pathogens. Soil and organic matter always harbour potential pathogenic organisms. Pathogens may develop and cause diseases during ambient storage at relatively high humidity or during germination. Potential damage is thus dependent on storage and germination conditions.
   (c) Pests. Insect eggs, larvae or adults mixed with seed and debris can cause the same damage as pathogens during storage and sowing.
3. *Price impact of impurities.* Where the price is set by the seed producer and there is no minimum standard on purity, the cheapest is always to base the price on pure seed and then reduce it according depending on the actual purity.
4. *Sowing conditions.* Machine sowing may be complicated by inert matter which gets stuck in mechanical parts, e.g. fruit parts and small stones. Sowing precision is also hampered by various types of non-seed material.
5. *Quantities distributed.* Small quantities of high quality seed lots would be expected to be absolutely clean, while a certain amount of impurities may be tolerated by bulk treating.

In seed trade, seed cleaning is sometimes more psychological than practically necessary. A product of clean material is more easily sold than a 'dirty' product. It is hard to believe that a seed lot consisting of half seed and half soil and fruits can be of better quality than a clean lot. Where purity is documented, e.g. via a seed testing certificate, a high figure looks more attractive. As in most other marketing, the image sells.

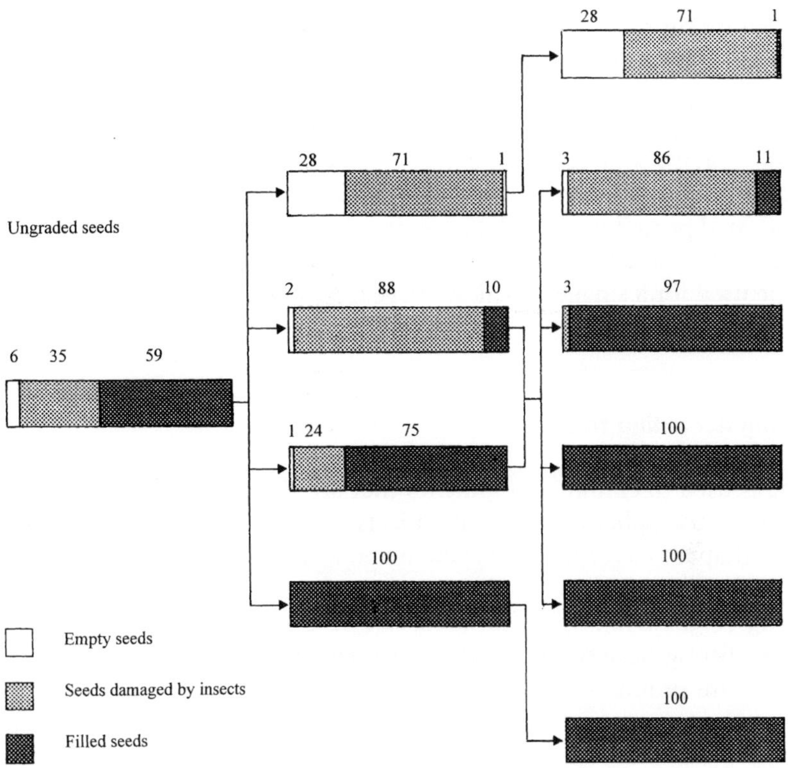

Fig. 3.20. Example of cleaning effect on purity of a seed lot. After the first cleaning the seed lot is separated into four fractions of increasing purity. One part consists of 100% pure seed, another part consists of almost pure debris and damaged plus infested seed. The two intermediate fractions consist of a mixture of seed and impurities. These fractions need a further cleaning with new adjustment of the same procedure (e.g. different air speed during winnowing) or separation by another method (specific gravity, indented cylinder, etc.)

Accessories for seed cleaning range from simple handheld sieves, baskets and cloth frames to advanced combined machines in which the fruits can be fed in one end and the cleaned seed collected at the other. Because of the diversity of forest seed types and thus the requirement for cleaning, it is usually preferred to have relatively simple equipment types which can easily be adjusted to different types of seed.

Seed lots which are very difficult to clean to high purity, e.g. if they contain a large fraction of empty or insect-infested seeds with very similar appearance as healthy seeds, can often be cleaned more efficiently by initially grading the seed lot, usually according to size: once the major portion of debris has been removed by sifting, the seed lot is divided into two, three or more size classes which are then cleaned by one of the other methods. When the individual size classes have been cleaned they are poured together again into one seed lot. This initial grading avoids size differences interfering with other parameters, e.g. specific gravity, and separation consequently becomes much easier in the succeeding cleaning procedures. Often the initial separation makes more complicated procedures like flotation, incubation, drying and separation (IDS) or pressure–vacuum (PREVAC; Sect. 3.8) redundant, or these methods need only be used for a smaller fraction of seed.

3.7.1 Cleaning According to Size

Sifting is used to eliminate impurities that are significantly larger and smaller than the seed. Spherical (round) objects will pass an opening in sieves or screens that are larger than their diameter. Asymmetrical objects may pass an opening larger than their smaller diameter when their small diameter faces the opening (Fig. 3.21a). Spherical seed can be cleaned to quite high purity by selecting fitting sieve types, but it is not very effective for flat or winged seeds.

Sieves are produced in a wide range of material, quality and sizes. Simple wire-mesh screens with different mask widths are readily available at most hardware stores. Laboratory screens consist of a series of six to eight, 20-cm-diameter screens with different holes or mesh sizes. Patented types used, for instance, for soil analysis are unnecessarily accurate (and expensive) for seed cleaning. Locally manufactured types are often available and just as good for seed cleaning. Small screens can be used for small seed lots or samples or, more operationally, for small seeded species like eucalypts and *Anthocephalus*. Examples of mesh size for some Australian species are given by Gunn (2001), e.g. from 1 to 4 mm for eucalypts and from 3 to 12 mm for acacias.

The simplest screening series consist of two sieves (Fig. 3.21b). The upper screen has a mesh size larger than the seeds; it will retain large material like fruits and twig fragments. The lower one, with a mesh size smaller than the seed, retains the seeds while smaller debris passes through. The holes are adjusted to retain the smallest viable seeds. Shaking or sliding the seeds over the screens will make them pass through. Sometimes several screens with gradually decreasing mesh or hole sizes may be used and the seeds graded according to size. The grading may be maintained during subsequent cleaning. In some instances small seeds are deliberately discharged.

Special seed cleaning screens with many types, shapes and sizes of holes are produced. They are made from iron sheets, plastic or wood. Seed cleaning uses a combination of different types of screens.

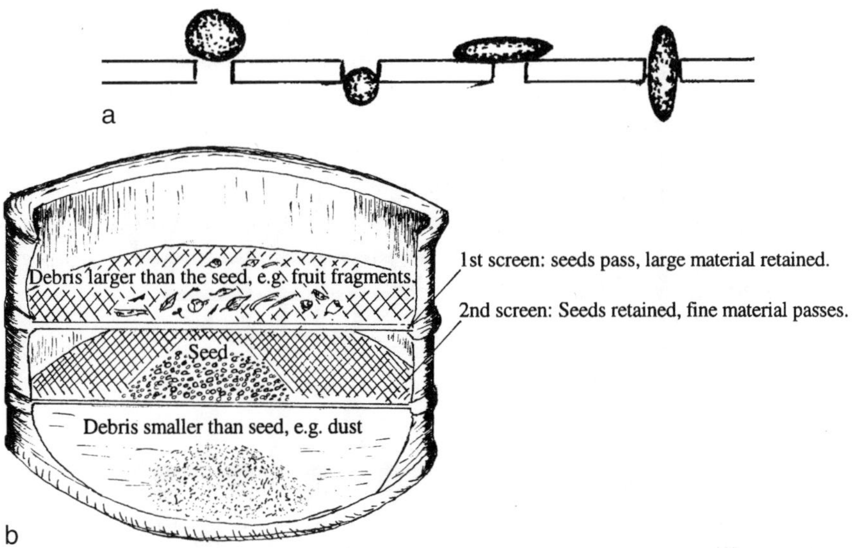

Fig. 3.21. Cleaning by sifting. **a** Seeds pass through openings depending on size and orientation. **b** Laboratory screen cleaning small seeded species. Example: Opening size, upper screen 2.5 mm for *Eucalyptus globulus*; 1.5 mm for *Eucalyptus obliqua*, *Eucalyptus delegatensis* and *Eucalyptus sieberi*; lower screen: 0.5+−0.7 mm as appropriate. (Redrawn from Forestry Commission 1994)

Mechanical seed cleaners use replaceable screens with different hole sizes and shapes (Fig. 3.22). Some smaller laboratory seed cleaners may be supplied with more than 100 different screens. Larger industrial cleaners are normally supplied with a smaller number of screens according to the main species processed.

The appropriate screen type is found by the following method:

1. Place a stack of screens with the correct hole type on top of each other, with the largest opening on top and then decreasing to the smallest opening at the bottom.
2. Pour the seed sample onto the upper screen and shake gently to let seed and debris pass through holes larger than their diameter.
3. Disassemble the stack of screens and examine the best separation. Choose the appropriate screen size(s).

Intermediate size seed and debris may block ('blind') the holes in the screen as intermediate size fractions get stuck in the opening, i.e. seeds and particles too

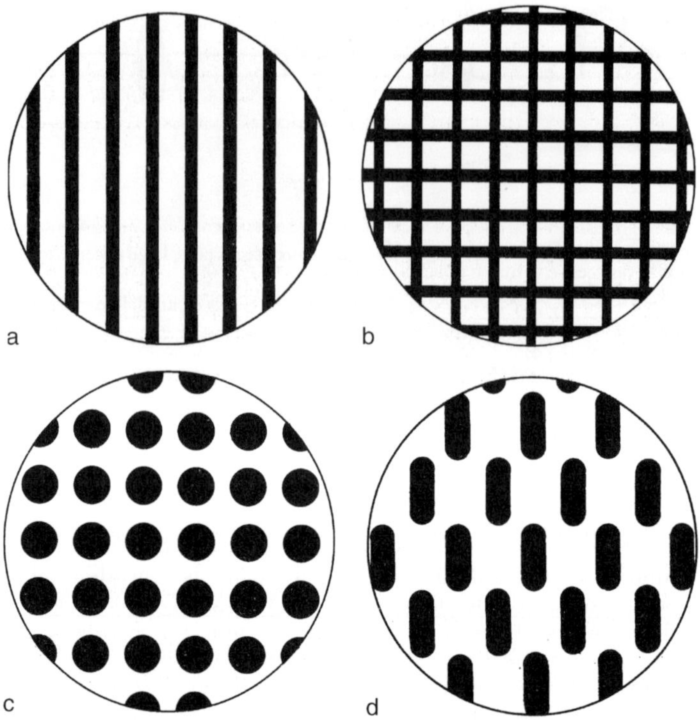

Fig. 3.22. Screens with different hole types used for different seed types in mechanical seed cleaners. **a** Grid type used mainly for precleaning, for example, branchlets and leaves from large seed. **b** Wire-mesh type; this screen has a relatively large opening area compared with metal sheets (**c**, **d**) and is thus faster in use than these. However, the wire mesh more easily gets blocked by material, especially if it has small opening sizes. **c** Metal sheet with round holes, especially used for round seed and for removing large debris (precleaning). **d** Metal sheet with oblong holes. Used, for example, for oblong seeds or for separating oblong debris like leaves, fruit stalks, branchlets and fruit parts. Screens with oblong holes are normally oriented with the holes following the direction of the seed flow (longitudinally)

large to pass through the opening and too small to be left above the screen. The screens may be cleared by regular brushing. Some mechanical cleaners use rubber balls placed on the screens: the balls tend to push down or break material getting stuck in the holes. A more efficient method is to place the rubber balls on wire-mesh screens with a large mesh size under the functional screens. The vibrating movements during operation will make the balls jump up against the screen above and push up material which blocks the holes (Fig. 3.23).

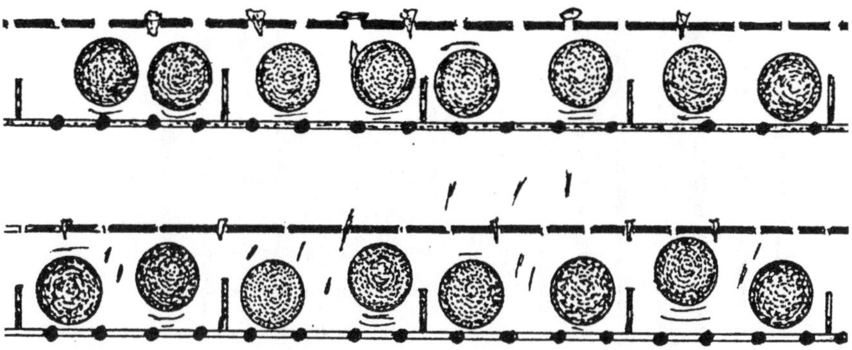

Fig. 3.23. Cleaning of blocked holes during mechanical sifting. Wire-mesh screens with rubber balls are placed under the functional screens. During operation the balls jump and push up any material stuck in the holes

3.7.2
Cleaning According to Form, Sieves and the Indented Cylinder

Long seeds prevail in grasses but are much less common in trees. Oblong-type debris such as pieces of branches and fruit stalks is, on the other hand, common. Debris that differs from spherical seed in shape and length, e.g. twig pieces, pine needles and the like, can be separated by using screen types with oblong rather than round holes. Figure 3.22 shows different opening shapes in screens. In general, round holes are used when the items to be separated differ in width (width is the greater diameter of the cross-section of the non-symmetrical seed); oblong holes are used when separation is according to thickness, i.e. the smaller diameter.

A widely used cleaning machine for crop seed is the indented cylinder. Separation is here according to length and is based on the difference in the centre of gravity of short and long seeds. The indented cylinder (Fig. 3.24) consists of a cylinder with numerous indentions in its inner surface, revolving round a sloping axle. Above the axle along its length is a fixed sloping trough. During operation, seed to be sorted is fed in at the upper end of the cylinder and slowly moves downwards to the lower end. As the cylinder revolves slowly, seeds fitting into the pockets are carried upwards. When indented in a short pocket in a horizontal position, short seeds remain there, while long seeds fall out.

There are two types of adjustments, viz. slope of cylinder, which determines the flow of seed, and pocket sizes, which change according to seed size. Change of pocket size requires change of the cylinder. Spare cylinders are relatively expensive and the machine is mostly used for large quantities.

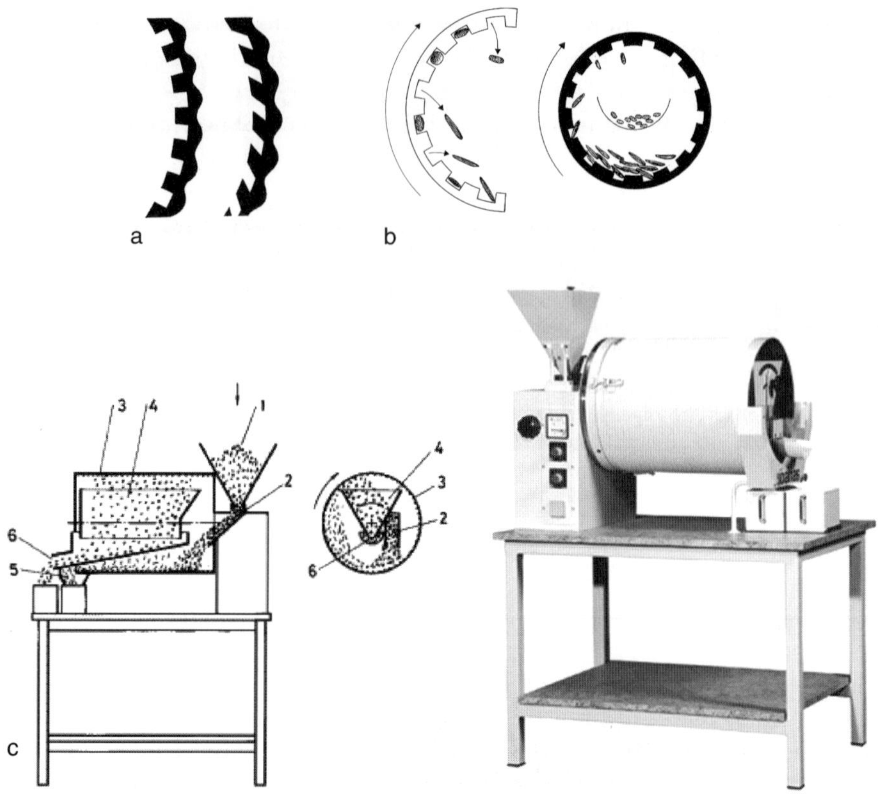

Fig. 3.24. Cleaning by an indented cylinder. **a** Different types of indents. Cylinders with asymmetrical indents (*right*) have proven more effective in separation than the type with symmetrical holes (*left*). **b** Separation of short and long seeds. Short seeds are carried upwards into the trough, long seeds fall back into the cylinder. **c** A laboratory indented cylinder. (Source: http://www.westrup.com)

3.7.3 Cleaning According to Gravity and Form – Winnowing and Blowing

Light material such as dust, chaff, leaves and wings can be separated from seed, and seed from heavier material such as wood, stones and soil particles by using their different specific gravity (weight-to-volume) and surface-to-volume ratios. An object with a large specific gravity, e.g. a stone, or a small surface-to-volume ratio, e.g. a spherical object, will fall faster vertically and be moved a shorter distance by a horizontal air current than an object with a smaller specific gravity, e.g. wood, or a large surface-to-volume ratio, e.g. a leaf.

Specific-gravity separation is also used in liquid media, where the material is separated according to its ability to float or sink in a liquid medium.

Winnowing is a simple, traditional, widely used and effective cleaning method. Natural wind is used. Rice farmers use two winnowing methods. In one method the seeds to be cleaned are held in a large-diameter, flat basket. The seeds are repeatedly thrown up in the air and gripped again; when seeds are in the air the wind will blow away light debris. The other method is to slowly pour seeds from a certain height into a pile (Fig. 3.25a). The wind will blow away any light material, while the heavier seed fall. The two methods are labour-intensive and dependent on natural wind. Mechanical seed cleaners use the same principles (Fig. 3.25b).

Fans or propellers can create an artificial air current. The air speed determines the degree of displacement: the stronger the speed, the more matter can be displaced. In practice, air speed is regulated to utilise a certain displacement distance, e.g. 2–3 m. Winnowing sorts seed into a gradient with heavy particles

a

Fig. 3.25. Applications of the winnowing system. **a** Traditional winnowing is used for rice cleaning in rural areas. The grain is thrown up in the air or slowly poured into a pile, in both cases using natural wind to blow away light matter.

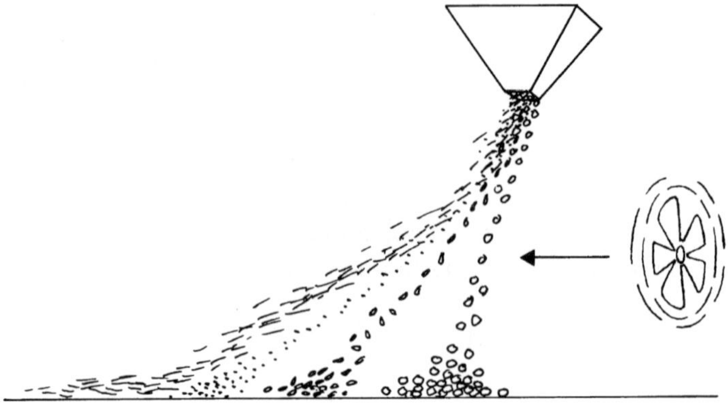

Fig. 3.25. (Continued) **b** A mechanical fan or propeller provides a stable air current, which can be regulated by the fan speed. Matter falling through the current of air will be displaced horizontally according to its weight and air resistance. The winnowing system is used in many seed cleaning devices

closest to the air source and the lightest particles farthest away. The intermediate fraction contains a mixture of light seed and debris with decreasing purity away from the air source. The air speed is increased when cleaning relatively heavy seed. Different fan types have different maximum capacities which depend on lamella type and direction, and fan speed. In practice, the right air speed at the outlet is regulated by air intake.

The simple principle of horizontal displacement by air flow was used by Chaplin (1985) for cleaning mahogany seeds (Fig. 3.26). The device also uses a specific-gravity factor by using a sloping table facing the air current.

The physical principle of air-current displacement is also used in seed blowers, which are mainly used for small seed types such as eucalypts and casuarinas. The seed lot to be cleaned is placed in a vertical cylinder connected to an electrically powered air current at the bottom (Fig. 3.27).

3.7.4 Cleaning According to Gravity – Specific-Gravity Separators

Specific gravity is measured as unit weight per unit volume (e.g. grams per cubic centimetre) – in common terms it is a measure of 'how heavy' something is compared with its volume. In gravity cleaning, differences in specific gravity are sorted according to how objects fall ('down'). By using a force in the opposite direction to gravity ('up'), the seed mixture will stratify. Seeds are separated on various forms of tilting decks. Friction resistance, which

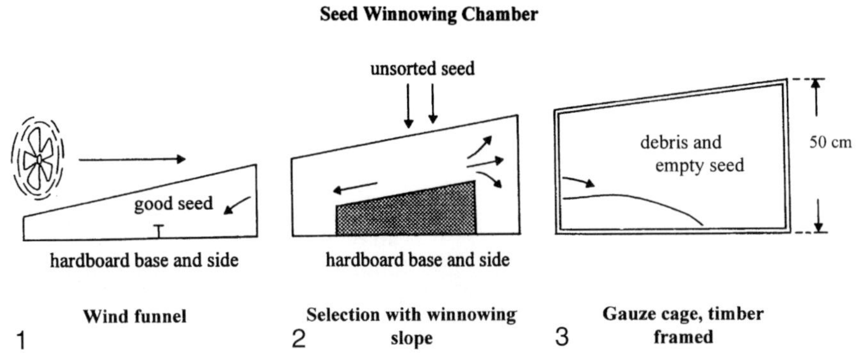

Fig. 3.26. Simple winnowing chamber constructed and used for cleaning *Swietenia* seeds. The chamber consists of **1** a wind funnel from where an air current is created by a fan, **2** a central section with a Formica top sloping against the air current and **3** a gauze cage where light debris is collected. The uncleaned seeds are fed from the top above the central slope and pass through the air current. Light matter is blown into the gauze cage, the seeds fall down on the slope and roll or slide down. Slope, height of drop and strength of air current can be adjusted according to seed type. (From Chaplin 1985)

normally prevents matter from moving on a slight slope, is overcome by shaking or blowing.

3.7.4.1 Oscillating Table

The separator consists of a slightly inclined table with zigzag partitions along its length (Fig. 3.28). Seeds placed on the deck will tend to move downwards according to gravity; however, during the sideways oscillation, objects will be struck by the partitions. These strikes will tend to move the objects upwards. Separation of the seeds is based upon the balance between these two forces. For light seeds the striking impact of the partitions overcomes gravity and hence moves them upwards. For heavy seeds the striking impact is insufficient to overcome gravity and the seeds slide or roll downwards. The sloping of the table and the sideways movement can be adjusted for different seed types: varying the tilt will influence the gravity force; varying the oscillation speed influences the strike. The seed surface will, to a certain degree, also influence cleaning. A rough surface with high friction will slow down movement and such seeds tend to follow the flow of heavy seeds.

Fig. 3.27. Separation by blowing. South Dakota seed blower. The upwards air current will displace all light material like chaff and wings to the top while the heavier seed are collected at the bottom. The cylinder can be emptied in sections so that the midsection, which typically contains a mixture of seed and debris, can be recleaned
(Source: http://www.seedburo.com)

3.7.4.2
Vibrator Separator

In the vibrator separator the main physical attribute is surface characteristics. A rough deck surface will tend to grip a rough surface of the seed. Vibration of the sloping deck will thus move seeds upwards, while smooth objects slide down according to gravity. Seeds tend to stratify according to their surface characteristics at the outlet end of the table (Fig. 3.29). There are several adjustment options, which help to create a fine balance and thus often a very efficient separation (Jensen 1987):

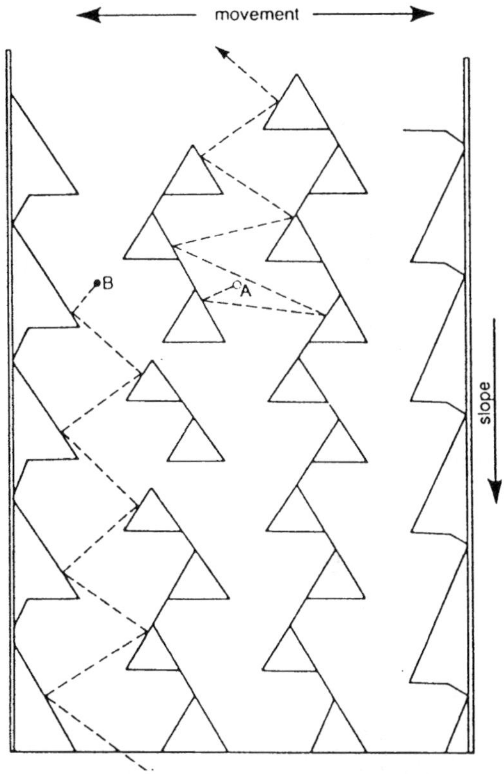

Fig. 3.28. Principle of the oscillating table. A light object **a** is hit by the zigzag movement of the partitions of the oscillating table and moves upwards. A heavy object **b** is hit by the first partition but because of the gravity force it fails to be hit by the next partition and consequently moves downwards. (From Jensen 1987)

Fig. 3.29. Vibrator separator. The cleaner is used for small quantities of seed. Adjustment for seed and debris types is done by using different deck types with different friction. Fine adjustment is done by adjusting the slope of the deck

1. *The deck surface using different material with different roughness.* For example cloth or linen as the smoothest material and sandpaper as the roughest material.
2. *Speed of deck vibration.* The faster the speed, the more frequent the particles are struck.
3. *Degree of side and end tilt.* Steeper slope means increasing the gravity force. There are two tilts: sideways tilt is the largest which influences the balance between gravity and friction, i.e. the actual separation; end tilt influences how fast seeds move from the hopper to the outlet, i.e. for how long separation forces work.
4. *The feeding rate from the hopper.* The rate is adjusted to create a smooth flow such that seeds have moved when new seeds are fed.
5. *The arrangement of the outlet gates.* The deck separation creates a gradient between different surface structures. The outlet gates are placed in a way of optimal separation of clean seed and impurities.

The vibrator separator is a versatile machine which can be used for fine cleaning of a large number of seed types and is thus very popular with seed cleaners.

3.7.4.3 Pneumatic Table Separator or Specific-Gravity Table

The separator consists of a slightly inclined table with a porous surface, e.g. a woven linen cloth, connected to an air blower (Fig. 3.30). The air blows from the bottom and 'lifts up' the seeds and debris on an 'air cushion'. The seed mix will tend to stratify vertically on this air cushion with the heavy seeds at the bottom and the light seed above. Vibrating or oscillating movements of the inclined table inflict separation. The heavy seeds at the bottom will be hit by the surface of the table and thereby move upwards. Light seeds and particles are not struck and will 'float' over the edge of the table at the lower end.

There are various adjustments:

1. *The pressure of the air stream.* This determines the stratification height, i.e. which fraction of the seed is struck by the deck and which fraction escapes.
2. *The surface of the deck.* This ranges from gauze to rough linen. The range is limited as the surface also must allow sufficient airflow.
3. *The speed of deck vibration.* As for the gravity separator, the greater the speed, the more frequent the particles are struck and the more they will move.

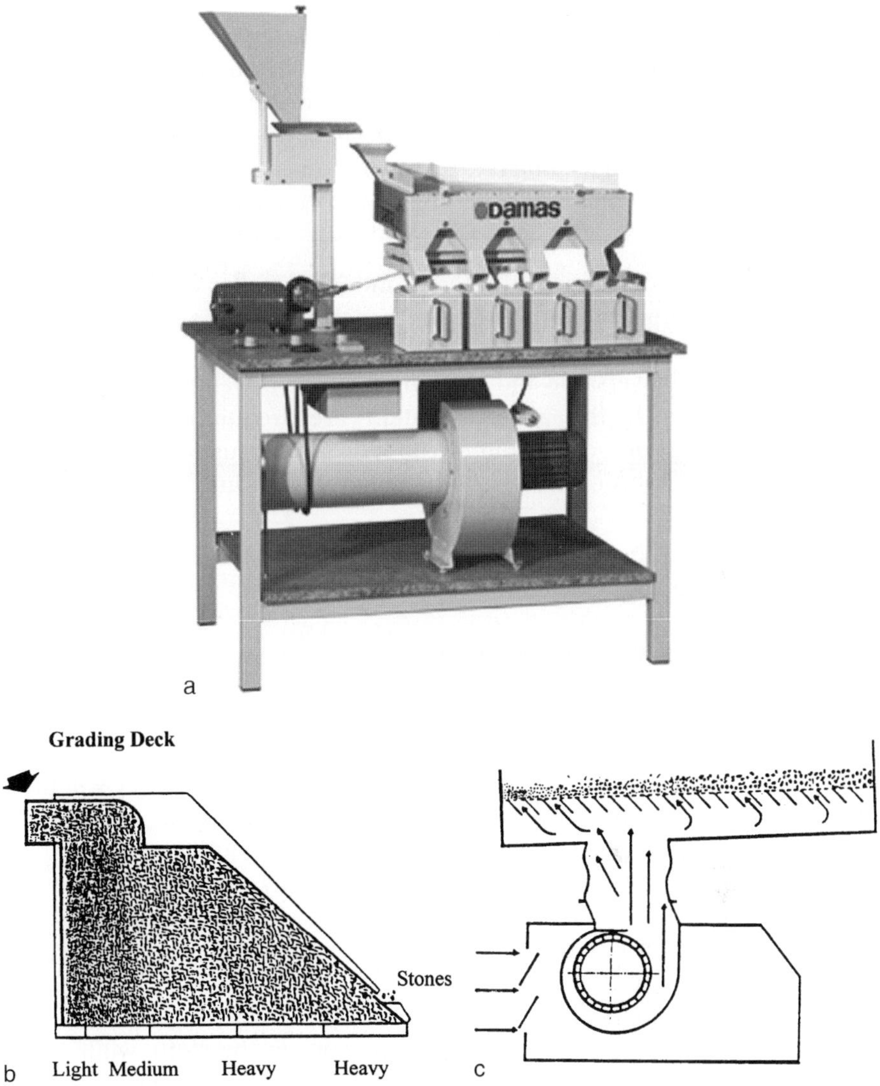

Fig. 3.30. Principle of separation by the pneumatic table separator. **a** Damas separator. **b** The separator seen from above. The seed is fed onto the table at the *arrow*. Vibration of the inclined table makes the heavier seeds move to one end of the table and the lighter seeds or debris to the other. **c** Transverse section of the separator showing how the seeds 'float on an air cushion' with the heavier seeds under the lighter seeds. (**a** From http://www.damas.dk, with permission, **b**, **c** from Jensen 1987)

4. *The arrangement of the outlet gates.* The deck separation creates a gradient between different surface structures. The outlet gates are placed in a way of optimal separation of clean seed and impurities.
5. *The inclination of the deck in two directions and the feeding rate.* The impact of tilt and feeding rate on separation and flow is the same as in the gravity separator.

The size of the deck varies in different models: smaller ones have a maximum length of some 25 cm; the larger models have a maximum length of about 125 cm. In general, the larger the deck, the easier it is to adjust. The pneumatic table is a versatile and efficient machine if well maintained; however, problems with the airflow system are frequently encountered. Mechanical damage or wearing out of rubber gaskets causes leaking and thus uneven pressure. Decks cannot easily be repaired without causing difference in the airflow.

3.7.5 Cleaning According to Form and Surface

Differences in friction and the centre of gravity can make separation efficient. Friction, which refers to the surface structure, influences how objects will slide; the centre of gravity, which refers to the height, influences how they will tilt or roll. For example, an object with high surface friction, e.g. a leaf, can stay on a steep slope, while an object with low friction, e.g. smooth paper, will slide down. Further, an object with a low centre of gravity may stay on the same slope, while an object with a high centre of gravity will roll down. Since spherical objects have both low friction and a high centre of gravity, they will roll down a slope with a relatively small angle.

Friction cleaning in its simplest form is carried out by letting the seeds move on a sloping cloth frame. Flat objects will remain on the cloth, while round objects roll down and are collected at the bottom, e.g. through a special outlet (Fig. 3.31b).

A mechanical derivation of this system feeds seed onto a sloping rotating cloth. Round seeds roll down the cloth and are collected in one fraction, flat seeds are carried up the slope into another fraction (Fig. 3.31a). Feeding speed is adjusted so that the seeds can be carried away or roll freely. The rotation speed must be adjusted so that the seeds move smoothly without jumping. The slope is adjusted so that the most effective separation is achieved.

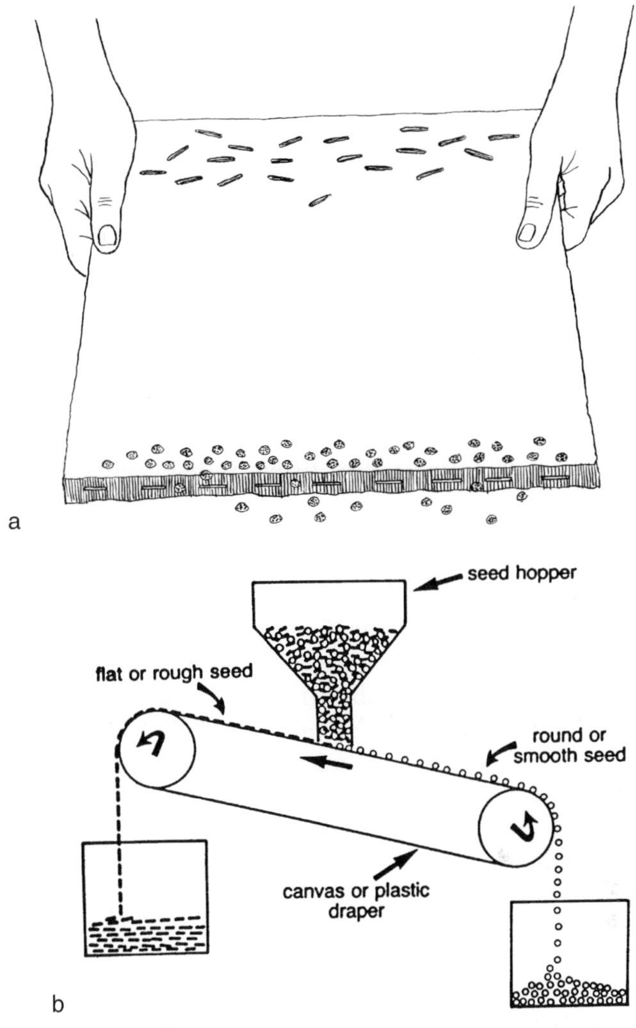

Fig. 3.31. Methods of separating seed according to surface friction and centres of gravity. **a** Sloping cloth frame; round or smooth seeds roll or slide down the frame, while flat and rough seeds and debris remain on the top part of the frame. **b** The rotating belt carries small flat or rough particles and seeds upwards, while round or smooth seeds and heavy particles roll down. (**a** P. Andersen, **b** from Jensen 1987)

3.7.6 Cleaning According to Specific Gravity – Flotation

Flotation also uses specific gravity, but in the sense of the difference between the specific gravity of a liquid and that of the seeds and inert matter. Matter placed in a liquid medium will float if its specific gravity or density (weight-to-volume ratio) is smaller than that of the liquid, and will sink if it has a higher specific gravity than the liquid. Most seeds have a density slightly below 1.0 g/cm^3 when dry and slightly above 1.0 g/cm^3 when imbibed. That implies that they will tend to float in water (specific gravity of 1) when dry and sink when imbibed. During processing, flotation is used, for example, to separate pulp and seed during depulping (Sect. 3.5.2). The flotation technique is especially applicable if a very small specific gravity difference between sound seed and inert matter makes separation by other means difficult. Such a small difference is found, for example, between healthy seed and seed with small insect infestation or shivered embryos (Bonner 2003). Flotation in pure water is easiest. However, for relatively heavy or relatively light seeds, different liquids with different densities are used (Table 3.3). Mixing two liquids with different densities, e.g. *n*-pentane and ethanol, makes a solution with intermediate density; adding a soluble compound to a liquid (e.g. salt to water) normally increases its density. It is thus possible to adjust the density of the liquid quite precisely for optimal separation. The liquid should have a density between that of full and empty seeds or debris. Under these conditions full seeds will sink and empty or damaged seed and light debris will float. An example of the use of flotation in a low-density medium is separation of *Araucaria cunninghamia* seed. The average density of the seeds is around 0.75 g/cm^3, with filled seeds slightly heavier than empty (embryoless) seeds – both filled and empty seeds will thus float in water. The density of a mixture of 95% ethanol and *n*-pentane

Table 3.3. Specific gravity of some liquids used for separation by flotation

Medium	Specific gravity
Pure water	1.0
Absolute alcohol (ethanol)	0.791
95% alcohol (ethanol)	0.806
50% ethanol	0.90
Diethyl ether	0.714
Petroleum ether	0.657
n-Pentane	0.626
Mixture of 95% ethanol and *n*-pentane, 3:1	0.76
Mixture of 95% ethanol and *n*-pentane, 12:13	0.71
Linseed oil	0.93

falls in-between that of the density of filled and empty seeds; filled seeds will sink and empty seeds float and the mixture is thus suitable for separating the two fractions by flotation (Haines and Gould 1983).

Some flotation media have a negative effect on seed viability. Potential damage depends on whether the liquid is absorbed and reaches the embryo, or whether it only affects the seed coat. This in turns depends on seed-coat structure and the period of exposure. Relatively hard coated species may thus tolerate short exposure to a poisonous medium, while less hard seeds are likely to absorb the liquid more quickly and thus suffer poisoning. Alcohols are poisonous to seed embryos, but harmless as long as they are only in contact with fruit or seed coats. Different types of alcohol have different effects: ethanol has been shown to have a negative effect on storability of some pine species (Barnett 1971). In a study by Simak (1973), he found that absolute alcohol had no negative effect on germination, while lower concentrations could apparently damage the seed. Short-term and long-term effects on viability of other organic flotation media have been documented by Hodgson (1977).

A further development of the flotation method is used in various types of absorption separation methods. These methods are used for separation of dead and empty seeds from healthy ones by using the character of different absorption and desorption rates for the different seed conditions (Sect. 3.8).

3.8 Seed Grading and Upgrading

Seed grading is a further development of seed cleaning aiming at achieving high viability by eliminating low-quality, yet intact pure seed. Such seed may be empty seed, immature seed, damaged or dead seed or seed developed after self-fertilisation. In the latter case the removal also serves to improve the *genetic quality* of the seed lot. Separation during grading is based on some correlation between physical character of the seeds and their viability/vigour. For example, assuming that there is a correlation between seed size and vigour, the deliberate removal of small seeds from the seed lot will increase the seed lot quality. Such a positive correlation between seed size and seedling size/vigour has been documented for several species. Often large seeds tend to germinate faster and produce larger and more vigorous seedlings than small seeds of the same species. Swift and uniform germination and seedling establishment is generally an advantage both for nursery and field establishment. The character of quick germination and early growth is not necessarily correlated with ultimate yield and tends to get smaller and may disappear after one or more growth seasons. (Dunlap and Barnett 1984; Fowells 1953; Griffin 1972; Sorensen and Campbell 1993). Seed size is also influenced by inheritance and in some instances it has been shown that elimination of the lightest seeds may influence the genetic composition of the seed lot by selectively removing seeds of families with relatively small seeds (Hellum 1976).

Seed grading may in practice be an extension of the seed cleaning process because the small and light seeds are removed together with chaff and other impurities. Methods of grading must, however, be adjusted much more precisely since the physical difference between seeds within a species is likely to be much less than between seeds of different species or between seeds and inert matter. Two very exact methods of separation have been used primarily for pines but are applicable to other species.

When seeds are poured into water (specific gravity 1.0), they will initially float, but mature viable seeds will absorb water and sink after some time, from a few minutes to several hours. Empty, immature or damaged seeds and other light material may remain floating and can be skimmed off after an appropriate period of time. Some empty and damaged seeds absorb water at the same rate as sound seeds and will sink accordingly. However, during a subsequent redrying, from a few minutes to a few hours, empty and damaged seeds and seeds with low viability tend to lose water faster during drying than full viable seeds. During a second flotation healthy seeds sink, while immature or damaged seeds float (Fig. 3.32). The seed lot can thus be sorted according to probable seed viability (Bergsten and Sundberg 1990). Since the sound seeds have imbibed during the separation process, they must be dried again before storage.

An inverse flotation separation method is used in the PREVAC method to separate seeds with mechanical damage from sound seeds. Dry (non-imbibed) seeds are exposed to low pressure (vacuum) for 1–20 min while lying in water. When the pressure is released, mechanically damaged seeds, e.g. with cracks or part of the seed coat missing, absorb water more quickly that undamaged seeds. During subsequent flotation, damaged seeds then tend to sink, while undamaged seeds tend to float (Bergsten and Wirklund 1987). (Note that the flotation principle here is the opposite of that described above in which sound seeds sink.)

A combination of the density method and the absorption method is used in Australia for the separation of live seeds of *Eucalyptus pilularis* from chaff and dead seeds. The seeds are initially preimbibed in water for 2–4 days, then separated in a sugar solution (ATSC 1996).

Removing dead and empty seeds will increase the germination percentage of a seed lot. Upgrading is therefore sometimes used for seed lots with low viability. In nursery operation, grading according to class is sometimes carried out to ensure a uniform germination speed and seedling growth within some specific grading classes. A uniform seed size facilitates sowing with sowing machines and a uniform germination and seedling growth rate will imply fewer cullings (Creemer 1990).

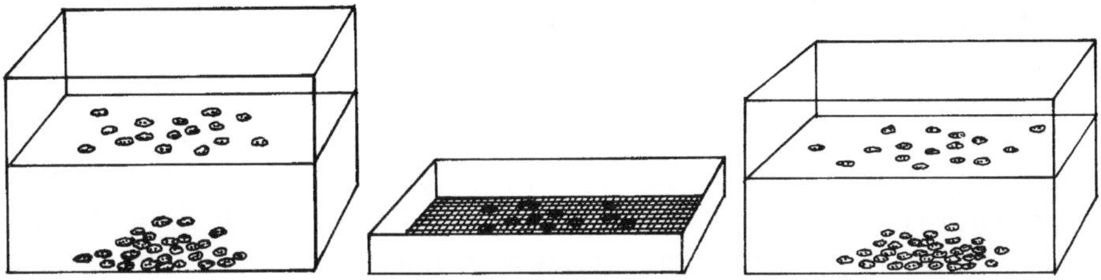

a b c

Fig. 3.32. Principle and procedure of separation by incubation **a**, drying **b** and separation **c**, also called IDS. The method can be very effective for separating sound and healthy viable seeds from dead and empty seeds. Its major drawback is that unintentional germination is difficult to control during incubation. IDS is therefore mainly used as a presowing separation to eliminate seeds that have lost viability during storage

3.9 Adjusting Moisture Content for Storage

In practice, moisture content regulation almost always means drying. Dry fruit species are extracted dry and the level of drying pertains to the potential storage period. In fleshy fruits immediate drying may also be necessary to prevent early germination. Germination in these species is, to a large extent, prevented by inhibitory compounds in the fruit flesh. Once the fruit pulp has been removed, the seeds may be able to germinate if the moisture content is high – which is usually the case after wet extraction using water. Drying should thus take place as soon as possible after extraction.

For orthodox seeds it generally holds that the lower the moisture content, the longer the seed can be stored; however, in some areas and with limited technical facilities it can be difficult to dry seeds. The target is here 'a safe level' for the storage period necessary. A moisture content of 6–8% is appropriate and safe for most orthodox species for at least about 2 years' storage. Appropriate means that viability will not decline significantly during this period. If this is the case it also means that there is little gain in further drying if the planned storage period is less than 2 years. If seeds are normally collected every year, mostly sown during the first coming season and only a limited amount of seed is carried over to the next year, drying to the aforementioned moisture content is fully adequate. Expected storage period and storage conditions are thus related.

Moisture content also relates to storage temperature. Both low moisture content and low temperature prolong the potential storage period. But in general it is much cheaper to dry than to cool down seeds.

Recalcitrant seed, which prevails among humid-zone climax forest species, does not tolerate much desiccation and must be stored with a high moisture content for the shortest possible time. A large intermediate group (semiorthodox, semirecalcitrant or 'intermediate' seeds) show varying levels of tolerance. While orthodox seed is either dried further or maintained at the moisture level achieved during processing, recalcitrant and intermediate seed may need remoistening for safe storage. The moisture content for these species is adjusted to the minimum for safe storage. The actual moisture content of the seed lot is measured before storage either by a calibrated moisture meter or by the International Seed Testing Association oven method (Chap. 6).

Seed drying is, in principle, the same as drying for extraction of seed in dry dehiscent fruits. Outdoor drying, sun-drying or a drying kiln may be applied. A few considerations are, however, of special relevance:

1. The volume of seed is significantly smaller than the volume of fruits – in pods and capsules typically less than 1%. Drying capacity is thus less critical for seed drying.
2. The desiccation rate differs with the difference between air humidity (relative humidity) and seed moisture content. The larger the difference, the faster the rate, or, in more common terms, desiccation is initially fast but the rate slows down as the moisture content approaches the equilibrium moisture content at a given relative humidity.
3. Relatively moist seed can be damaged by high temperature. Fruit cover protect the seeds, but heat is easily transferred through moist tissue over a short diameter.
4. Seed which is to be finally dried for storage is supposed to be clean. While a certain level of contamination may be acceptable during extraction, greater hygiene is required during seed drying.

Moisture content should be followed during seed drying to guide the procedure. Unfortunately, oven-drying methods take a longer time and if carried out according to standard methods may require some precision equipment, such as an oven and precision scale. Less precise methods may be used during current drying. Seed moisture meters are widely used for agricultural crops during harvest and processing (Box 3.5). These moisture meters consist of a container, about 0.25–0.5 l, and an electric device that measures conductivity of the material poured into it. The moisture content can be read after a few seconds. Various companies produce seed moisture meters under various brand names. Seed moisture meters have a few drawbacks:

1. The moisture meters are calibrated for a short-range crop seed only. Calibration tables or graphs must be drawn for tree seed on the basis of exact comparative measurement by the oven-drying method.
2. The devices are only useful for relatively small seed of 'crop size', i.e. anything smaller than a 'maize seed'.
3. The moisture meters are not precise in a number of species of Leguminosae because of their hard seed nature. If moisture meters are used for these species, seed must be ground.

Box 3.5

Seed moisture meters

Quick moisture meters are much used for agricultural seed to determine the best time for harvesting (Fig. 3.33a). The devices can, with appropriate calibration, be used for some species of forest seed. Moisture meters use electric properties of seed to compute moisture content.

Moisture meters can be used for either the entire seed or homogeneous ground material. However, since the devices are mainly used as field equipment, grinding may be inconvenient and the device is, in practice, mainly used for relatively small seeds like those of pines, which can be used directly. The principle of moisture meters is that a standard volume of seed is poured into the measuring chamber – moisture content is read directly on a digital display for a selected agricultural crop (wheat, millet, barley, etc.). Since there is no standard calibration for forest seed, a conversion factor, table or graph must be obtained for the particular forest species in question (Fig. 3.33b). The conversion is established by a series of tests in which the meter reading for a given selected agricultural species is plotted against the moisture content measured by the International Seed Testing Association (ISTA) oven-dry method. Calibration should encompass the range of moisture contents normally encountered for the species, i.e. typically 6–25% for orthodox seed. Where the ISTA value plotted against the meter reading turns out to be a straight line, the calibration factor is calculated as the slope of the line. This factor is then used for all readings for the particular species. In some cases the line appears different for different moisture content levels. Subsequent readings must here use different conversion factors depending on moisture content.

Where applicable, seed moisture meters can be very useful to guide drying. Unfortunately the meters are not useful for very small seeds like those of eucalypts (where the measuring sample may be the whole seed lot) and large seeds, which cannot flow into the chamber, or where material from only a few ground seeds fills up the chamber. Hard-coated seeds of leguminous species tend to absorb or desorb moisture so slowly that moisture meters give a large error. This leaves out most of the common tree species and moisture meters thus have a limited applicability specieswise.

(*Continued*)

Box 3.5

a

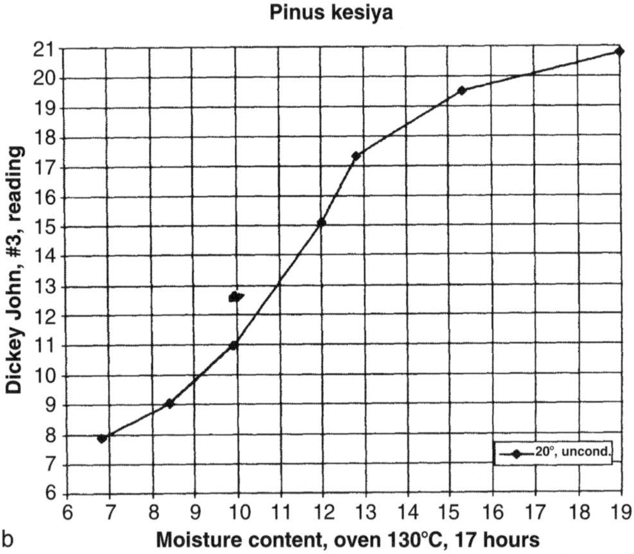

b

Fig. 3.33. **a** Two types of moisture meters. **b** Calibration graph for a seed moisture meter

3.10 Seed Moisture and Principles of Seed Drying

The rate of seed drying is determined by the physical relationship between seed moisture, temperature and relative humidity (Stubsgaard and Poulsen 1995).

3.10.1 Temperature and Humidity

The maximum amount of water that can be contained in atmospheric air depends on temperature: the higher the temperature, the more water the air can contain. When the air contains this maximum amount of water vapour at a given temperature, it is said to be saturated. The maximum amount of water at a given temperature between −10 and 60°C is depicted by the lower curve of Fig. 3.34. Air containing less than the maximum amount of water at a given temperature is not saturated. The actual amount of water is expressed as the relative humidity, i.e. the actual water content as a percentage of that of saturated air at the same temperature. For example, if air at 20 C contains 10 g water/kg dry air where its capacity (saturated air) is 15 g/kg dry air, its relative humidity is 10/15 × 100% = 67%. Figure 3.34 shows the relationship between saturated air (relative humidity 100%), temperature and relative humidity.

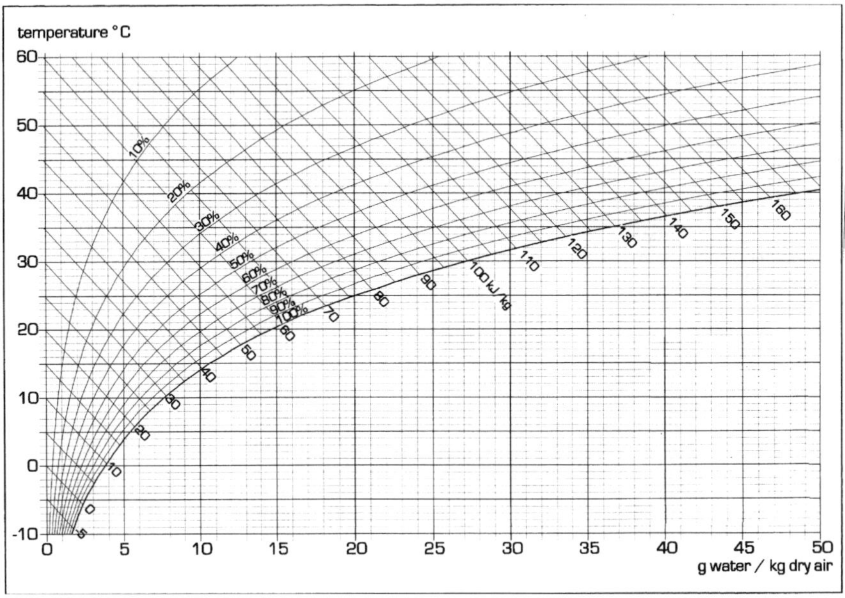

Fig. 3.34. The relation between temperature and air humidity. Increasing the temperature at a given absolute humidity (grams of water per kilogram of dry air) reduces the relative humidity. A relative humidity of 100% is the saturation curve or dew point curve. Temperature decline at 100% relative humidity causes water to condense (dew). Diagonal lines indicate the amount of energy given in kilojoules per kilogram of water at a given temperature and humidity. (From Stubsgaard and Poulsen 1995)

If the temperature of air goes up or down and the air contains the same amount of water vapour, its relative humidity changes accordingly. For example, if air at a given relative humidity (e.g. 70%) is warmed up (e.g. from 20 to 35°C), the relative humidity drops (in the example to 30%). The opposite occurs if the temperature of the air drops (e.g. at night): relative humidity increases. If the initial relative humidity was high or the drop in temperature large, the air may reach the saturation point, where the relative humidity is 100%. This is also called the dew point since a further drop in temperature will cause condensation of the water vapour into dew droplets.

3.10.2 Seed Moisture and Relative Humidity

Water in seeds (measured as moisture content; Chap. 6) tends to be in equilibrium with atmospheric water (measured by its relative humidity) surrounding the seed. If the air is dry and the seed moist, water will tend to move from the seed to the air; the seed dries and the surrounding air becomes more humid. If the air is humid and the seed dry, water will tend to move in the opposite direction; hence, the seed gains moisture.

The larger the difference between the relative humidity and the equivalent seed moisture at the same temperature, the quicker the water movement will take place towards equilibrium. Consequently, the lower the relative humidity of drying air, the quicker the seed (or fruit) will dry. A warm air current with low relative humidity is thus the most effective for drying.

The equilibrium exists immediately around the seed. If the air around the seed is replaced by ventilation, a new equilibrium will be established with the new air now surrounding the seed. The faster the humid air is removed and replaced with dry air, the quicker the seed will dry. Therefore, air circulation by natural wind or artificial ventilation promotes drying.

The actual moisture content in equilibrium with air humidity at a given temperature depends on the species. Examples of equilibrium moisture content are shown in Fig. 3.35.

3.10.3 Seed Moisture and Temperature

Temperature influences seed moisture in two ways: partly via the previously described relation to relative humidity; partly directly by evaporation. As the temperature increases, liquid water from the seed will evaporate.

Absorption and desorption of water are influenced by the seed or fruit size, and the structure of the fruit or seed coat. Small seeds and fruits absorb or

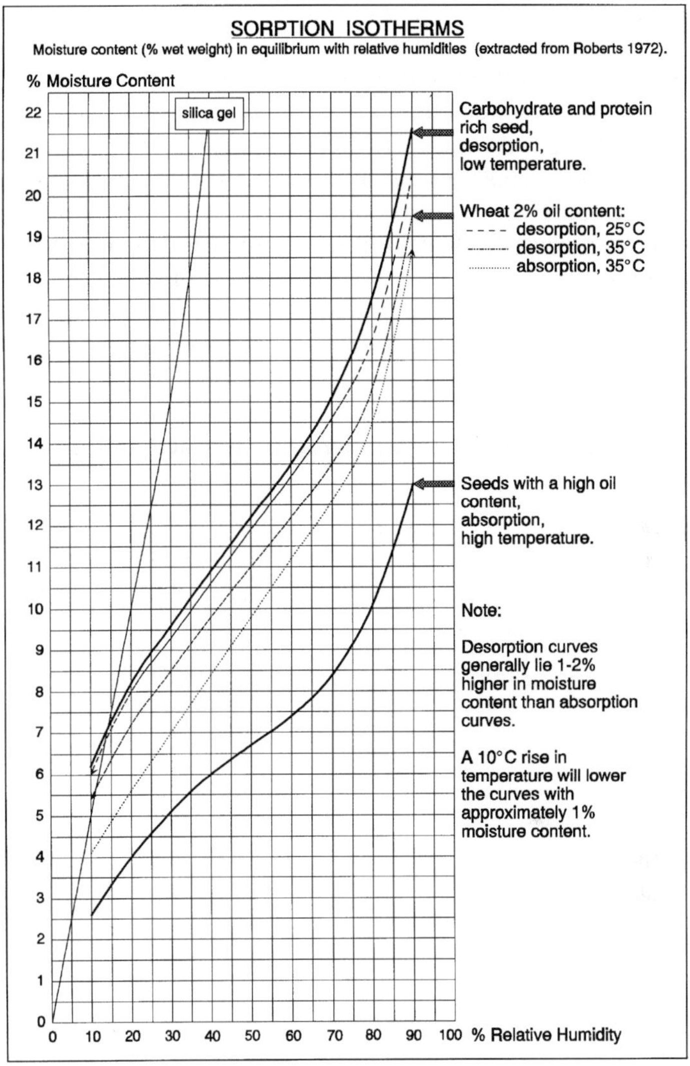

Fig. 3.35. Equilibrium moisture content for different types of seed. (From Stubsgaard and Poulsen 1995)

desorb water faster than larger ones because the surface area is large relative to the volume, and the distance of migration of water is shorter. The anatomy of the seed (or fruit) determines how fast water can migrate from the interior to the outside during drying. A thick or dense structure is likely to restrict water movement. Tompsett (1987) found that seeds of *Dipterocarpus intricatus* dried to 7% moisture content in a week, while seeds of *Dipterocarpus obtusifolius* retained 30% moisture content after 5 weeks in the same drying-room environment. The extreme case is in legumes where the seed coat becomes impermeable to water. As the cells shrink during drying, water movement becomes constricted. That can cause the so-called case hardening of cones where the inner part of the cones and the seeds remain moist because of too fast drying. The outer cone cells have collapsed and form a physical barrier to further desiccation.

With progressive drying the forces resisting desiccation of the cells increase. As the moisture content decreases, the remaining water is 'bound' to the cell constituents and macromolecules in the cells and becomes practically immobile (Bewley and Black 1994). Drying to low moisture content is consequently difficult and high temperature and dehumidified air may be necessary.

The absorption and desorption curves of Fig. 3.35 differ. That means that while the seeds relatively easily lose water at high temperature and low relative humidity, absorption is much slower. In other words: seeds are often more likely to lose water during dry conditions than to regain it under humid conditions. In legumes a special structure of the hilum, the hilar valve, regulates drying. The function of that structure is to allow water to leave the seed, while water is unlikely to enter (Hyde 1954; Dell 1980; Chen and Fu 1984). Hence, the seeds tend to establish equilibrium with the driest atmosphere they have been exposed to.

The type of storage tissue in the seed also influences moisture content. Nutrients are stored in seeds mainly as sugars, starch, protein and fat (oil). Simple sugars prevail in some extremely recalcitrant seeds but are rare in orthodox seeds. The four components differ in their water affinity, sugar being the most hygroscopic (binding most water), followed by protein, starch and oil in decreasing order. Hence, at the same relative humidity, seeds with a high oil content will contain less moisture than seeds with a low oil (and high protein or starch) content. Because of the differences in anatomical structure and storage tissue of seeds the equilibrium moisture content differs between species.

Certain chemicals have the ability to absorb moisture from the air at relatively low relative humidity. One of the most common ones is silica gel, the equilibrium moisture content of which is shown in Fig. 3.35. As silica gel absorbs moisture, the relative humidity of the surrounding air decreases. Seeds kept in a closed container together with silica gel will thus obtain moisture content in equilibrium with the air, dehumidified by the silica gel. Storing seeds with silica gel in order to keep them dry is practically applicable to small seed lots, e.g. those stored in glass jars.

3.11 Potential Seed Damage During Processing

Seed processing aims to achieve a balance between maximising effectivity (extraction, cleaning, protection against deterioration) and damage to the seeds. In practice, processing always implies a risk of damage or injury to some seeds. Damage may occur in various ways:

1. *Mechanical damage.* Usually on the seed coats but occasionally on the embryos with well-developed seed cotyledons. Generally, spherical seeds and small seeds tend to suffer less damage than elongated or irregularly shaped seeds (Bewley and Black 1994).
2. *Heat damage.* Often occurring by exposure to high kiln temperatures for extracting seeds from cones, or deliberate burning for removal of fruit or seed hairs. Fatal high temperatures can also occur during fermentation of fruit pulp. Moist seeds are more prone to heat damage than dry seeds, and recalcitrant seeds are, accordingly, sensitive to heat damage.
3. *Chemical damage.* Sometimes occurring during separation by flotation in organic liquids. Other potential sources are fungicides.
4. *Water damage.* Prolonged submersion in water, e.g. to soften the fruit pulp, may hamper respiration of the seeds. Prolonged soaking may also cause imbibition and initiate germination in seeds with no dormancy. The rate of seed drying is particularly crucial in species with desiccation intolerance.

The severity of the damage depends on the extraction/handling procedure and the seed type:

1. The more fragile the seed, the more sensitive it is to damage. Seeds with thin seed coats or large cotyledons without or with little enclosing endosperm are easily damaged by some processing methods.
2. The more frantic the process, the higher the potential damage. Threshing and beating, e.g. of indehiscent pods imply a potential risk of breaking the embryo. Especially sensitive is the attachment site of the cotyledons to the embryonic axis (More 1972). Mild impact to seed coats can have a beneficial influence on germination by breaking physical dormancy (Chap. 5).

Studies on the mechanical effect of seed processing on seed quality have mainly concentrated on conifers. In comparison with other species, conifers require quite a lot of handling and processing in order to extract the seeds, and their seeds are often subsequently dewinged to ease handling. At the same time, seeds of many conifers are fragile and easily damaged by handling. Handling damage while the seeds were still enclosed in the cones has been reported for *Abies* spp. Throwing sacks of cones to the ground during collection was sufficient to cause quality reduction (Edwards 1981).

Damage during dewinging may occur when seeds are tumbled, e.g. together with debris. Great damage has been encountered in seeds of *Abies lasiocarpa* where up to 50% of the seeds were lost. Excessive dewinging resulted in dull dusty seed coats, susceptible to mould and resulting in weak seedlings (Edwards 1981).

Temperature damage to conifer seeds during extraction in kilns has been widely reported. Most sensitive are immature and moist seeds, and temperature damage is accordingly most likely to occur during the early phases. Potential damage also depends on the length of exposure. Wang et al. (1992) found a substantial loss of viability of *Pinus contorta* var. *latifolia* seed by exposing the cones to a scorching temperature of 220°C for more than 1.5 min. Shorter exposure did not impair viability, possibly because the temperature inside the seed did not rise to a fatal level for the embryo.

A few investigations suggest that seed damage may occur during chemical treatment. Fumigation with carbon disulphide or hydrocyanic acid for killing *Megastigmus* spp. in conifers affected viability (Sweeney et al. 1991). Some alcohols used for flotation had a negative effect on the viability of *Pinus* seeds; others caused no loss in viability (Hodgson 1977).

There is little documentation on mechanical damage to forest seed during processing, but some parallels may be drawn from experience with agricultural seeds. In maize (*Zea mays* L.) various degrees of damage to the seed coat, endosperm and embryo could be detected in 89% of the seeds after processing (Jahufer and Borovoi 1992). The injuries affected germination, seedling development, susceptibility to diseases, plant growth and development and grain yield. The germination rate and seedling quality were influenced by the location of the damage and the embryo, with especially the central part being the more sensitive.

Minor damage to seeds during processing may not immediately affect viability but may cause reduced seedling vigour and misshapen seedlings (Moore 1972). Damage also affects storage potential since injured or deeply bruised areas may serve as centres for infection (Bewley and Black 1994; Brandenburg 1983; Moore 1972; Veira et al. 1994). This may partly be caused by an accelerated progressive deterioration (Chap. 4) or interaction with other deteriorating factors, e.g. increased susceptibility to fungal infection through cracks in the seed coat. Injuries to or near delicate parts of the embryo are prone to both primary and secondary deterioration.

While heat damage is most likely to affect moist seeds, dry seeds seem more susceptible to mechanical damage (Moore 1972); therefore, it must be advised that seeds should only be moderately dried before mechanical treatment, e.g. extraction, dewinging and cleaning.

The desiccation rate has been shown to have crucial effects on desiccation tolerance in desiccation-sensitive seed (Sacande et al. 2004). Fast drying appears to be less damaging than slow drying. An explanation has been suggested that during slow drying seeds spend a longer time with intermediate water content, which appears to be more damaging than both higher and lower moisture content (Pammenter and Berjak 1999; Peran et al. 2004).

3.12 Safety Precautions During Processing

As with seed collection, processing implies both general and specific safety hazards. Processing staff should be familiar with these potential risks and observe appropriate precautions:

1. *Fire danger.* Dry fruit parts, resin and dust released during processing of dry fruits can easily catch fire and therefore pose a fire hazard (Morandini 1962). Use of artificial heat or other electric appliances during extraction increases the danger. Dust may catch fire when coming into direct contact with glow wires or the like. Therefore, heat sources should be safely shielded and dust removed regularly during processing. Water and/or fire extinguishers should be readily available at the seed-processing unit.
2. *Respiratory, eye and skin irritations.* During processing, floral parts, fungal spores, dry pulp and other fine particles become suspended in the air and form what is commonly known as dust. Some species, e.g. acacias, are known to release especially large amounts of dust when threshed.
   Dry dust causes a general irritation of eyes, nose and skin, with resulting itchiness, coughing and sneezing. For most people this is merely annoying, but for some people, some dust elements cause allergic reactions. Dust problems can be minimised by appropriate ventilation, possibly by outdoor handling. Threshing machines and other equipment which release large amounts of dust should be provided with extractors, e.g. powerful vacuum cleaners placed as close to the dust source as possible. Staff working with species or equipment with particular dust problems should be provided with dust masks (Fig. 3.36) and possibly also dust glasses. Regular vacuum cleaning of processing rooms can reduce the dust problem.

Softening the fruit flesh of sugar palm (*Arenga pinnata*) by soaking prior to depulping requires caution. Decomposing fruits develop a fluid causing intense itching and burning whenever it comes into contact with the skin. Also contact with the seed coat can cause skin irritation (Masano 1990). Processing of seeds of *Platanus* spp. and several species of the family Boraginaceae are known to produce similar skin irritations. Rubber gloves must be used when handling these fruits.

3. *Mechanical equipment.* The risk of accidents with mechanical equipment such as threshers and grinders can be greatly reduced by safe construction and maintenance of the equipment and appropriate training and instruction of the operators. Potentially dangerous mechanical or electrical parts (rotating devices, cords, etc.) should be shielded with screens. Screens should be mounted in the front of inlets to, for example, threshers, and operators should observe a safe distance. Emergency switches should be positioned near the place of operation so that machines can easily be stopped in the case of an accident.

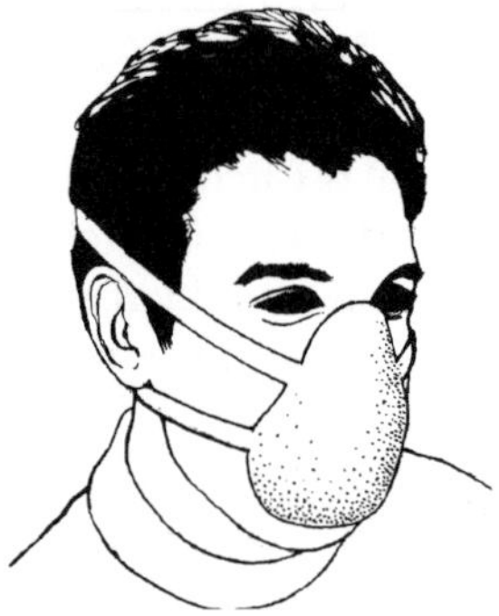

Fig. 3.36. Disposable dust mask covering mouth and nose, used during fruit and seed processing where excessive dust is released. (P. Andersen)

4. *Poisonous fruit pulp.* Some fruits like *Strycnus* spp. have poisonous pulp, fatal to humans and livestock. Removed pulp and water used for extraction must be discharged and disposed of safely.

3.13 Maintaining Identity During Processing

During processing the fruits and later the seeds pass through a number of processes, they are unloaded and loaded into different containers and processing equipment, and are often handled by a number of people. The risk of losing or accidentally mixing labels is obviously high, especially when handling a number of minor samples of the same species, e.g. single tree collections or provenance collections. A system must be created to minimise the risk of losing seed identity. Handling of labels is, in many cases, as important as handling of the seed itself. Simple routine procedures are recommended. Even if some members of the staff are not able to read the labels, they should still be able to maintain the routines. Some points are summarised below:

- Two labels should always follow the seed lot during collection. One is placed outside the container, one is put inside together with the seeds. The labels should be written with water-repellent ink; the labels should be resistant to some degree of moisture.
- Labels that are no longer valid should be discarded to avoid later confusion, e.g. if new labels are written because the old ones become difficult to read, or if several seed lots are mixed.
- When fruits or seeds are poured into, e.g. trays, depulping or cleaning machines where the label cannot be kept with the fruits or seeds, or where it would be easily lost by wetting or blowing away, the labels should be clipped or stuck to the processing equipment. Once the particular processing part has been concluded, the label must be placed with the processed seeds.
- Partly processed seeds are preferably put into the same containers again. After reduction of the major bulk (e.g. after extraction) fewer, smaller or different types of containers may be used. The new containers must be labelled, and redundant labels discarded.
- If part of the seed is fully processed and another part needs additional processing, the two parts must be separated and labelled individually, e.g. A, B, C, ...
- Discarded labels should be torn or removed completely from the processing site (not just thrown on the floor) in order that they will not later be confused with valid labels.

A second point in maintaining identity relates to the risk of physically contaminating the seed lots. If the seeds are to be used for trials, contamination may completely distort the results. It is rarely possible to clean a seed lot for seeds of the same species, and separation of seeds of some species with very similar seeds may also be impossible. Therefore, contamination is often irreversible.

The risks of contamination during seed processing are many. Light seeds may blow from one seed lot to another; perforations in containers or trays may cause seeds to slip from one container or tray to the next if stacked; seed may be stuck in storage containers or processing equipment, especially tiny seeds like eucalypts, *Anthocephalus* or casuarinas. Hygiene routines must be followed:

1. The same containers are used before and after part-processing.
2. Emptied containers are thoroughly cleaned before they are used for any other seed lot. Bags are turned inside out to be cleaned in stitching and corners.
3. Processing equipment is thoroughly cleaned after each process. Brushing and the use of compressed air or a strong water current is often necessary for appropriate cleaning.

Seed Storage 4

4.1 Introduction

Seed storage normally refers to any prolonged safekeeping of seed material, which is beyond a mere delay of the processing or distribution chain, i.e. storage would have a purpose in itself. However, there is no sharp distinction between 'keeping' seed for a while until it can be distributed or planted and storing it for the same purpose. The aim, whether keeping or storing seed, is to maintain viability.

Seed supply systems generally aim at minimising the storage requirement. Seed procurement may be demand-driven in the sense that the seed supplier collects seed he knows is needed, based on experience and/or actual orders. If this works, a large part of the seed may not need any storage. However, seed storage is an essential part of a seed procurement system, the main purpose of which is the following:

1. *To secure the supply of good-quality seed for a planting programme whenever needed.* Storage is necessary if there is a timely delay from collection until practical sowing. This is normally the case in seasonal climates with a relatively short planting season. Many species produce seed (or good seed crops) at long intervals, ranging from a few years to many years. To ensure seed supply during the period between two good seed crops, a seed stock should be established (Wang 1975).
2. *To rationalise seed collection.* Collection from far-away sources is costly but may be rationalised if only carried out occasionally. Expeditions would thus typically collect surplus seed to cover several years' supply rather than to undertake collection every year.
3. *Opportunistic collection.* Species or sources with occasional high production may justify collection beyond actual demand. Surplus seed can be stored temporarily until sold or distributed.
4. *Business stock.* The seed business has a shopping aspect. Some customers would like to see what they buy when they buy it. And most customers want seed when they order it.

An operational seed store thus primarily serves as a buffer between production and demand in the same way as any other production storage: seeds are stored during periods of seed production and shipped to nurseries or other recipients when required. Storage facilities thus have a regular turnover. The quantity and duration of seed to be stored depends on the supply–demand balance, the storage physiology of particular species and the cost and other constraints by keeping seed in storage.

Seed is living plant material and many seeds are designed to survive for long periods of unfavourable germination conditions. In some types of 'orthodox' seeds there is practically no limitation in storability in the sense that there is no significant decline in viability within the operational or potential storage duration (say 10–15 years) (Hong and Ellis 2002). By improving storage conditions, e.g. reducing temperature and moisture content, storage potential for orthodox species can be improved. Storage is thus primarily determined by economic considerations of the advantage of keeping a buffer stock.

A different storage physiology is found in so-called recalcitrant seeds, which are characterised by their intolerance to conditions normally conducive to seed storage, i.e. dry and cool. Because of their low tolerance to desiccation, they are also called 'desiccation-sensitive' or 'desiccation-intolerant'. Storage of desiccation-sensitive seeds is limited by their inert seed physiology. However, within this limit it has been possible, at least for some species, to prolong storability significantly by adjusting drying rate, drying to lowest safe moisture content (LSMC) and storing at a relatively low temperature (Sacande et al. 2004).

Seeds are concentrated packages of genetic material designed to grow into mature trees. If vegetative populations are threatened and disappearing, conservation of genes as a reserve population is sometimes applicable (Linington 2003). Gene banks can have different roles. In some cases they can be used to conserve a last bit of genetic material from a vanishing natural source. In other cases they are used as reserve populations of genetic material, which are not worth actually growing but containing potential that may be useful in future tree improvement. In these so-called gene banks, seeds (and sometimes other propagation material) are stored for long periods at very low moisture content and temperature (cryopreservation) (Marzalina and Normah 2002; Smith et al. 2003). The techniques applied for storage at ultralow temperatures are quite different from those for conventional seed storage. As they are outside the scope of operational forestry, they will only be mentioned briefly in this chapter.

4.2 Storability and Metabolism

Storability or storage potential refers to the inert or inherited ability of species to maintain viability for a certain period under 'ideal' conditions. Storage potential is closely connected to the ability to develop and maintain a condition of physiological inactivity during storage. The 'ideal' conditions are conditions that reduce physiological activity, generally low temperature and water content. Non-ideal conditions, i.e. high moisture content and temperature, are consequently accompanied by physiological activity. This can lead to one of two results: seeds germinate or they deteriorate. The conditions between physiological dormancy and germination represent a range: conditions that are too dry or too cold to initiate germination, yet too moist and too warm to stop physiological activity are generally unfavourable storage conditions.

Loss of viability is a progressive process that may be caused only by internal ageing or deteriorating events, e.g. denaturation of cell components. For orthodox seed, these events are slowed down by desiccation and cooling. In recalcitrant seed, desiccation beyond a critical level (critical moisture content) or LSMC, which is different from species to species) causes irreversible damage to cells. Moisture in connection with higher temperature has both direct and indirect effects. The direct effects are that moisture causes physiological activities (respiration) which can cause accumulation of toxic metabolites, which in turn cause deterioration (Pammenter and Berjak 1999). Indirectly increased moisture and temperature are conducive to growth and activities of pest and pathogens, which in turn damage the seed (Agarwal and Sinclair 1987).

The general rule for orthodox seed is to dry to the lowest (practically) possible moisture content and then cool it down to the lowest (practically) possible temperature. A combination of say 5% moisture content and –5°C will practically stop metabolism and thus greatly reduce internal ageing, and also block most external damaging factors such as insects, mites and fungi (Fig. 4.3).

Recalcitrant seed imposes a real storage problem. Metabolism cannot be 'switched off' and the best storage conditions are thus the best combination between maintaining enough physiological activity to keep seeds alive and yet avoid germination, an 'idling' metabolism. Desiccation sensitivity and intolerance to low temperature do in practice exclude pest and pathogen management via moisture and temperature control for these seeds. In fact, in terms of pest and pathogen management the lowest physiological activity may be very poor conditions, as such conditions may reduce the inert resistance of seeds. Slow-germinating plants are thus known to be more susceptible to fungal attack than fast-growing ones.

4.3 Classification of Storage Physiology

The traditional classification of seeds according to their storage physiology in 'orthodox' and 'recalcitrant' (Roberts 1973a) is a linguistic curiosity. Orthodox means the normal, the right, how it should be. Recalcitrant means the difficult, problematic or against what is rational or sensible. This type of classification, which has become commonly used in seed handling literature, takes its point of departure in the orthodox type as the 'normal' and the rest as deviating from this norm. The terminology became even more confusing with the term intermediate', originally referring to a definite category of desiccation-tolerant but temperature-sensitive species (Ellis et al. 1990; Table 4.1). The 'intermediate' group is now mostly used for a more continuous or variable storage behaviour in-between that of orthodox and recalcitrant. Most recent literature also prefers the more illustrative term 'desiccation-sensitive' or 'desiccation-intolerant' for the term 'recalcitrant' (Pammenter and Berjak 1999; Sacande et al. 2004[1]). In the same way 'desiccation-tolerant' is becoming more used for orthodox.

Orthodox seeds are seeds that tolerate drying and subsequent cooling and storage at low temperature. Drying takes place as a part of maturation ('maturation drying') and may continue after dispersal. 'Dry' can be anything between 2 and 10% moisture content Very low moisture contents are, however, primarily found in seeds with relatively large non-embryonic material (fruit or seed coat with natural low water affinity). Most orthodox seeds tolerate drying to at least 5% moisture content (Berjak and Pammenter 2002). However, under a certain lower limit of moisture content, orthodox seeds may experience an indirect 'desiccation damage', e.g.:

1. Susceptibility to mechanical damage as structural water disappears and embryos become more 'fragile'.
2. Temperature damage during drying. Although very dry seeds are quite resistant to high temperatures during drying, seeds do not survive, e.g. oven drying used during testing. Air drying to very low moisture content requires high temperature. Cold desiccation, e.g. using silica gel or 'freeze drying', is effective only down to 'normal' 3–5% moisture content.
3. Imbibition damage. In particular, a slow imbibition rate appears to be potentially damaging for very dry seeds (Walters et al. 2001; Peran et al. 2004).

Table 4.1. Physiological storage classes as related to temperature and moisture content. 'Low' and 'high' are relative concepts and are not directly comparable. For example, low moisture content for intermediate seed is, for example, 10–12%; low moisture content for orthodox seed is 4–7%

	Orthodox Seed	Intermediate Seed	Temperate recalcitrant seed	Tropical recalcitrant seed
Storage moisture content	Low	Low	High	High
Storage temperature	Low	High	Low	High

[1] This reference contains a number of relevant papers; where statements or findings occur in several papers, reference is made to the compilation (Sacande et al. 2004).

Temperature tolerance is linked to moisture content in the way that the lower the moisture content, the lower usually the temperature tolerance. Seeds with 10% moisture content can be damaged by subzero temperature, while seeds dried to 4–5% moisture content may be frozen to −20°C or even lower (Gamene et al. 2004; Omondi 2004; Gamene and Eriksen 2004). Sporadic exceptions exist: Merritt et al. (2003) found, for instance, that both germination and seedling vigour were reduced for two Western Australian species after storage for 18 months at −18°C compared with storage at 23°C. However, within the practically applied range of moisture content and temperatures, any lowering of either temperature or moisture content will prolong viability for orthodox seed. Harrington (1972) suggested a rule of thumb that every 1% reduction in moisture content or every 5.6°C reduction in temperature will approximately double storage life. The rule is claimed to be valid for moisture reduction down to 4–5%, depending on the species. More exact models have been elaborated for predicting storage life under different sets of storage conditions for different species (Roberts 1973a; Ellis 1986, 1988). The linearity of ageing predicted by the mathematical equations is, however, limited to a range of 'typical' storage conditions (Hong and Ellis 2002). Orthodox seed is by far the most common, making up the prevailing storage behaviour in more than 90% of all seed plants (Tweddle et al. 2003). Orthodoxy dominates in all dry and seasonal environments and is prevalent in pioneer species in humid climates (Farnsworth 2000). But also almost 50% of late successional species in humid areas have orthodox seed (Tweddle et al. 2003).

Recalcitrant or desiccation-sensitive seed has loosely been identified from species which do not follow this 'normal' behaviour, i.e. species which maintain a high moisture content at maturity (often more than 30–50%) and undergo very little maturation drying (Berjak and Pammenter 1996, 2002). The seeds are sensitive to desiccation below a certain level (LSMC *sensu* Tompsett 1992). The LSMC differs between species but often ranges somewhere between 12 and 30% (Tompsett 1992) for desiccation-sensitive seed. Owing to desiccation sensitivity, the seeds cannot be dried to a level where they switch off metabolism and they rapidly lose viability under any kind of storage condition. Their inherited storage potential is thus generally low, although some species can be kept for several months in an imbibed stage or cool with reduced moisture content. A number of other characteristics of the two groups are listed in Table 4.2.

The recalcitrant group contains a wide variation in terms of temperature and desiccation tolerance and storability under various conditions. Desiccation sensitivity is most frequent in humid ecosystems. The mangrove family Rhizophoraceae contains species with extreme desiccation-sensitive and short-lived seed. In fact the species contain no true seed stage as development, maturation and germination are continuous events, which occur while the seeds are still attached to the mother tree (Farnsworth 2000). The phenomenon is

Table 4.2. Summary of some features of orthodox and desiccation-sensitive (recalcitrant) seed

	Orthodox	Desiccation sensitive
Natural occurrence	Dominating strategy in arid and semiarid environments, and pioneers in humid climates. Also prevalent in temperate and tropical high-altitude species	Prevalent in warm humid climates, especially climax forest species of tropical rain forests and mangroves. Also some temperate and a few dry-zone species
Families and genera where the particular storage behaviour prevails	Most families, e.g. prevailing in Myrtaceae, Leguminosae, Pinaceae, Casuarinaceae	Dipterocarpaceae, Rhizophoraceae, Meliaceae, *Artocarpus, Araucaria, Madhuca, Triplochiton, Vitellaria, Agathis, Syzygium, Quercus.*
Seed moisture content and temperature during storage	Tolerant to desiccation and low temperatures. Conventional storing 5–7% moisture content and 0–5°C. Cryopreservation 2–4% moisture content and −15 to −20°C	Intolerant to desiccation and low temperatures (except some temperate recalcitrant species). Tolerance level dependent on species, normally minimum of 20–35% moisture content and 12–15°C for tropical species
Potential storage period	With optimal storage conditions several years for most species; for some several decades	From a few days for extremely recalcitrant species to several months for more tolerant ones
Seed characters	Small to medium-sized seed, often with a hard seed-coat	Usually medium-sized to large and heavy seeds; this is partly attributed to a high moisture content
Maturation characters	Accumulation of dry weight ceases before maturation. Decline of moisture content typically to 6–10% at maturity. Little variation between individual seeds	Accumulation of dry weight up to the time of seed dispersal. Little or no maturation drying, moisture content at maturity 30–70% with large variation between individual seeds
Dormancy	Often occurs	Absent or weak. Maturation and germination often more or less continuous
Metabolism at maturity	Not metabolically active when shed	Metabolically active when shed

called vivipary or precocious germination (Sect. 6.2). *Avicennia* spp., another mangrove species, also show extreme desiccation sensitivity, although the species are usually not viviparous (Le Tam et al. 2004). Most Dipterocarpaceae are recalcitrant. Some *Vatica* spp. lose viability in few days and so do many other species in this family. *Dipterocarpus alatus* is comparatively desiccation tolerant (a short-lived orthodox) and *Dipterocarpus imbricatus* is somewhere in-between. Recent research has shown that high moisture content at the time of collection does not necessarily mean intolerance to desiccation. Several species from moist forest, which are usually dispersed with high moisture content (more than 45–50%) and germinate readily, have shown desiccation tolerance to less than 10%, which in turn makes at least some months' storage possible (Sacande et al. 2004; cf. Fig. 4.6). Although desiccation sensitivity is not common in dry ecosystems, it does occur. Danthu et al. (2000) found, for instance, desiccation-sensitive seeds in four species from the Sahelian–Sudanean area.

Tropical recalcitrant seeds are normally sensitive to low temperature. Chilling damage often occurs at 15–20°C (Sacande et al. 2004). However, recalcitrant species of *Cordia* and *Vitex* in Kenya tolerate storage temperatures of 2°C (Schaefer 1991), and Bonner (1996b) reported low temperature tolerance in recalcitrant subtropical *Citrus* spp. Desiccation-sensitive species tolerant to low temperatures, even slight frost, are common in temperate areas and in tropical highland. Recalcitrant cold-tolerant species are common in temperate genera *Fagus*, *Quercus*, *Lithocarpus*, *Castanea* and *Coryllus*, but also occur in the mainly tropical high-altitude species of *Illicium verum*, *Cinnamomum cassia* and *Michelia mediocris* (Pritchard et al. 2004; Kha et al. 2004). Temperate recalcitrant species are sometimes regarded as a distinct and definite category (Table 4.1). Desiccation intolerance is most common in wetland or flooded environments where it prevails in many different families and genera (Farnsworth 2000). Some species of dry and seasonal areas also have desiccation-sensitive seeds (Tweddle et al. 2003; Pammenter and Berjak 2000).

A storage behaviour showing sensitivity to low temperature even with relatively low moisture content was first described for coffee and was later found for a handful of species. The category was originally classified as 'intermediate' (Ellis et al. 1990). This completes the theoretical model as depicted in Table 4.1.

Research on a number of species has suggested that the storage physiology from orthodox to recalcitrant contains many intermediate stages. Dickie and Smith (1995) found that the critical moisture content, below which viability was impaired, was 5 and 7% for *Agathis australis* and *Agathis macrophylla*, respectively. These species were classified 'suborthodox'.

The whole spectrum from orthodox to recalcitrant seems to cover an open-endedness and continuum across species (Pammenter and Berjak 1999, 2000; Figs. 4.1, 4.6). At one end of the scale, seeds are extremely orthodox, the viability of which under optimal conditions will last for decades or centuries (Farrent et al. 1988). At the other end, there are extremely recalcitrant seeds, which lose viability in a few days no matter how they are stored. Although the LSMC covers a wide range of critical levels, some authors suggest that there are discrete levels of critical water potential among desiccation-sensitive seeds (Sun and Liang 2001).

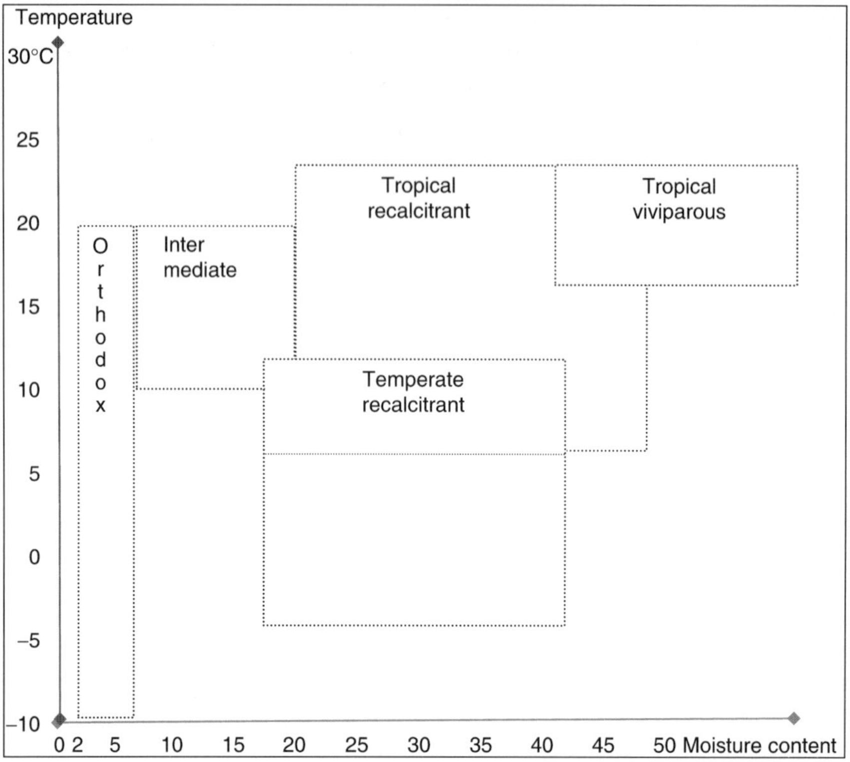

Fig. 4.1. Model of continuity of seed-storage physiology. Orthodox seed includes the largest number of species (more than 90%) and is a relatively distinct and well-defined group

4.4 Ecophysiological Role of Storage

With the few exceptions of viviparous seed, seeds go through a physiologically inactive (quiescent) period from maturity to germination. The period of quiescence may be short, if the seed falls in an environment conducive to germination, or long if dispersal and germination are delayed. Dormant seed (Chap. 5) has an additional mechanism to delay germination in order to increase the chances for seedling survival. Seeds deposited at a site unfavourable to germination may stay dormant for a period waiting for conditions to improve. If conditions remain unfavourable to initiate germination, seed ageing and possible predation will gradually eliminate seed. For recalcitrant seed, the survival period is short – often a matter of days. For long-living orthodox seed, e.g. seed of some legumes, subsequent seed deposits may accumulate and with time build up quite large soil seed banks (Auld 1986; Holmes et al. 1987; Leck et al. 1989; Doran et al. 1983; Cochard and Jackes 2005).

4.5 Seed Longevity

Archaeological excavations suggest that an orthodox crop seed can remain viable for hundreds of years in a dry climate. This may be true for orthodox tree seed as well, but actual long-term storage records of forest seed are rare and are measured in decades rather than centuries. In *Sophora chrysophylla*, Norton et al. (2002) found a high viability (84%) of 24–40-year-old seed. The potential storage period, seed longevity, is determined by an interaction between genetic storage potential, the physiological conditions and the storage conditions. Storage conditions only refer to those with an impact on physiological ageing and not, e.g. predation or instant destruction.

1. *Genetic.* Storage potential is inherited. The two main genetic groups are the orthodox and the desiccation-sensitive. Although there are species and even provenance variations in orthodox seeds, most orthodox species are quite similar (Bonner et al. 1994). The prototype mechanism of orthodoxy is to pack storage material as tightly as possible and then switch off metabolism completely. This feature gives room for little variation. When variation between different orthodox seeds is observed, it is mostly caused by variation in the extent of age damage to cell components and their subsequent repair and turnover. This is manifested in genetic variation of storage potential. For orthodox seed Ellis and Roberts (1980) suggest that within a species there may be more than a sevenfold genotypic variation in seed longevity. Seed lot variation for storability has been documented from different land races (Lauridsen and Souvannavong 1993), different provenances (Emmanuel and Dharmaswamy 1991) different mother trees (Oloo et al. 1996) and different clones (Chaisurisri et al. 1993).
   Variation is much higher when looking at individual seeds in a seed lot as represented by a normal viability graph: a few seeds lose viability relatively early in storage, while the last few may stay viable for a long time. If, for example, viability declines from 100 to say 99% in 3 months, and is 1% after 5 years, it means that the longevity of 1% of the seed lot was only 3 months, while another 1% remained viable for 5 years, i.e a 40-fold difference within a seed lot.

Species with large provenance variation also often show large individual variation. For example, in neem, *Azadirachta indica*, variation ranges from near recalcitrant behaviour for humid Southeast Asian provenances to near orthodox for dry East African land races (Lauridsen and Souvannavong 1993). In Kenya within-provenance variation was shown between ten mother trees of neem. They showed a decline in viability after 4 months from 74 to 60% for the most orthodox seed lot, and from 79 to 4% for the most desiccation-sensitive seeds.

2. *Physiological conditions.* Seeds collected at an early maturity stage generally have a shorter storability than seeds picked at full maturity, even if the initial germination was the same (TeKrony 2003; Seeber and Agpaoa 1976). Storability thus develops later than the ability to germinate. The physiological cause of reduced storability may be ascribed to failure to accomplish essential stages of late maturation events, e.g. incomplete embryo development, inadequate protection from desiccation or inadequate formation of storage proteins or chemical compounds necessary for storability (Hong and Ellis 1990). For example, in *Taxus brevifolia* the embryo grows in size right up to the stage of full maturity, and only fully mature seeds tolerate desiccation to a level necessary for storage (Vertucci et al. 1996). However, as stated in Chap. 3, seeds collected early may be after-ripened to attain full maturity, including normal storability. The developmental stage is especially evident and important in recalcitrant seed. Firstly, because dry weight continues to accumulate up to the time of seed maturity, so seeds collected just before natural shedding may be underdeveloped. Secondly, because the processes of maturation and germination are more or less continuous. If germination does not occur, deterioration proceeds rapidly, making late collection equally unsuitable (Berjak and Pammenter 1996). Marambe et al. (1998) found that seeds from immature fruits of neem (*Azadirachta indica*) stored better (viability decline from 80 to 69%) than seeds from ripe fruit (from 80 to 42%) after 12 weeks' storage at 4°C and 8% moisture content. A probable explanation is that immature seeds after-ripen during the beginning of the storage period, while mature ones are quicker to enter the stage of ageing. Any physiological damage happening prior to storage will affect storability. Such damage could happen during processing, e.g. mechanical or temperature damage (Moore 1972). In practice, the physiological conditions of seed are measured as a high initial germination before storage: Seed lots with high initial viability have a higher longevity in storage than that of seed with low initial viability.

The progression of natural ageing with resultant loss of viability typically follows a sigmoid pattern as indicated in Fig. 4.2. Loss of viability is initially slow, followed by a period of rapid decline. The higher the viability when the seed lot enters into storage, the longer the seed will remain viable under a given storage environment.

Prestorage conditions may strongly influence the response to storage conditions. 'Vigorous', high-quality seed of most species store surprisingly well even under relatively adverse conditions, while badly deteriorated seeds store poorly even though conditions are quite favourable' (Delouche et al. 1973). This is a product of both the progressive ageing and the role of fungi and microorganisms during storage.

3. *Storage conditions.* In general, any storage condition that will reduce physiological activity without physiologically damaging the seed is ideal. Storage conditions closely interact with the physiological conditions of seed. Dry, sound, undamaged seed of orthodox seed may store well under ambient conditions (Box. 4.1). Damaged and infected seed may easily deteriorate under such conditions but maintain a relatively high viability under good storage conditions. For example, minor damage to the seed coats that may serve as entry points for fungal attack hampers storage under storage conditions where fungi are active, i.e. more than 5–7% moisture content, while such damage may be of no harm at lower moisture content. Temperature and humidity are the most important factors in seed storage and, compared with the atmosphere, which also plays a role, the easiest to regulate – reduction of either of them improves storability. Humidity interacts with seed moisture content (Chap. 3). Non-dormant seeds may germinate if their moisture content is above 30%. Rapid deterioration by microorganisms can occur if the moisture content is 14–30%, and seeds with a moisture content above 14–20% respire and metabolise actively. Metabolising seeds may be damaged by accumulation of toxic metabolites or heat if improperly ventilated. Certain seed insects are active at a moisture content of less than 10%, and damage by fungi may occur down to 4–5% (Bewley and Black 1994). For all the above reasons, it follows that the higher the storage moisture content, the more rapid will be the deterioration of the seed. Figure 4.3 illustrates this. Thus, it follows that storage conditions for orthodox seed are outside the 'critical area' of physiological damage, while 'safe' storage conditions for recalcitrant seed coincide with the conditions in which both seed metabolism and activities of insects and fungi prevail.

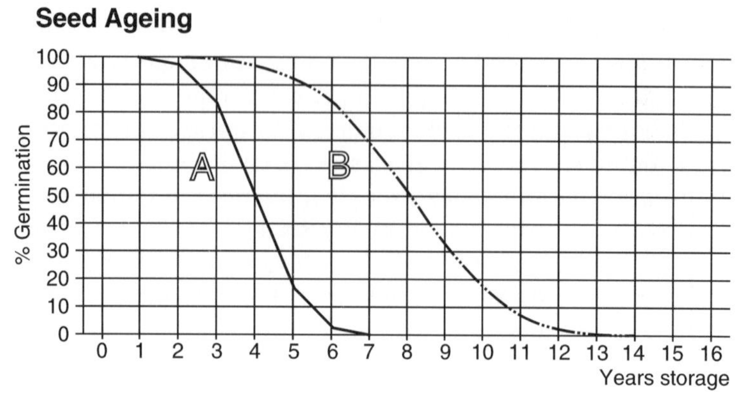

Fig. 4.2. Survival or viability curves indicating decline in the percentage of viable seeds in a seed lot during a storage period. Survival curves of seed lots of a given species tend to follow the same pattern under a given set of storage conditions; hence, curves *A* and *B* could be two species under similar storage conditions, or two seed lots of the same species exposed to different storage conditions. Compare with Fig. 4.5

Box 4.1

Storage conditions

Storage conditions should be designed to prolong the viability of seeds by reducing or limiting any factor that impairs viability. The general storage conditions should therefore aim at:

1. Reducing the metabolism of seeds
2. Keeping insects, fungi and other pathogens away
3. Reducing general seed ageing

The general prescriptions for seed storage are thus:

- Store seeds at the lowest possible temperature that will not damage the seeds.
- Store seeds with the lowest possible moisture content that will not damage the seeds.
- Eliminate as many pathogens as possible before storage.
- Protect seeds from pathogens during storage.
- Store in the dark.
- Store orthodox and intermediate seeds with low moisture content in airtight containers.
- Store recalcitrant seeds in material permeable to gases but that retains of moisture.

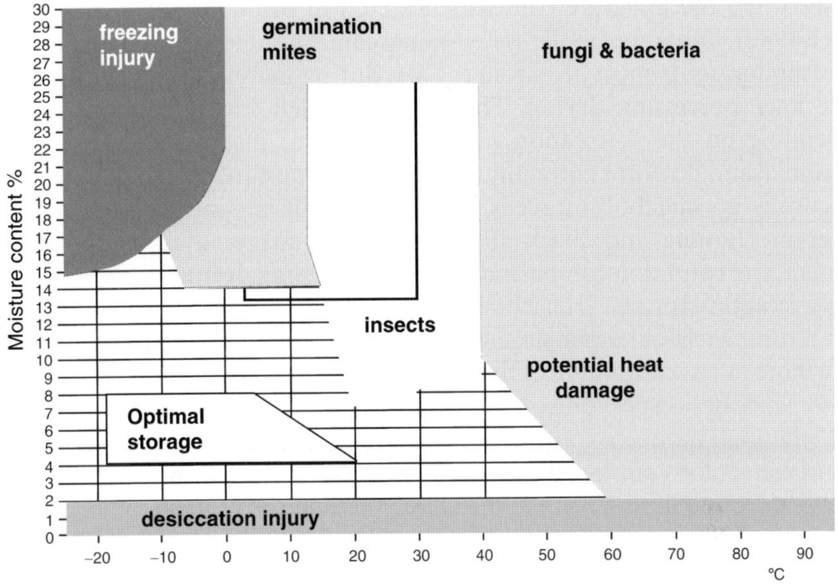

Fig. 4.3. Storage conditions of seed. 'Safe storage' is the storage condition in which viability is maintained as long as possible. Safe storage conditions for orthodox seed are low temperature and humidity in which both physiological activity of seeds are low and where conditions for infecting and infesting organisms are poor. (Redrawn from Roberts 1972)

4.6 Seed Ageing, A Physiological Background

Ageing in living organisms refers to the gradual exhaustion of life processes, which are necessary to maintain adequate physiological activity to survive and compete, which in turn include, for example, the ability to rejuvenate and regenerate tissue. Seed ageing has a slightly different character because it occurs in inactive living material. Compared with ageing in living plants, seed ageing represents primarily deterioration of the apparatus, rather than a wearing out or senescence of the organism.

4.6.1 Desiccation and Metabolism

Maturation drying, which occurs in practically all seeds, including recalcitrant ones, causes reduction of metabolic activity as water is an important component in metabolism. Respiration requires relatively high amounts of water and ceases when the moisture content is lowered to below 14–20% (Bewley and Black 1982). Other metabolic systems may continue at lower moisture content. In orthodox seeds most 'free' water is lost during maturation drying and possible later processing drying. The little water left in the seeds (e.g. 4–6% depending on the desiccation rate) is 'bound' to macromolecules, i.e. it is immobile and does not enter into chemical reactions. In dry seeds there is thus practically no metabolism; seeds are alive without any measurable life manifestation (Bewley and Black 1994). For orthodox seed of high moisture content, the condition can be 'induced' by processing drying.

Desiccation-sensitive (recalcitrant) seed does not possess the ability to enter into a completely quiescent stage without metabolism. Here germination events are a more or less continuum of the maturation processes (Berjak and Pammenter 1996). Since desiccation causes damage, the moisture content must be kept high, and with high moisture content seeds remain metabolically active. However, the *rate* of metabolism can usually be significantly reduced by storing seeds at reduced temperature and drying them to the LSMC (Sacande et al. 2004).

4.6.2 Physiological Changes During Ageing

Ageing denotes the progression of deteriorating events that take place within the seed and which ultimately lead to the death of the seed (Roberts 1972). The term 'progression' suggests that ageing takes place over a prolonged period, during which cytological and biochemical deterioration accumulate. Ageing does, accordingly, not include momentary loss of viability owing to instant damage, e.g. by temperature or mechanical impact. Insect predation is in this connection not considered as ageing, while fungal infection, being a more progressive process of deterioration, may be part of or closely linked to the ageing process (Roberts 1972). Seed ageing is always deteriorating and the ultimate effect is a decline in viability. However, it is sometimes observed that stored or accelerated aged seed shows increased germination compared with fresh seed (Masilinami et al. 2002). This can be ascribed to an after-ripening effect or a break of seed dormancy during storage of the accelerated ageing treatment (Chap. 3).

Physiological ageing is influenced by both internal and external factors and the interaction between the two. Internal factors include, for example, damage to cell membranes and organelles, denaturation of enzymes and oil, and accumulation of toxic metabolites. The direct influence of external factors is primarily fungal infection; indirectly the external factors (temperature and moisture) accelerate or reduce the rate of intrinsic factors (Fig. 4.4). Dry and cool orthodox seeds have no metabolism and thus no accumulation of metabolites and no fungi, so the influence of extrinsic factors is negligible. For a detailed discussion on the cytological and biochemical factors in seed ageing,

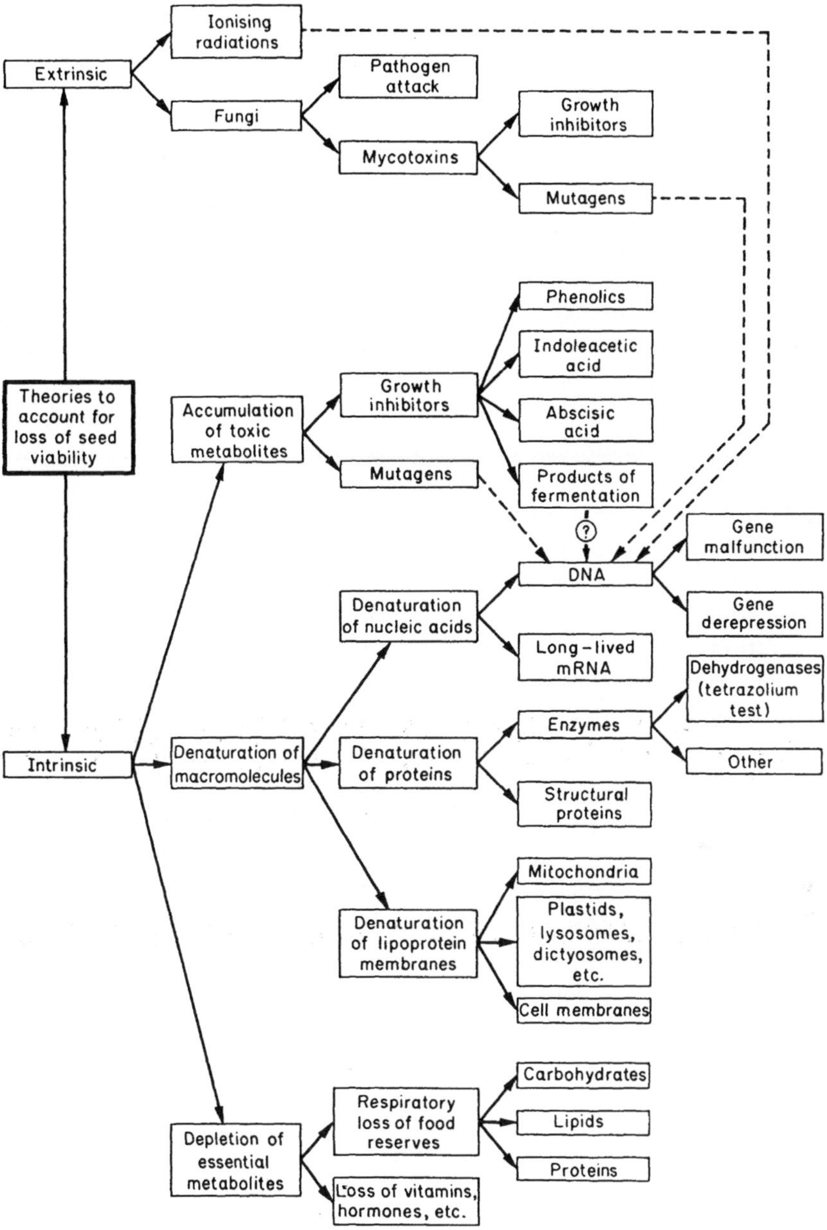

Fig. 4.4. Summary of factors and events of deterioration leading to seed ageing. (From Roberts 1972)

reference is made to Roberts (1972, 1973b, c), Bewley and Black (1982, 1994), Heydecker (1972), Wilson and McDonald (1986), Berjak and Villiers (1972a–d) and Pammenter and Berjak (1999).

As long as physiological processes prevail, possible damage to cells and organelles is repaired. When the processes are switched off, damage will accumulate. For orthodox seed, deterioration will thus start when the moisture content becomes low, i.e. at maturity. However, much cytological and biochemical damage is never expressed as it is readily repaired during the first stages of germination, the 'lag phase', when seeds have imbibed and germination processes are initiated (Chap. 6). Ageing initially manifests itself as higher sensitivity to stress factors. Aged seeds germinate under a narrower set of germination conditions ('optimal conditions') and are more prone to stress and damage during germination than fresh seeds (Neya et al. 2004). Aged seeds are said to have less 'vigour' (see further discussion in Chap. 7). The more severe and the more progressed the damage, the more difficult the repair, and the higher the risk that damage temporarily or permanently hampers seed quality. A theoretical critical 'point' is when deterioration has progressed to a stage of irreversible damage (Berjak and Villiers 1972a). At this stage, viability of the seed lot declines, which in turn is an indication that in some seeds the deterioration has progressed beyond the point where seeds can germinate.

Storage conditions have a direct connection to the ageing events in following manners:

1. Temperature influences biochemical processes. The biological optimum is at ambient temperature. In practice, the lower the temperature, the slower the process and thus the slower the deterioration. Low temperature (below 4–10 C) further inactivates most seed insects and storage fungi. However, very low temperature can be damaging for moist seed, e.g. by ice-crystal formation at subzero temperature, and in tropical recalcitrant seed, low temperature reduces essential metabolic processes (Sacande et al. 2004; Prichard et al. 2004). For temperature-sensitive seed, chilling injuries start at about 20°C (see further later). High temperature in connection with high moisture content accelerates ageing (Bewley and Black 1982).
2. Moisture content has various direct and indirect effects. Water in plants is a transport and dissolvent medium, a temperature regulator, a structure component and an essential molecule in numerous biochemical reactions. Most biochemical and cytological events take place in a watery environment. Some types of deterioration take place at high moisture content only, e.g. accumulation of toxic metabolites, denaturation of enzymes and fungal deterioration.

3. Oxygen is a component in virtually all biochemical processes. Seeds which are stored under conditions where aerobic metabolism is necessary (respiration in moist seed including recalcitrant seed) will die quickly if seeds are deprived of oxygen. In dry seeds where metabolic oxygen is not needed, oxygen has mostly a negative effect. For example, denaturation of cell constituents (membranes, enzymes, DNA) only occurs under aerobic conditions (Roberts 1972, 1973b, c). Accordingly, high oxygen pressure promotes and low pressure represses this type of deterioration. Low oxygen pressure is achieved in vacuum storage and storage in CO_2 (see later). In addition to preventing denaturation, low oxygen pressure also prevents insects, fungi and other aerobic microorganism problems at temperatures where they are potentially active.
4. Light could have an indirect influence on storage by preventing fungi, as most fungi prefer darkness. Roberts (1972) suspects ionising radiation influences seed ageing in nature. Under artificial storage conditions, light probably has no influence. In species with photodormancy, dark storage could be used to prevent germination at high moisture content (Vasquez-Yanes and Orozco-Segovia 1996). In practice, however, photodormant seeds are mostly small, orthodox seeds, which are more easily preserved by other means.

4.6.3 Longevity Models

Viability tests carried out at regular intervals during storage of a number of species and for a number of storage conditions have shown that the survival curve typically forms a sigmoid pattern as depicted in Fig. 4.5. The shape and steepness of the curve in relation to a storage time scale depend on both seed lot (species, provenance, maturity stage, moisture content, etc.) and storage conditions (temperature and humidity). The survival curves tend to show two distinct types.Type A shows a prolonged period of relative stability with very little loss of viability – a plateau phase frequently found in orthodox seed of good quality and under good storage conditions. Type B, where the initial 'plateau' is lacking and which immediately enters the rate of cumulative mortality, shows the typical pattern of deteriorated orthodox and recalcitrant seed (Bernald-Lugo and Leopold 1998). Seed lots of the same species with similar initial quality and stored under the same set of storage conditions tend to show the same pattern of decline in viability over the storage period, i.e. similar viability curves. Different storage conditions usually alter the viability

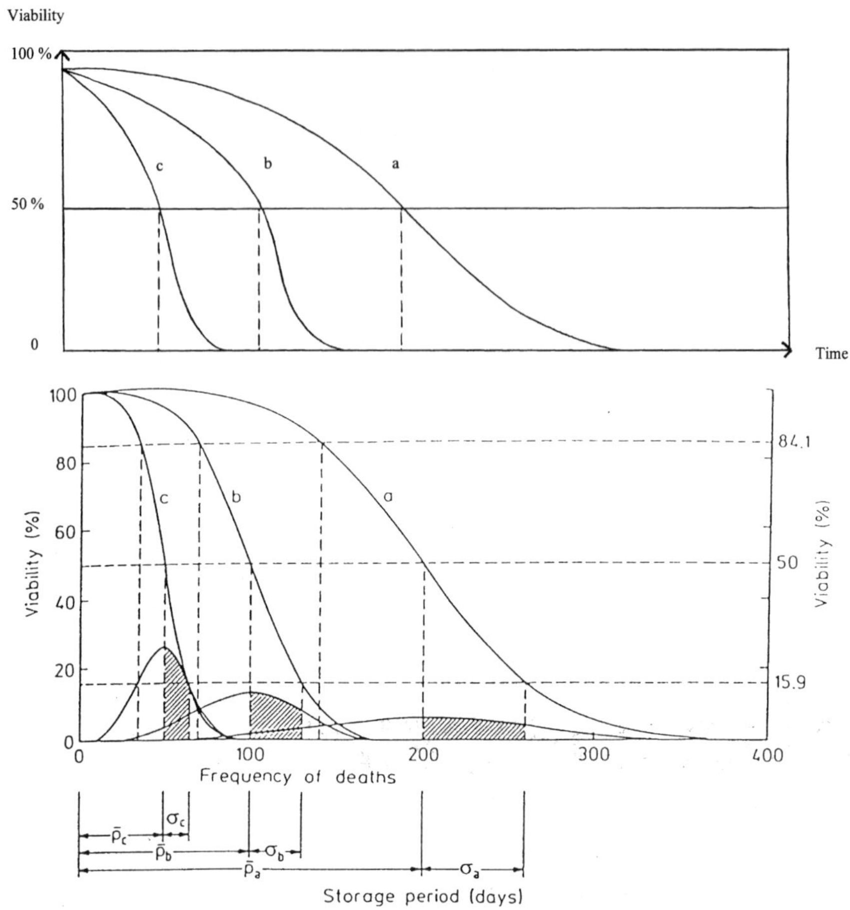

Fig. 4.5. Viability curves for three different storage conditions. *a* represents the best storage condition and *b* and *c* represent less favourable storage conditions. Mean viability periods are found on the time scale by noting the distance from zero to the points where the viability curves intersect with the 50% viability level. On the *right* is the 'frequency of death', which has a normal distribution. (From Roberts 1973a)

curves: the poorer the storage condition, the steeper the slope of the curve (Fig. 4.5).

Viability curves can be used for comparing different storage conditions or different seed lots under similar storage conditions. Developed further, viability curves can be used for predicting the storage life of a seed lot.

The storage conditions differ significantly between the two main groups, orthodox and recalcitrant seeds, and the two groups are considered separately.

Appropriate humidity control and preventing absorption of moisture by airtight storage is particularly important in the ever-humid tropics or during rainy seasons. Cloth bags were thus found inferior to airtight containers during moist-season storage in Nepal (Napier and Robbins 1989). As the relative humidity increases with decreasing temperature, airtight storage is normally mandatory for cold storage.

4.7.2 Temperature

Low temperature prolongs the storage life of seed. Temperature has a direct impact on the ageing processes and an indirect influence via fungal and insect activity. Cool storage means reduced temperature compared with ambient conditions. Ambient temperature under tropical conditions can be anything between 30 and 35°C in tropical lowland to less than 10–15°C in the cold season in the subtropics or in highland conditions. Both seed insects and fungi are active under ambient temperature, and ambient storage always implies a risk of damage by fungi and insects. Precautions where cold conditions are not available may be, for example, fumigation or application of a pesticide, or both (see later).

Cold storage is mandatory if seeds are prone to lose viability at ambient temperature, i.e. for short-term storage of sensitive seeds and any long-term storage. In the Philippines, seeds of *Pinus merkusii* are reported to lose viability within 4 months when stored at ambient temperature. At 2°C they can be stored without significant loss in viability for up to 14 months (Seeber and Agpaoa 1976); hence, any storage beyond a few months of this species must be under reduced temperature. At least two species of eucalypts, *Eucalyptus deglupta* and *Eucalyptus microtheca*, have short viability under ambient conditions and must be stored at low temperature (3–5°C) to maintain viability beyond 2 years (Boland et al. 1980; Table 4.3). Generally, the lower the temperature, the longer the viability provided seeds have low moisture content (with the few exceptions mentioned before for overdried seed). Most orthodox seeds maintain viability for decades under storage temperatures of −10 to −15°C. Such low temperature may, however, only be economic in special cases. Where the availability of cold-storage facilities is limited, only seeds that are likely to lose viability significantly during the potential storage period are stored under cold conditions.

4.7.3 Storage Atmosphere

Since orthodox seeds are preferably dried to moisture contents where they are no longer metabolically active, they do not require oxygen for respiration.

Table 4.3. Effect of cold storage on viability for some orthodox seeds. Temperature interacts with, for example, moisture content. Many dry, orthodox seeds will show little or no decline in viability during a 24-month period. Regarding specieswise information on viability under various storage conditions reference is made to Kaul (1979), Gunn (2001), ATSC (1995) and the seed database of Royal Botanic Gardens, Kew (http://www.kew.org)

	Initial conditions		Viability after 24°months' storage	
Species	Moisture content	Viability	Ambient conditions	Cold storage
Khaya senegalensis	2.5–5	100	98	98 (−4 and 5°C)
Khaya ivorensis	≈6	100	Not available	44 (2°C)
Swietenia macrophylla	2.5–3.5	100	90	85 (5°C)
Cordia alliodora	<10%	≈ 60	2.5	38 (5°C)
Chukrasia tabularis	Not available	Not available	29	59 (4°C)
Chukrasia velutina	Not available	Not available	69	72 (79)
Acacia mellifera	5	99	99	100
Pinus merkusii	≈6	80	40	80
Eucalyptus deglupta[a]	Not available	100	3 (air conditioned)	37 (2–5°C)
Eucalyptus microtheca[a]	Not available	100	20 (air conditioned)	72 (2–5°C)
Casuarina equisetifolia[a]	Not available	100	44 (air conditioned)	100 (2–5°C)
Grevillea robusta[a]	Not available	100	100 (air conditioned)	98 (2–5°C)
Tectona grandis[b]	6.5	NA	21	57 (4°C)

[a] Tested after 5 years' storage (Gunn 2001)
[b] Tested after more than 3 years' storage

Bewley and Black (1994) found that reduction of O_2 pressure, e.g. by replacing O_2 with N_2 or CO_2, had little effect on seed longevity as long as temperature and moisture content were kept low. However, as seed insects and microorganisms respire and hence need O_2 at a moisture content where the seeds themselves do not, replacement of the seed atmosphere with CO_2 is a common, effective and safe method of seed treatment (Sect. 4.11.1).

4.8 Storage of Desiccation-Sensitive and Intermediate Seeds

As the group desiccation-sensitive species has been defined as species whose seed cannot be dried and stored, the heading 'storage of desiccation sensitive seed' is already problematic. The simple connection is that since the seeds cannot be dried, they cannot be stored (Hong and Ellis 2002). However, a lot of research and tests have been completed during the last 10 years, the International Plant Genetic Resources Institute (IPGRI)/Danida Forest Seed Centre (DFSC) project being the most comprehensive (Sacande et al. 2004). The results have shown that the feature of recalcitrance is represented by a large group of species with different desiccation sensitivity (Fig. 4.6). The storage behaviour ranges from the extremely recalcitrant and viviparous seeds of some lowland rainforest and mangrove species to seeds that tolerate a substantial reduction of their maturity moisture content. Viability is generally short and certainly shorter than that of orthodox seed. However, up to 3 years' storage has been reported both for temperate (Suszka and Tylkowski 1982) and tropical species (Corbineau and Come 1988).

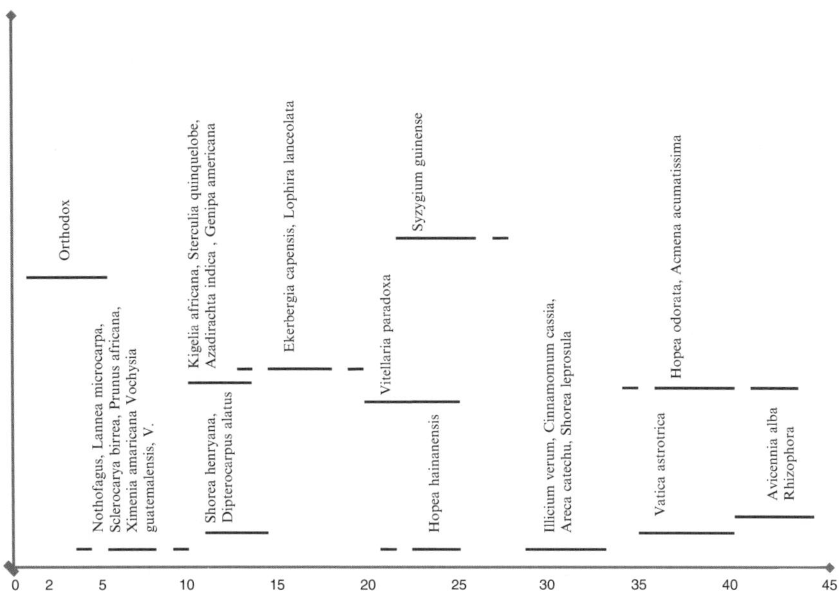

Fig. 4.6. Model of desiccation tolerance of some selected species from the International Plant Genetic Resources Institute/Danida Forest Seed Centre research series (Sacande et al. 2004). The lowest safe moisture content (*LSMC*) is the moisture content below which germination tests show a strong decline. Critical moisture content is in some tests defined as the level where viability is reduced to 50% (Salomao 2004). For practical purposes the LSMC should be considered 'open-ended', a relative level below which damage is likely to occur. There will be some seeds that suffer desiccation damage above that critical level and some seeds that survive desiccation below that critical level. The LSMC covers a continuous range but some LSMCs are more frequent than others, suggesting that there are some discrete levels of critical moisture contents (Sun and Liang 2001)

Because recalcitrant seeds are not designed to become quiescent, storage conditions must allow metabolic processes to proceed but at the lowest possible level. Pest and pathogens are physiologically active under the range of storage conditions suitable for recalcitrant seed, so special precautions are usually needed to control infestation, in particular with fungi.

Storage conditions should basically aim at the following (King and Roberts 1979):

- Prevent desiccation
- Prevent germination
- Control activities of pests and pathogens
- Maintain conditions for minimum physiological activities, e.g. adequate oxygen supply

In practice these requirements leave very narrow limits for storage conditions, which are a balance between allowing necessary base metabolism, and at the same time limiting metabolism to carry the process towards germination. Storage of desiccation-sensitive seeds is always short term. But progress on processing and storage methods, which keep seed alive from collection to sowing, can overcome crucial bottlenecks in both delivery systems and seasonality. Some considerations should also be given to the 'least tolerable germination percentage'. The ageing process in recalcitrant seed is of a different nature from that in orthodox seed because metabolising seeds maintain the active physiological repair and turnover mechanism during storage. Ageing and ultimate loss of viability is thus less likely to be due to accumulation of irreversible damage to the cytological mechanisms and DNA damage. Accumulation of toxic metabolites is likely to play a larger role. If that is the case, any seed that germinates may develop into a good seedling, which further implies that there is no lowest tolerable germination percentage other than the one set by convenience of handling. Where orthodox seed lots may be discharged when germination reaches a low 50% of the initial germination for fear of permanent vigour reduction, such rules may be applied less strictly to recalcitrant seed.

Reduction of moisture content has a dual purpose, viz. to reduce metabolism and to prevent germination. In species where desiccation tolerance is insufficient to prevent germination, any dormancy is an advantage for storage. In temperate and some highland recalcitrant species, e.g. *Quercus*, *Lithocarpus* and *Aesculus*, there is some temperature dormancy, which prevents germination of non-pre-treated seed. Unfortunately this is rare in tropical species. Vasquez-Yanes and Orozco-Segovia (1996) found that photodormant seeds of four rain forest pioneer species stored better in an imbibed state under dark conditions than at any stage of reduced moisture content. However, this is probably a rare observation since photodormancy is normally confined to pioneer regeneration strategies and most pioneers have orthodox seeds.

Attempts to prevent germination by management of dormancy has largely been unsuccessful. King and Roberts (1979) applied the natural germination inhibitor abscisic acid but failed to prolong viability. Schaefer (1990a) stored recalcitrant *Prunus africana* seeds without depulping to keep the natural germination inhibitors, but the viability of the seeds was significantly lower than that of extracted seeds. It should be noted that progress on the storage behaviour of *Prunus africana* suggests that the species is fairly desiccation tolerant (Were et al. 2004).

Seeds that are extremely desiccation sensitive, non-dormant and short-lived can neither be dried nor stored imbibed, and the only way to maintaining viability is to allow germination to proceed (Sect. 4.8.6).

4.8.1 Moisture Content and Desiccation Rate

The LSMC is the moisture content below which desiccation damage occurs. Because the aim is to reduce metabolism as much as possible, the LSMC is generally the target moisture content for seed storage.

The LSMC or desiccation tolerance varies considerably between species. Some extremely desiccation sensitive seeds may be damaged by drying to less than 50–60% moisture content (mangrove species *Avicennia*; Farrent et al. 1988). However, most recalcitrant seed can be dried to 12–17% moisture content and stored at least for several months (Hong and Ellis 2002), and intermediate seed to somewhere between that of orthodox and that of recalcitrant (Sacande et al. 2004).

Most Dipterocarpaceae are desiccation-sensitive. *Shorea, Parashorea, Hopea, Cotylelobium* and *Vatica* spp. tend to be more recalcitrant (LSMC 30–40%), while some *Dipterocarpus* contain some more desiccation-tolerant species with suborthodox to intermediate storage behaviour (Fig. 4.7). Seeds of *Dipterocarpus intricatus, Dipterocarpus alatus* and *Dipterocarpus tubeculatus* can be dried to 10, 17 and 12%, respectively, without great damage, and have prolonged storability when the moisture content is reduced within the range 6–20% (Tompsett 1992[2]).

The desiccation rate, i.e. the speed with which seeds are dried, has been shown to have a crucial effect on viability. Of more than 50 species studied in

[2] These seeds are categorized as 'orthodox' in the terminology of Tompsett as he does not include the term 'intermediate'.

a

b

Fig. 4.7. Recalcitrant/desiccation-sensitive seeds. **a** *Illicium verum* (star anis) is a highland species which tolerates low temperature. **b** *Vatica subglabra* is a typical lowland species with highly desiccation sensitive seed. Recalcitrance is a prevailing character in dipterocarps, although the family also contains species with quite desiccation-tolerant seed

the IPGRI/DFSC project, it was shown either that the desiccation rate was of no importance or that the seeds maintained viability better after fast drying than after slow drying. (Sacande et al. 2004). Most desiccation-sensitive seeds tend to keep better either at the LSMC or fully imbibed, while intermediate moisture content is less suitable (Walters et al. 2001). The benefit of fast dehydration and subsequent hydration may be connected to the increased ageing sensitivity at intermediate moisture level (Peran et al. 2004). Progress on knowledge about the desiccation rate and the LSMC has helped prolong the storage life of several species; some species have been reclassified from recalcitrant to intermediate or even orthodox. Desiccation sensitivity may vary over time, e.g. it may increase during storage (Farrent et al. 1997). It is thus better to reduce the moisture content during the initial processing than to permit seeds to dry during storage.

4.8.2 Temperature

Chilling injury depends on species, moisture content and possible duration of chilling. For sensitive species, chilling injury may occur below 20°C; some lowland species are tolerant to low temperatures (2–5°C), while most temperate and highland species tolerate slight frost. Intermediate seeds also show an increased storability at lower temperature, although chilling damage is a risk in some species.

On the basis of the IPGRI/DFSC research series, Pritchard et al. (2004) recommend 15°C storage for most tropical lowland recalcitrant seed, while the temperature for highland species can usually be reduced to 0–5°C.

For tropical seed, chilling injury is closely connected to the lack of quiescence. Temperatures below the physiological range block metabolism and thus the necessary life processes. Potential chilling damage is also closely connected to moisture content. Seed with a high moisture content is most prone to low-temperature damage, which in turn means that the lower the moisture content, the less the risk of chilling injury (Hong and Ellis 2002).

The lowest germination temperature is often close to the chilling limit, and temperature reduction is thus generally unsuitable to prevent germination. Corbineau and Come (1988) studied storage of four recalcitrant species from Thailand under a range of temperature and moisture regimes. Germination was high under wet storage even at low temperatures (for *Hopea odorata* and *Mangifera indica* even at 5°C, viz. 95 and 40%, respectively), but the germinated seeds soon died at low temperatures. Exposure to higher temperatures to avoid chilling injury increased both germination (to 100%) and growth rate. Only *Symphonia globulifera* did not germinate at low temperature and could be stored at 15°C. Constant temperature during storage is better than fluctuating temperature to maintain viability of desiccation-sensitive seed (Seeber and Agpaoa 1976).

Despite the general intolerance to low temperature, cryopreservation of excised embryos of recalcitrant seeds at ultralow temperatures has been successful for several species (Krishnapillay and Engelmann 1996; Marzalina and Normah 2002).

4.8.3
Storage Atmosphere and Media

Respiring seeds suffer from anoxia if deprived of oxygen. Short fumigation in CO_2 may be applicable to kill seed insects, but the seeds must subsequently be stored in an atmosphere with oxygen (ATSC 1995; Gunn 2001). Otherwise adequate ventilation is necessary both to prevent heating and anoxia, and to remove toxic metabolic gases (CO_2) (Tompsett 1992).

Gunny, cotton or hessian bags allows sufficient ventilation but tend to be overgrown with mould if the seeds have a very high moisture content (above 20%). Thin polyethylene material, 0.1–0.25-mm thick, is permeable enough to prevent excessive moisture loss, yet allow some ventilation (Bonner 1996; King and Roberts 1979). However, such thin material easily tears and most people use loosely folded plastic bags for storing seed with a high moisture content (Panochit et al. 1984).

Storage media with some moisture-retention capacity to prevent desiccation have been found suitable for some species. Song et al. (1984, quoted in Tompsett 1992) stored seeds of *Hopea hainanensis* in moist coconut dust, and perlite has been used successfully for a number of recalcitrant species. Schaefer (1990b) stored seeds of *Podocarpus milanjianus* and *Prunus africana* in cold moist sawdust, which also helped to reduce fungal infection.

4.8.4
Seed Treatment

Storage conditions are within the physiological range of pathogens including insects and fungi. Insects are most efficiently controlled by prestorage treatment, e.g. CO_2 fumigation or short immersion in cold or warm water. Up to 10 days' fumigation (depending on metabolism) is used for Australian recalcitrant species. For seeds sensitive to CO_2 fumigation, 24-h immersion in cold water is used (ATSC 1995). For *Citrus* spp. 50°C for 10 min has been recommended (King and Roberts 1979).

Fungal development is reduced by adequate ventilation or by storing in, for example, sawdust. Where application of fungicides is necessary, they may be applied by immersing the seeds into a solution, or by dry treatment. In the latter case the seeds must be surface-dry.

4.8.5
Hydration–Dehydration

The storage life of some recalcitrant and intermediate seeds can be prolonged by a midstorage hydration–dehydration treatment. The treatment apparently activates the innate repair mechanism during the hydration stage. The method was originally developed in India for prolonging the storage life of bamboo (*Dendrocalamus strictus*) seed (Sur et al. 1988). In a study of *Ailanthus excelsa* (Simaroubaceae), seeds stored for 3 months at a moisture content of 14% were hydrated (soaked) for 44 h in various liquid media, reaching a moisture content of approximately 62%. The seeds were then redried at ambient temperature (33°C) for 72 h, back to the original 14% moisture content The improvement in germinability after another 2 months' storage depended on the soaking media: soaking in water doubled the germination rate compared with that for the untreated control (from 13 to 26%); soaking in 10^{-4} M Na_2HPO_4 improved germination to 44% (Ponnuswamy et al. 1991).

4.8.6 Storage of Germinants

Natural vivipary/precocious germination occurs mainly in the mangrove family Rhizophoraceae (genera *Rhizophora, Ceriops* and *Bruguiera*) where no true seed stage exists; the dispersal unit is a seedling (Fig. 4.8). Spontaneous vivipary is prevalent in several other mangrove species and in lowland rain forest species. Since recalcitrant seeds often germinate immediately after seed fall, collection from the ground almost inevitably implies collection of already germinated seed. Where viability is low even under the best storage conditions and germination cannot be prevented in practice, storage of germinants is applicable. The main disadvantage of handling germinants rather than seed is that germinants are more fragile and thus easily prone to damage during, e.g., handling and planting. Keeping the temperature down reduces the germination speed.

4.9 Seed Store Units

Seed stores should, as far as possible, fulfil the requirement listed earlier to maintain seed viability as long as required. This pertains mainly to the moisture and temperature regime. Seed moisture can be managed by airtight storage. Temperature is the overriding factor in all seed store considerations. Partly because it is the only manageable factor after processing, partly because it involves potentially high running costs if it implies use of artificial energy for cooling. Choosing the right location, position and construction of a storeroom is a permanent cost reduction of a seed supply system. For practical purposes, seed stores should be manageable and easy to use, i.e. to put in seed and take it out again when required. Storerooms should be

Fig. 4.8. Viviparous seeds of the mangrove genus *Rhizophora*. Abscission from the tree and thus dispersal is delayed until the seeds have germinated and developed into seedlings while they are still attached to the tree

easy to clean and have intact sides – seeds are food for both insects and rodents, and mice and rats in seed stores are both a nuisance and potentially destructive.

4.9.1
Physical Setting of Storerooms

For the overall reason of practicalities of transport, administration and distribution, it is usually sensible to locate seed stores near to processing units and sales offices. However, cold storage is one of the expensive operations in seed supply, and there are instances where an alternative location should be considered, e.g. long-term storage units. High temperature is generally more damaging for highland species than low temperature for lowland species, so storing highland species at a lowland location is not advisable. The conditions of seed processing and storage differ in one factor, viz. temperature: where processing for orthodox seed is most effective under warm condition, storage is better at lower temperature. It should be recalled that for orthodox seed the storage life roughly doubles as the temperature declines by 5.6°C (Harrington 1972), and where artificial cooling is to be applied to bring down temperature further, reduction in the outside temperature will save the energy cost of cooling. Under dry conditions 5°C is equivalent to about 500–800-m increased altitude. Cooling down a poorly isolated cold room from say 30–25°C to 20–15°C (by 10°C) requires about 50–100 $kW/h/m^3$ annually. With an energy price of $0.12 per kilowatt-hour, the annual energy cost per cubic metre is $12–25, or $200–400 for a 'standard' 16-m^3 storeroom. The figure can be higher or lower depending on isolation type, but an energy saving of 50% is reasonable in highland locations, and locating the seed store at higher altitude may in some instances be applicable in hilly areas. Since relative humidity also tends to increase with increasing altitude, seed to be stored at the lowest possible humidity should be stored in hermetic containers before transfer from a lowland processing unit to an uphill seed store. Consideration of uphill location of seed stores mainly applies where there are no cooling facilities or where storerooms cannot be properly isolated. Well-isolated cool rooms have low heat transfer and thus small energy loss (Box 4.2).

4.9.2 Storeroom Capacity

Storeroom capacities should fit the need. In cold stores, waste of space is waste of energy for cooling. The need for storage capacity changes during the year. Large quantities of seed come in during the main seed harvest, which, depending on seasonality, can be a very concentrated period. With a regular and sensible turnover most of the seed store will be emptied around the time when the nursery season starts. A certain base volume will persist, which is the seed carried over from one year to another, i.e. the 'buffer'. The total seed store capacity should be for the largest quantity of seed that is likely to be stored at any time (Linington 2003).

The physiological requirement for storage and the difference in turnover rates require a certain breakdown of storerooms with different conditions and capacities. At least two to four different storage conditions are appropriate at a

Box 4.2

Energy in cold stores

Cold rooms must be continuously cooled because heat is transmitted from the outside through the walls of the room.

The temperature inside a cold room tends to be in equilibrium with the outside temperature because heat is transmitted through the walls of the room and is lost by ventilation. Therefore, cold rooms must be continuously cooled to compensate for the energy transmission. Major heat transmission is through the walls and is proportional to the total wall area (ceiling, floor and walls) and the temperature difference between the inside and the outside of the room. A large wall area and a large temperature difference increase the loss of energy per unit time. Heat transfer through the wall depends on its insulation, which in turn is determined by the material and the wall thickness: transmission is slow through a thick wall of insulating material like polyurethane, and fast through a thin wall of wood. The exact heat transmission through a wall is calculated as follows:

$$P = S \times \lambda \times \Delta t,$$

where P is the heat transmission in kilojoules per hour, S is the wall area in square metres, Δt is the temperature difference between the inside and the outside of the wall and λ is the specific heat transmission of the wall material measured in watts per metre degrees Celsius, Table 4.4.

If heat transmission is likely to be different through different sections of the walls (e.g. floor, ceiling and walls), the values for the individual sides should be calculated separately.

Heat transmission also occurs through ventilation, e.g. through the doors during handling. A value of 5 times the volume of the empty chamber is used in this calculation (which is half of that suggested by FAO 1984 for general cold stores, but is justified here because there is less frequent opening of seed stores than, for example, food stores).

The quantity of heat per cubic metre of air exchanged may be taken as 2 kJ/m3 °C (FAO 1984). Heat transmission through ventilation is then calculated as

$$R = 5 \times V \times 2 \times (\theta_e - \theta_i),$$

where R is heat transmission in kilojoules per /24-h period, V is the volume of the empty storeroom in cubic metres, θ_e is the external temperature and θ_i is the internal temperature.

Example of use of the formula for calculating heat transmission: The total energy transmission for a cold store of 16 m^3 with dimensions 2 m × 2 m × 4 m, and cooled down from 30 to 10°C is calculated. The walls consist of 40-mm cold-store panels with a heat transmission value of $\lambda = 0.93$ kJ/h m^2/m °C. The wall area is 40 m^2; $\Delta t = 20$°C.
Heat loss through the walls in 24 h is 24 h × 40 m^2× 20°C × 0.93 kJ/h m^2 °C = 17,456 kJ.
Heat loss by ventilation is 5 × 16 m^3 × 2 kJ/m^3 °C × 20°C = 3,200 kJ.
Total heat transmission per day is 17,456 kJ + 3,200 kJ = 21,056 kJ.

Table 4.4. Heat transmission values of some frequently used isolation materials

Material	Specific heat transmission, λ (W/m °C)	Material	Specific heat transmission, λ (W/m °C)
Air	0.024	Kapok insulation	0.034
Dense bricks	1.31	Paper	0.05
Concrete	0.9-2	Rock wool	0.045
Cork	0.044	Saw dust	0.06
Glass wool	0.04	Straw insulation	0.09
Wood	0.15	Styrofoam	0.01

Source: http://www.engineeringtoolbox.com
The heat transmission is also called thermal conductivity, sometimes called *k* values, indicated as watts per meter-kelvin.
Note that the lower the figure, the better the isolation. The thicker the material, the higher the *m* value and the better the isolation.

standard seed suppliers' unit, viz. (1) ambient temperature, (2) air-conditioned room (one or two different temperatures) and (3) cold storage (one or two levels). Seeds are stored under one of these conditions depending on species requirement and potential storage period. For example, at the Australian Tree Seed Centre four levels of storage temperature are used: (1)air-conditioned 23 – 25 C, 35% relative humidity; (2) air-conditioned 16– 14°C, 60% relative humidity; (3) cool room 3–5°C, approximately 90% relative humidity; and (4) freezer −15 to −14°C. Deep freezing is only used for long-term storage for conservation and storage trials (ATSC 1995).

The storeroom requirement for each storage condition is estimated separately. The room capacity is obviously measured in volume and not in weight as we usually use in seed handling. Most seed has a specific gravity of approximately 0.5–0.4, i.e. there is about 1.25–2 l per kilogram of seed. With some addition for container space and unutilised space, a volume of some 2.5–3 l per kilogram of seed is a reasonable calculation. This would be larger in storerooms where space is needed to move and handle seed within the store, while it would be smaller in cooled cabinets like refrigerators and freezers, where seeds are handled from the outside space. Freezers and refrigerators can usually be utilised to most of their full volume capacity (Fig. 4.9). One hundred kilograms of seed tightly packed in plastic bags in a refrigerator takes up 150–200 l; packed in hard storage containers it fills 250–300 l, and with handling and shelf space it may take up as much as 500–600 l on average.

Fig. 4.9. Ordinary household refrigerator used for seed storage. Refrigerators can hold large quantities of small seeds or be used as small-capacity 'buffer' storage for carryover seeds during seasons where larger storerooms are empty. The energy requirement and the efficiency in temperature distribution vary between different types

4.9.3 Cold Stores

Cold stores are storerooms where the temperature is artificially brought down below ambient temperature. Cold storage is used for a number of products, in particular food, and storerooms in all sizes and capacities are available. Volume and cooling capacity are the main factors to consider in relation to the choice of cooling device. However, the energy efficiency varies by more than a factor 2 between different brands of, for example, refrigerators, which in an economic context means that choosing the most energy efficient appliance reduces the

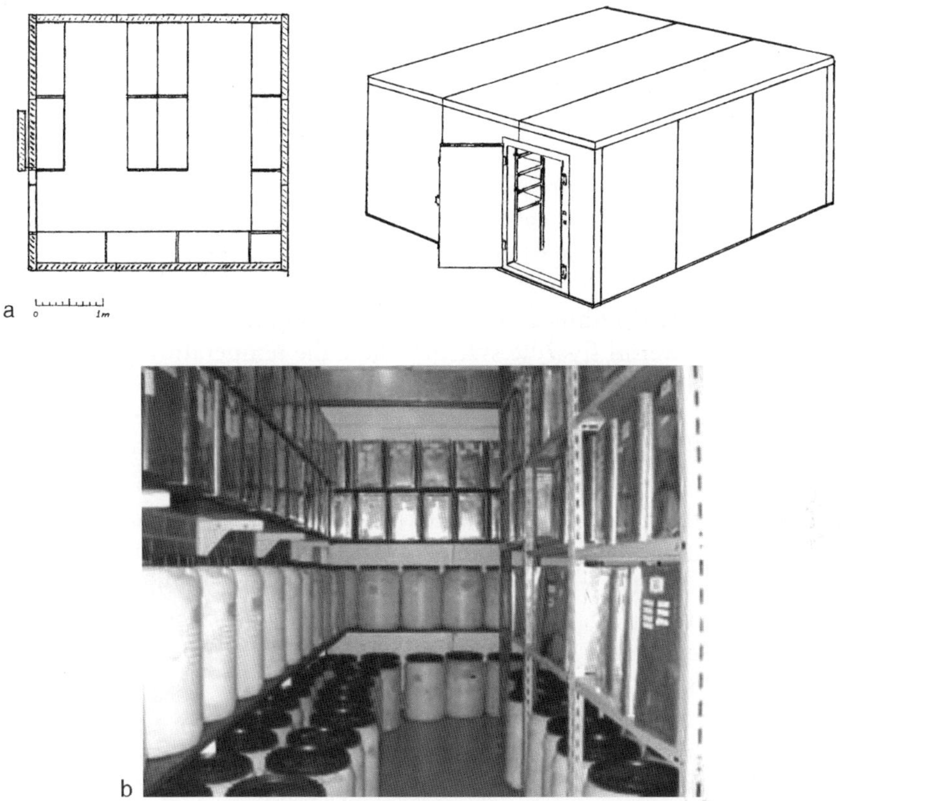

Fig. 4.10. 'Walk-in' cold stores are used in large seed supply units such as national or regional seed centres where large storage capacity is needed. Predesigned cold stores consist of special isolated panels for wall material and doors. Seeds are stored in storage containers that fit the shelf design so as to economise the room capacity. **a:** Design of a storeroom. **b:** Seed store in the Australian Tree Seed Centre. (*a* from Stubsgaard 1992, *b* courtesy of B. Gunn, Australian Tree Seed Centre)

energy cost by 50% or allows cold storage of twice the amount of seed for the same price.

Large ordinary household refrigerators have a capacity of up to 200 l. A 200-l refrigerator may hold more than 100 l of seed, which constitutes a large quantity of *Eucalyptus*, *Casuarina* or other small seeds, but is usually insufficient for larger-seeded species. The temperature of refrigerators can usually be adjusted between 5 and 10°C. Most refrigerators are designed for household food and thus are provided with small freezers for ice cubes and drawers and shelves for special items. It is usually convenient to remove all these accessories if the device is used for seed storage only. Temperature distribution should be considered: the best refrigerators have ventilators or several cooling devices.

Deep freezers with capacities from 50 to more than 400 l are available and are frequently used for subzero temperature storage (–15 to –20°C) of relatively small seeded species or research samples. Refrigerators and freezers should not be placed in closed rooms where seeds are stored at ambient temperature or in cold rooms, since their operation generates heat which will warm up the room they are placed in.

Larger cold stores are isolated rooms with cooling devices (Fig. 4.10). Ready-made cold stores consist of highly isolating wall elements and a powerful cooler. Capacities range from 16 to more than 100 m^3. Cooling systems range from ordinary air conditioners, where the temperature is lowered typically to 20–25°C, to powerful freezing systems where the temperature can be lowered to below 0°C (FAO 1984; Linington 2003). Cold rooms can be established in any appropriately sized and placed room provided with ordinary room air conditions. Isolation efficiency is crucial to reduce heat transmission and thus reduce electric energy consumption. Isolation efficiency is a product of specific material and thickness. is for instance rock-wool and Styrofoam, for example, are good isolating materials (Table 4.4). Cooling to safe temperature will normally help keep seeds viable during storage; however, prolonged exposure to adverse conditions, e.g. during transport, can easily accelerate deterioration. For sensitive material it is thus advisable to keep the temperature low during transport (Box 4.3).

If seeds are stored in non-airtight containers, e.g. to allow some respiration, air humidity should, as far as possible, be regulated. Humid climates during rainy seasons are conducive to fungi – mould will grow everywhere. Electric dehumidifiers can be used to reduce the relative humidity to 10–15% (Linington 2003)

Box 4.3

Mobile cooling vans

Cooling vehicles are available and much used for transport of easily decomposable food items. Cooling vans can be used for transport of easily deteriorating seed, e.g. during longer collection expeditions under adverse conditions or during transport of recalcitrant seed from cooled seed stores to distant seed users. As more cooled goods are being transported from distant locations, more such transport facilities are becoming available, for example, for difficult seed transport. Small cooling units for installation in ordinary cars are becoming available for ordinary use and may contribute to overcome bottlenecks in distant seed supply. Killing sensitive seeds and microsymbionts by exposure to adverse conditions in ordinary transport vehicles can take a short time; applying the 'luxury' equipment of a cooling unit can make a difference.

4.9.4 Some Cost–Benefit Considerations for Seed Stores

Energy use for cold rooms is a considerable additional seed procurement cost, which is in turn added to the seed selling price. The price of cooling is generally proportional to the seed volume, so the size of the seed influences storage economy. One litre of *Swietenia macrophylla* seed contains about 200 seeds, while the same volume for *Anthocephalus chinensis* seed contains 1,500,000 seeds. One and a half million *Anthocephalus chinensis* seeds fill a corner of a refrigerator, while the same quantity of mahogany seeds requires about 7.5 m^3, or a considerable part of a storeroom. Despite this the mahogany seeds could produce a better return for cold storage because they are short-lived and may suffer great mortality in ambient storage, while *Anthocephalus* seeds hardly show any difference during short-term storage. It is not economic to store seeds under expensive cold storage if it does not significantly prolong storability compared with ambient conditions.

Cold storage is necessary for short-lived species and in particular where short life is combined with periodicity in seed production. However, for such species, seeds may be fractioned into several portions where seed that will be distributed in the first year may be stored in a simple air-conditioned room and seeds to be kept for several years are stored at low temperature. This strategy is used, for example, for *Araucaria cunninghamii* in northeast Australia. Good seed crops of this species occur about once every 6–10 years. The seeds are large and short-lived under ambient conditions. Long-term storage requires deep freezing down to −14 C. Seeds with a shorter storage period are stored in ordinary cold storage (Keys et al. 1996).

When seeds are disposed of either through a season or during a prolonged period between two bumper crops, storeroom capacity becomes redundant. It is practical and economical if separate storerooms or sections of storerooms can be closed as seeds are disposed of. Alternatively smaller quantities may be moved to large refrigerators and the main storeroom switched off.

4.10 Storage Containers

The explosion in the plastic and glass bottle industry has made the selection of seed containers almost unlimited (Fig. 4.11). Storage containers have two main functions, viz. to help maintain viability and to facilitate packing. Suitable materials and containers for long-term storage of orthodox seed in storerooms have the following properties:

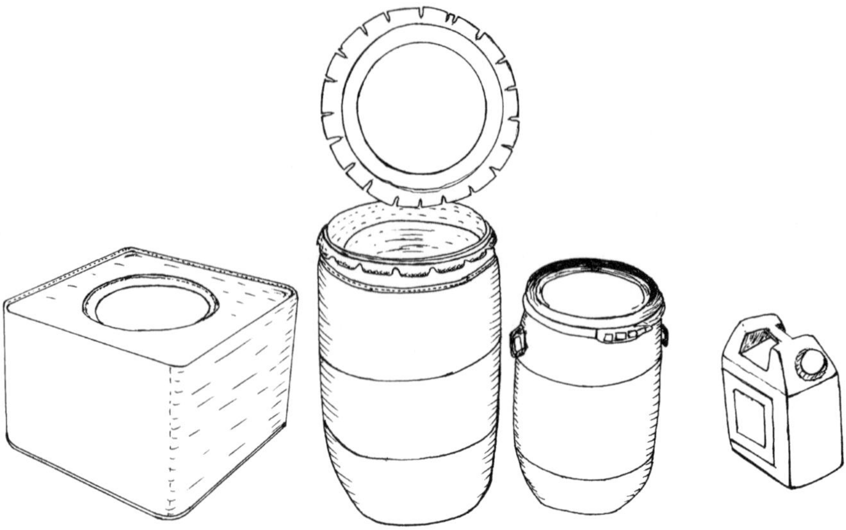

Fig. 4.11. Traditional storage containers for seed. Manufacture and extension of plastic material in all sizes and forms has made it easy to find suitable designs of the required size and form. The 'old-fashioned' barrels, jerry cans and glasses are still much used because they are of suitable size and have undergone few changes

1. *Airtight.* Airtight fittings prevent absorption of water from humid air in storerooms. In hard containers, airtightness is achieved by using lids with gaskets of rubber or other material. Such lids will provide a tight fitting as long as the gasket is soft and flexible. Old, worn, dry or damaged gaskets do not remain tight and should be replaced. Greasing can extend the lifetime of rubber. Standard-sized, locally available container types in which the rubber gaskets can be replaced are therefore preferred. Soft containers such as plastic bags may be kept airtight if closed with special sipper-closing or sealed by melting. High-moisture seed (recalcitrant) should not in stored in airtight containers.
2. *Strong material.* Mechanical damage and tearing imply the risk of spilling seed and hampering viability. Strong storage containers can be reused many times. Metal tins are occasionally used but can rust and should therefore be internally protected by an anticorrosive covering. Polyethylene bags are not reused but strong material is necessary if complete airtightness is wanted. Thinner material has perforations.
3. *Easy to fill, empty and clean.* Containers designed for liquid materials are often less suitable because they usually have narrow openings. It is practical if the opening is large enough to get a hand inside when cleaning.

4. *Of suitable volume.* Filled containers provide the best storage environment and space utilisation. It is practical to use a container of a size which can be emptied fast during disposal. Large containers should preferably be used for large seeds and seeds with rapid turnover. If small portions of seeds are likely to be taken out regularly, e.g. for trials, it is appropriate to split up the seed lot before storage and store the seeds in smaller portions in the containers, for example ten bags of 50 g rather than one bag of 500 g.

4.11 Storage Pests and Pathogens

Physiological activity by all living organisms requires a certain minimum of free water; therefore, a moisture content below the physiological minimum of seeds is also too low for activity of pests and pathogens[3]. This makes, at least theoretically, management of pests and diseases of orthodox seeds relatively simple: conditions which promote general seed longevity, i.e. low temperature and moisture content, also lower the activity of storage pests and pathogens. Properly dried and packed seeds are thus safe as long as they are kept in storage. However, many types of pests and pathogens have adapted to very dry conditions: they are active at very low humidity and have dormant stages in which they can remain alive and inactive under storage conditions, yet be a potential infective source when conditions improve.

All seeds have some innate protection against pests and diseases, e.g. hard pericarp or seed coat, and/or chemical protection. Protection is strongest in fresh, healthy and mature seeds and weakest in aged, damaged and immature seeds. The strongest prevention against pests and pathogens is thus to collect the sound and mature seeds. The second prevention is to keep them under conditions with the least possible chance of development of pest and pathogens.

The distinction between cause and effect of microorganisms in seed ageing is often two-way and self-accelerating. Once microorganisms infect, they promote their own environment, for example, by increasing the moisture content by respiration. Microorganisms also cause a similar breakdown of cell constituents as occurs during natural ageing.

[3] A pathogen is a disease-causing microorganism, for example bacteria, viruses or fungi. A pest is a macroorganism associated with predation (consuming part of the seed), for example insects and mites.

Recalcitrant seeds pose a problem because they must be stored at a temperature and moisture content conducive to insect and fungal development (Fig. 4.3). Infecting organisms are often the major cause of deterioration for seeds stored at high temperature and moisture content.

However, pest and pathogen control is applicable under certain conditions:

1. Where storage conditions are conducive to pests and diseases, e.g. humid, ambient conditions where seeds have not been properly dried and/or moisture absorption can take place because seeds are not stored airtight.
2. Where there is a risk that pests and disease will develop after storage before germination. This risk is accentuated if the initial pathogen infection is high, if presowing conditions are conducive to development and if there is a long time lapse between distribution and sowing.
3. Where seeds do not tolerate a storage environment that prevents pest and pathogen activity. This is typically the case for desiccation-sensitive seeds.

Some seed pests and pathogens are host-specific in the sense that they are closely associated with one or a few plant species. Others can infect a wide range of species and may even attack other plant parts. Some seed-borne fungi do not cause any damage to the seed itself but only to seedlings or larger plants. In this case, seeds are a 'vehicle' for dispersal of the pathogen rather than a food source, so-called seed-borne pathogens.

Pests and diseases must be controlled during seed handling, both to prevent the infected seed itself being destroyed and to prevent the pest being spread to other seeds during handling. Also, in the case of seed-transmitted diseases, to prevent them from spreading to plants in the nursery or further afield. The latter is especially important where seeds are shipped into an area in which the pathogen is not found and where possible introduction could cause major loss.

The type and level of seed pest and pathogen control vary with infection rate, type of infecting organism and the likelihood that the organism may multiply and destroy seeds during storage. Preventive measures like early collection, swift processing, good hygiene and appropriate storage conditions are often sufficient to reduce the loss caused by both insect and fungal attack and to make chemical control redundant. In cases where pesticide treatment cannot be avoided, e.g. because of suboptimal storage conditions, recalcitrant seeds or for phytosanitary reasons, the use should be limited to that strictly necessary. Treatment should be applied with due consideration to possible impairment of seed viability, risk to labourers during handling and danger to the environment when disposed of. It should be noted that most chemicals are generic (target specific), i.e. insecticides generally have little, if any, effect on fungi and fungicides have little effect on insects.

Pathogens are disease-inducing organisms like bacteria, viruses and fungi. The last of these are by far the most important in seeds, although most fungi are non-pathogenic. Some pathogens are parasites in the sense of deriving their food from the host during a prolonged period. Often, however, injury to the host occurs because the pathogen releases enzymes or toxins detrimental to the host organism. The deterioration caused by pathogens is often manifested as poor or unhealthy performance (reduced 'vigour'; Chap. 7) of the seed or seedling; therefore, the term 'disease-inducing' is used. Insects and pathogens may be carried on, in or with the seeds, all of which are referred to as seed-borne. Organisms which cause no harm to the seed itself but only use the seed as a vehicle of dispersal (in practice only pathogens) are called seed-transmitted. The distinction between infection and infestation is somewhat blurred. In the strict sense, infection refers to the situation where the foreign organism lives inside the host (endoparasitism), while infestation is the invasion by exterior organisms (ectoparasites).

4.11.1 Seed-Storage Insects

Insect larvae that have infested seeds in the field may continue their predation in storage. However, only species that are able to breed and reinfest seeds in storage can be considered true storage pests. Most insects are unable to do so either because they cannot complete their life cycle under dry storage conditions (e.g. the adult insect cannot survive and mate under storage conditions) or because the new generation is unable to penetrate the seed coat during infestation (Fig. 4.12).

Despite the hard seed coat of legumes, a small group of bruchids are able to reinfest seeds during storage and produce several generations until the whole seed lot is destroyed, provided the temperature is conducive to their survival and continuous activity (Fig. 4.13). The adults of these bruchids need no food intake for reproduction: the feeding is entirely by the larvae (Southgate 1983).

Some species, for example, of the genus *Caryedon* attack both young immature seed in the field and fully mature seed in storage (Singh and Bhandari 1988). In a study of *Caryedon serratus* infesting *Acacia nilotica* in the Sudan, the initial field infestation was from 10% on standing trees to 17% on the forest floor, but the infestation increased to 90% after 3 months' storage (El Atta 1993). Some estimation of the duration of the life cycle and thus the reinfestation rate can be made. In the above example, El Atta (1993) observed the

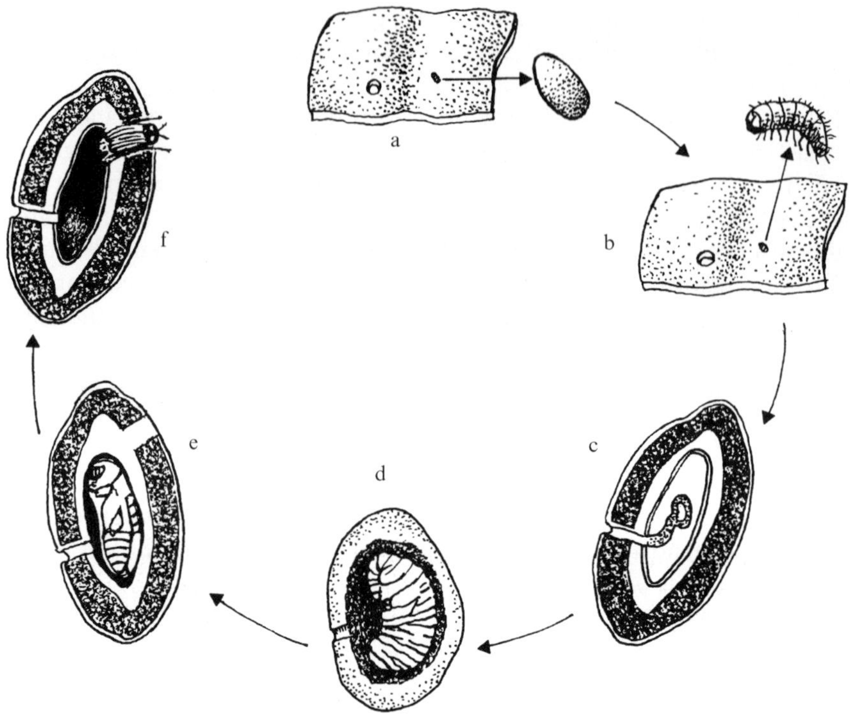

Fig. 4.12. Life cycle of bruchid beetle, here on *Prosopis* sp. **a** Eggs glued to pod surface or laid in cracks in a pod or in emergence holes of an adult bruchid (round holes). **b** Entry holes of the first-stage larvae that have burrowed through the pod wall and first-stage larva enlarged to show hairs, spines and legs which are modified for entering seeds. **c** Cross-section of pod and seed showing the burrow made by the entering first-stage larva. **d** Later-stage larva inside cavity chewed in seed. **e** Pupa inside larval feeding chamber. The larva penetrates the testa except for a thin 'window' before it pupates. **f** Adult emerging through hole prepared by the last-stage larva. (From Johnson 1983)

following lengths of the individual stages of *Acacia nilotica* infestation by *Caryedon serratus* under storage conditions:

1. Egg incubation period, i.e. from oviposition to hatching, 7–16 days
2. Larval feeding: four larval stages ('instars') with average durations of 12.4, 10.6, 11.5 and 7.2 days, respectively, were recognised, i.e. a total feeding period of approximately 42 days
3. Metamorphosis and pupal stage, 10–15 days
4. Emergence and mating, 1 day
5. Time from mating to new oviposition, 2–3 days

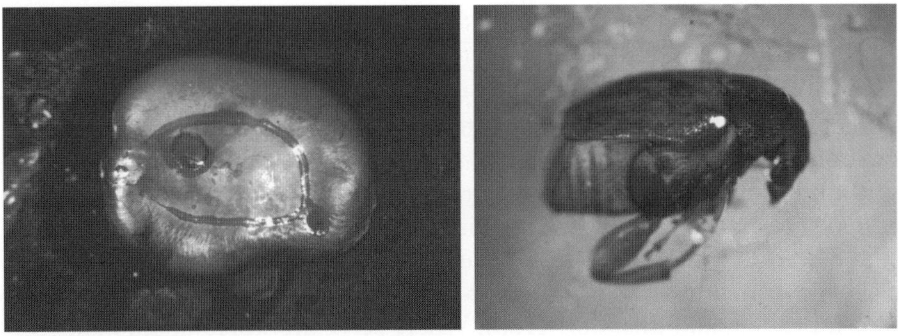

Fig. 4.13. Bruchid infestation of acacia seeds. Bruchids often attack immature seeds but under suitable conditions the insects can reinfest seed in storage and cause destruction of most of a seed lot

The total life cycle for this species under the given conditions is thus 63–76 days. The average number of eggs per female in the investigation was 93, of which some 80% hatched. However, as 12–25 eggs were laid per seed by this bruchid species, each female may have infested on average only five to six seeds. Hence, with a low initial infestation rate, several generations may be necessary for a total destruction of the seed lot. Storage under ambient conditions for less than one generation turnover (approximately 2 months in the example) may be acceptable if the initial infestation rate is low.

The life cycle of 2–2.5 months given in the example relates to optimal conditions for insect activity. A much longer life cycle would be expected where temperature and moisture content are low. The activity of insects during low temperatures varies with their specific environmental adaptations. Many tropical species show little activity below 15°C, while subtropical and high-altitude species may be active at temperatures down to a few degrees above 0°C. The moisture content may be a limiting factor for feeding, but in this respect there are also large variations. Although a moisture content below 10% is limiting for many field-infesting insects, others insects such as several bruchid species remain active at a lower moisture content. However, though the activity of some insects may not be stopped altogether under a given storage regime, any reduction of temperature and moisture content below the physiologically optimal will delay their development. For example, the duration of reproductive cycles may be doubled or tripled under conditions of reduced temperature. It has also been suggested that insects will lay more eggs in a dark and damp store than in a light and dry one (Singh and Bhandari 1988).

The best preventive measure against seed pests is early collection (i.e. to minimise field infestation) and appropriate processing including cleaning, which will eliminate infested seed (Tschinkel 1967; Southgate 1983).

4.11.1.1
Storage Conditions

Where seeds are insect-free when entering storage, they should remain so by excluding any exposure to later infestation. Seeds left over from a previous seed lot or insects (typically adults or pupae) left in cracks, corners, old containers, etc. may also carry a potential source of seed destruction for a new stored seed lot. Thorough cleaning and possible disinfection of storerooms, bags and containers are measures to prevent transmission from a previous seed lot (Singh and Bhandari 1988). Where seeds are stored at ambient temperature, mechanical barriers such as plastic sheets or insect nets may prevent the entrance of flying insects and hence infestation (Howe 1972). However, while, for example, polyethylene sheets may be effective in preventing insects entering into the bags, escaping insects may easily penetrate the sheet upon departure. Such exit holes may affect seed moisture conditions by permitting free gaseous exchange.

Deep freezing to, for example, –5 to –10°C will kill infested seed after some time. Where freezing is not applicable, the activities of insects can be greatly reduced by cooling. For most tropical species there is little activity under 5–7°C, but even a smaller temperature reduction will reduce both general activity and development. However, whenever the insects metabolise, they produce heat and water. They may thereby improve the microenvironment in their immediate vicinity, which in turn may promote their activity (Singh and Bhandari 1988).

4.11.1.2
Seed Treatment

Seed treatment is the application of remedies to actively eliminate infestation in cases where preventive measures cannot be applied or are insufficient. Seed treatment can be given as a single treatment that will eliminate all stages of the pest at the time of treatment but does not have any longer-lasting effect. Or it can be used as a long-lasting application, which will prevent new infestations.

Fumigation denotes application of a metabolic inhibitory or toxic alloy in gaseous form. The method is used in other connections to plant propagation, e.g. sterilisation of nursery soil. The advantages and disadvantages of fumigation, compared with other seed treatment methods, like insecticide powders, relate to their volatile nature:

1. The effect is usually quick since the gas will always reach the target organism, unless it is deeply hidden within the seed.
2. Insects absorb gases through the 'skin'. The more metabolic activity, the larger and quicker the effect. Accordingly, fumigation has a greater effect on adult stages and feeding larvae than on eggs and pupae; and the effect increases with temperature within the physiological limits.
3. The gases normally have no effect after they have escaped from the seeds.
4. Since the gases do not adhere to the seeds, no cleaning or preventive measures need to be taken after treatment, e.g. for export or during sowing.
5. The possible toxicity to humans is unfortunately easily overlooked because the gases are often invisible and without smell.
6. A prerequisite for application is that facilities and material impermeable to gases are available.

Several fumigants with proven effects on insect control are available. The most common ones are ethylene bromide, hydrocyanic gas, a mixture of carbon disulphide and carbon tetrachloride, phosphine and pirimiphs (Willan 1993). For bruchid beetles in *Acacia tortilis*, fumigation with carbon disulphide, aluminium phosphide or chlorosal (a mixture of three part ethylene chloride and one part carbon tetrachloride) has been used in India (Singh and Bhandari 1987). These fumigants are all toxic to humans and should be handled with utmost care, and only by authorised staff using safety protection. Further, most of the fumigants are phytotoxic, so prolonged exposure of the seed should be avoided (Singh and Bhandari 1988).

One non-toxic gas, CO_2, has been successfully used for seed treatment of many species of orthodox seeds and is described in detail here. Because CO_2 is harmless to dry seeds, the seeds can be stored with the gas for prolonged periods (Sary et al. 1993). CO_2 is a product of aerobic respiration. In a normal atmosphere it makes up some 0.03%, while in comparison O_2 makes up some 20%. CO_2 is non-toxic to insects, but when present in large concentrations relative to oxygen, it blocks some physiological pathways of the respiration system. Since dry seeds are not metabolically active at a low moisture content, CO_2 is harmless to dry seeds, whereas living insects are killed in an atmosphere of high CO_2 and low O_2 content. Adults and larvae have the highest rate of respiration and are easier to kill. Pupae have an active metabolism during their metamorphosis, but can also be dormant and therefore more resistant. The sealed bag must initially be stored at room temperature for about 8 weeks. This is to ensure a sufficiently high metabolic rate of the insects, which might otherwise be inhibited by low temperature. Seeds with high moisture content and hence metabolism, e.g. recalcitrant seeds, do not tolerate prolonged exposure to CO_2 fumigation. In Australia a maximum of 10 days' fumigation with CO_2 is recommended for recalcitrant seeds (ATSC 1995).

Some technical details of CO_2 application equipment are given in Box 4.4. If the bags are properly sealed, CO_2 will protect seeds against insects as long as the seeds are packed.

Box 4.4

Equipment and application method for CO_2 fumigation

Compressed CO_2 is commercially available in refillable metal bottles of about 6 kg or larger, as it is extensively used for fire extinguishing and thin plate welding, for example. Stored under pressure of approximately 40 atm (bars), CO_2 is a liquid. The pressure in the bottle decreases as the CO_2 is used. An adjustable pressure-reduction valve is therefore connected to the bottle to reduce the pressure of the outlet and allow a steady flow of the gas. The most convenient type is a flow meter in which the pressure of the outlet can be adjusted quite exactly, and which makes it easier to determine the time it takes to fill a bag with CO_2. A hose is fitted to the flow meter/reduction valve at one end and a blowing pistol for pressurised air at the other. All connections should be provided with tight-fitting gaskets and connections should be tightened to avoid leakage of the gas.

Seeds to be fumigated with CO_2 are placed in heat-sealable plastic bags with low permeability to CO_2. Laminated plastic material consisting of an outer layer of approximately 0.03 mm thick polyamide (nylon, low CO_2 permeability) and an inner layer of approximately 0.07 mm low-density polythene (ordinary plastic, heat-sealable) is suitable. An alternative laminate has aluminium foil instead of the outer polyamide layer. Bags with volumes of less than 4 l are preferred as larger bags are more difficult to fill and tend to puncture easily. CO_2 is invisible, and the amount of gas needed to fill a given bag size is estimated as the time with a given gas flow, which is initially checked by filling an empty bag. Note that the flow rate must be low, so it takes about 4–10 s to fill the bag. The blowing pistol end is inserted into the bottom of the seed bag. As CO_2 is denser than air it will replace the normal air between the seeds.

After filling the bag, the bag is closed with the help of an electric heat sealer. Special heat sealers make a broad tight seal and allow adjustment of temperature and sealing time to the particular material thickness. It is important that the temperature and time are adjusted so that the two sides melt together without melting holes in the material. During sealing the bag is kept upright and the sealing site kept clean (Sary et al. 1993). Seeds will initially absorb some of the CO_2 gas and the seeds will therefore after some time appear as if they were vacuum-packed.

(*Continued*)

Box 4.4

Equipment and application method for CO_2 fumigation—Cont'd.

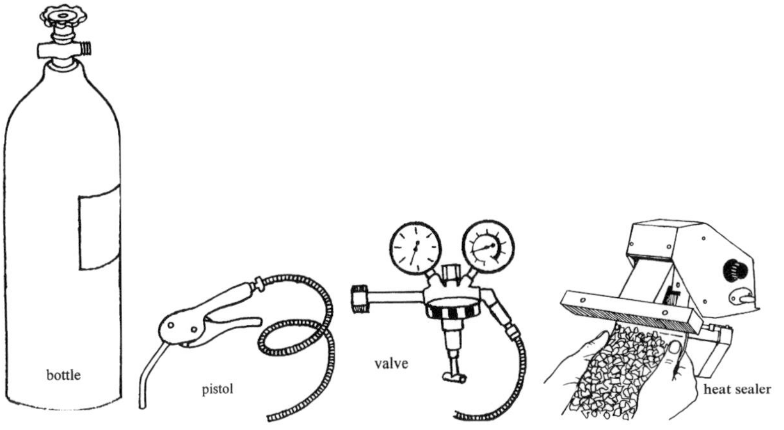

4.11.1.3 Insecticides

Insecticides may be used as an alternative to fumigation, or where a long-term effect is desired, e.g. if hidden and dormant stages may escape a short treatment and appear later during storage. Several insecticides are available. Most of them have been developed and are mainly used for agricultural seeds, e.g. so-called grain protectants. Application is normally in the form of dust where the seeds are mixed with the dry powdered chemical. Some insecticides have been banned or restricted, particularly in Western countries, for environmental reasons. This pertains in particular to the group of chemicals known as chlorinated hydrocarbons which contain, for example lindane, DDT, aldrin, endrin and chlordane. Most of these products are no longer produced and available in Europe and the USA but are (unfortunately) still produced and available in a number of tropical countries. Their use should generally be discouraged and an environmentally less dangerous product used where necessary.

Organophosphate insecticides are environmentally safer. They have a wide toxicity range; some are extremely poisonous to humans, while others are relatively harmless. Among the moderately toxic ones is phenitrothion, a relatively common seed insecticide known under various trade names such as Cytel and Folithion. In India, dusting of gunny bags with 5% Folithion dust for short-term storage and 10% Folithion dust for long-term storage was recommended for control of bruchid beetles in *Acacia tortilis* seed in storage (Singh and Bhandari 1987). Phorate and malathion are other organophosphate seed insecticides with relatively wide use (Singh and Bhandari 1988; Cremer 1990). Among the environmentally safest insecticides is pyrethrum; originally it was extracted from flower heads of the herb *Chrysanthemum cinerariaefolium*; now a synthetic equivalent to the flower extract is manufactured. Pyrethrum insecticides such as pyrethin dust mixed with seeds are used in, for example, India (Singh and Bhandari 1988) and Australia (Cremer 1990).

4.11.1.4
Biological Methods

Some plant species contain alleged insect-repellent compounds, which traditionally have been used for seed protection in storage (Table 4.5). Several of these species release a strong odour which is apparently avoided by insects; few of the biological remedies are toxic in normal doses. Apart from pyrethrum mentioned already, one of the most effective plants with insecticidal effect is neem (*Azadirachta indica*). Neem seeds have a particularly high concentration of the active compound azadiractin, and crushed seeds or oil are especially effective, but the chemical is present in all parts of the tree (Soon and Bottrell 1994). In Vietnam a local *Milletia* species, *Milletia ichtyochtona*, has been widely used as a disinfectant and insect repellent.

Table 4.5. Plants and plant parts used for insect control in storage, particularly for bruchids (Galop and Webley 1980, quoted in Johnson 1983)

Plant species	Plant part or extract used
Azadirachta indica	Neem kernels, seed oil, powdered leaves or bark
Chrysanthemum cinerariaefolium (*pyrethrum*)	Whole plants or flower heads
Capsicum	Pepper chillies
Cactus spp.	Stem powder
Anona reticulata	Custard apple seed powder
Mundulia sericca	Stem bark powder
Piper nigrum	Black pepper powder, extract
Madhura latifolia	Stem bark powder
Acorus calamus	Rhizome powder, oil
Thevetia nerifolia	Powder of drupes
Adhatoda vasica	Leaf powder
Ipomea cornea	Leaf powder
Derris elliptica	Oil
Pogostemon heyneanus	Pachouli oil
Nigella sativa	Black cumin oil
Phaseolus vulgaris	Bean oil
Allium cipa and *Allium sativum*	Oil

Another biological insect control method is through trapping, where the insects are attracted by pheromones (a group of female sex hormones which attract male insects). Practical application of the method has not been documented for seed insects. Also the use of seed-insect predators or parasitoids has a largely unused biological potential (Southgate 1983).

The main advantages of the biological methods are that they are likely to be non-toxic and hence safe for both labourers and the environment, and for plants/plant extracts that they are often locally available.

Some minerals such as diatomite, termite mound soil, kaolin and lime have alleged insect-repellent or insecticidal effects (Galop and Webley 1980, quoted in Johnson 1983).

4.11.2 Seed Fungi

Fungi are a diverse group of plant-infecting pathogens. Some fungi attack seeds directly, e.g. through cracks or damage to the seed coat, others infect only the germinating seedling. The latter are seed-transmitted (Neergaard 1979) and include the fungi known as 'damping off', which cause great damage in nursery seed beds (Gardner 1980; Kamara et al. 1981; Ivory and Spreight 1993).

Fungi multiply by spores, sometimes of different types. Spores are produced in vast numbers; they are tiny, long-lived and usually dispersed by wind, which can carry them over long distances. Once the spores have been deposited on a suitable substrate, provided temperature and humidity are appropriate, the spores may germinate and form minute threadlike filaments, or 'hyphae', that penetrate into the plant tissue. Hyphae and their aggregate network, mycelium, make up the vegetative stage of the fungus. Whereas the fungal spores are relatively resistant to adverse environments (e.g. drought), the hyphae normally grow only under high moisture conditions and warm temperatures. The hyphae penetrate between and within cells while absorbing organic material. However, for most pathogenic fungi the damage to the host plant is not so much caused by the depletion of nutrients as by damage to the cell caused by the release of enzymes and toxic metabolites by the infecting fungus (Halloin 1986; Vijayan and Rehill 1990). Some of the fungal exudates cause damage to the cell membranes, others inhibit vital life processes of the germinating seeds. A moderate infection may reduce germination energy and affect embryo development during germination, for example causing malformation or discolouration (inhibition of the chlorophyll synthesis) of the seedling (Christensen 1973; Halloin 1986). Infection of the radicle of germinating *Pinus* spp. by the fungi *Alternaria alternata*, *Aspergillus*, *Penicillium* and *Trichoderma* spp. either killed or caused temporary setback of the seedling.

However, some seedlings were able to fully recover from the attack (Rees and Phillips 1986).

Fungi may spread over short distances, e.g. from one seed to another in a seed lot, by the mycelium. As the fungus grows and completes its life cycle, it forms new spores which may be spread via equipment or the like to other seed lots. Many fungi multiply primarily vegetatively when conditions of moisture and temperature are optimal for growth, while spore formation becomes important under stressed conditions, e.g. when food supply is depleted or under suboptimal temperature and moisture conditions. Many fungi are able to survive long periods as spores. Fungal spores are often efficiently dispersed via wind, soil, impurities and/or equipment, and this together with an often very high resistance to adverse environments makes them almost omnipresent.

Storage fungi, commonly called moulds, are facultative saprophytes living on most dead organic materials. They can thus germinate and live on dead organic material such as dead flower parts, leaves or other impurities. There is, accordingly, little host specificity among storage fungi. Most species belong to the genera *Aspergillus*, *Penicillium*, *Rhizopus*, *Chaetomium* and *Mucor*. *Aspergillus* is by far the most common in seed store infections. *Aspergillus niger* and *Aspergillus flavus* attack seeds of a wide range of species (Mittal et al. 1990). *Penicillium* is more common in temperate than in tropical regions (Agarwal and Sinclair 1997), although it does occur frequently also in the tropics. Hong (1981) found, for instance, several *Penicillium* species on stored dipterocarp species in Malaysia.

Storage fungi require at least 10% moisture content (equivalent to about 70% relative humidity) for growth and development. Fungal attack is thus not common for dry orthodox seed. Desiccation-sensitive seed stored with high moisture content may continue to support a microflora of fungi usually only active under field conditions. For example, Mycock and Berjak (1990) found *Fusarium*, a typical field fungus, becoming dominant during storage for four recalcitrant species. In India, *Aspergillus niger* is a frequent storage fungus attacking seeds of *Shorea robusta*. This intermediate species can be stored at a LSMC of 12% (equivalent to about 75% relative humidity). The fungus attacks at this moisture content and the attack becomes increasingly worse at higher humidity (Singh et al. 1979). Seeds collected during the rainy season may contain more than 50% moisture, which is difficult to bring down to a safe level during conditions of high air humidity. Such seed will, accordingly, often be extremely prone to fungal attack.

Storage fungi may be active in a temperature range from 0 to 55°C, some even as low as –5°C; however, below 10°C the activity of most fungi is extremely low. Temperature reduction is the safest way of reducing fungal infection in seed stored with a relatively high moisture content (recalcitrant seed).

Whole, undamaged, intact seed coats are usually quite resistant to fungal attack as the fungi are normally unable to penetrate intact seed coats. An intact seed coat forms a usually effective physical and chemical barrier to infection

(Halloin 1986). Fungi therefore usually use cracks or damage to the seed coat as entry points (Halloin 1986). Also natural 'weak sites' of the seed coat may occasionally be attacked. A natural weak site is the chalazal region consisting of easily penetrable parenchyma tissue, through which fungi may invade (Christensen 1973). Germination of fungal spores requires high humidity. Since spores germinate on the surface of the seed, air humidity rather than seed moisture is the critical factor.

The severity of fungal attack is proportional to the initial infection rate (inoculation) plus the condition during growth and development. Seeds that are 'relatively' free of contaminating fungi and stored dry will thus never experience problems. Fungal problems can thus to a large extent be prevented by early collection, effective cleaning and good hygiene with equipment and containers. Sound healthy seeds are less likely to be infected by fungi than aged and damaged seeds. In this way there is a mutual cause effect: aged seeds are prone to fungal attack but fungi themselves contribute to ageing.

4.11.2.1
Fungal Treatment

In most cases preventive measures like ensuring appropriate time and method of collection, and appropriate processing and storage make chemical treatment redundant. However, where seeds are heavily infected with seed-borne pathogenic fungi and these are likely to cause damage during storage or germination, treatment may be indispensable. Further, where seeds are to be exported, treatment may be necessary for phytosanitary reasons (see later). A number of chemicals are available, some of which are listed below. The basic requirements for a seed-treatment chemical are, according to Agarwal and Sinclair (1997):

1. Effective under different agroclimatic conditions
2. Harmless to the seed and seedling, i.e. non-phytotoxic
3. Safe to operators during handling and sowing, and to wildlife
4. Not leave harmful residues in plants or in the soil
5. Compatible with other seed-treatment chemicals
6. Low price

It may in practice be impossible to find chemicals which fulfil all these requirements. For example, most seed-treatment chemicals are phytotoxic even when used in safe prescribed doses, but may still be economically beneficial by outweighing the detrimental effect of the pathogens. Potentially harmful effects to humans may in most cases be overcome by safety precautions during handling (Box 4.5).

Box 4.5

Safety precautions during handling of pesticides

Use of pesticides in seed handling should be limited to the absolutely necessary. Pesticides are poisonous, some more than others, and should be handled and disposed of with minimum harm to humans and the environment. Instructions for users (e.g. quantities to be used and possible dilution instructions) as well as safety precautions should be prescribed by the manufacturer. Any toxic chemicals should be provided with a label from the manufacturer indicating toxicity, e.g. in classes A, B, C, etc. The rules vary from one country to another. Highly toxic pesticides should only be handled by authorised personnel under observation of strict safety rules throughout handling. Some general rules and precautions are listed here:

1. Read instructions from the manufacturer carefully and handle the remedy accordingly.
2. Use the concentration prescribed by the manufacturer.
3. Never experiment by mixing different chemicals.
4. Prepare prescribed pesticide mixtures in a well-ventilated place.
5. Always use gloves during preparation and application; for liquid remedies, waterproof rubber gloves should be used.
6. Use masks and protective glasses when applying toxic fumigants and sprays.
7. Check and repair any leak from containers and equipment. Replace worn gaskets in equipment used for fumigation and spraying.
8. Do not leave pesticides unattended. Have a locked room or cabinet especially for pesticides and application equipment.
9. Dispose of any leftover remedy safely.
10. Be prepared for accidents; the universal emergency agent is water, which should always be available within reach.

To these points should be added that only personnel having received appropriate instruction and training should be allowed to handle pesticides.

Harmful environmental effects are subject to increasing concern, and in many countries a number of chemicals have been banned for environmental reasons. Mercury-containing chemicals were formerly common seed fungicides; now they are being replaced by more rapidly decomposable and less harmful products. Mercury-based fungicides can also be harmful to some seed, e.g. certain *Pinus* spp. (Willan 1993). When seed for export is treated with pesticides, the rules and legislation of the importing countries should be consulted. Failure to comply with such rules may cause import problems.

Most seed-treatment fungicides are targeted to a wide range of fungi and are likely to affect the total microflora and fauna on the seeds, including beneficial organisms such as mychorrhiza, rhizobia and *Frankia*. As a consequence, it is generally not possible to apply fungicides together with, for example, microsymbiont inoculants, e.g. by pelleting, and during any inoculant application seeds must be cleaned for possible adhering fungicides. The problem may in some instances be overcome by using an instant treatment like heat or surface sterilisation rather than pesticides with a long-term effect.

Fungicides applied before storage are normally targeted only at seed-borne and potential storage fungi. Another fungicidal treatment is sometimes applied just before sowing, targeted at the seed-borne and soil-borne fungi that may attack the germinating seed or seedlings in the nursery.

Surface Sterilisation

Fungi adhering to the surface of the seeds may be exterminated by instant exposure to a sterilising agent. Various types are available; the following are listed by Bonner et al. (1994):

1. Hydrogen peroxide (e.g. 30% for 20 min)
2. Sodium hypochlorite (10% solution of commercial bleach)
3. 75% ethanol[4]

Under laboratory conditions seed surfaces may be sterilised by a 0.1% solution of mercuric chloride ($HgCl_2$); because of its content of the heavy-metal mercury, the agent should be handled and disposed of especially carefully and safely.

Prolonged exposure of seed to all the above agents is harmful. Exposure time and concentration should be adjusted to the individual species to achieve the highest efficacy whilst avoiding phytotoxic side effects. After exposure the seed should be rinsed in water to remove possible residues of the chemical. Sterilising agents are effective for pathogens adhering to the seed surface and those present in superficial seed-coat crevices, but deep infecting fungi will normally escape the treatment. Surface sterilisation is widely used in experimental work on small seed lots but is impracticable on a larger scale.

Heat Treatment

Brief exposure to high temperature applied by dry air or submersion in hot water is applicable in cases where the fungus is heat-sensitive and the seed is heat-tolerant (Agarwal and Sinclair 1997). In temperate oak (*Quercus* spp.) a 2–2.5 h'submersion of seed in water at 40–45°C is used to control fungal infection of *Ciborea*. Such prolonged exposure must be carefully adjusted with regard to time and temperature since too long an exposure is likely to be harmful to the seed. Further, the heat treatment may leave the seed coat more vulnerable to invasion of other pathogenic fungi; therefore, a fungicidal treatment may still be necessary (I. Knudsen 1997, unpublished report).

[4] Should be pure alcohol and not denaturated alcohol as the latter may damage the seed.

Fumigation

Fumigation with methyl bromide is effective to control certain fungal pathogens. Other less widely used fumigants are hydrogen cyanide, carbon disulphide and aluminium sulphide.

Fungicides

Some of the most common fungicides are Bavistin-SD, Thyride, Ceresan, Brassicol, Thiram, Panoctine 35%, Orthocide, Dithane M-45, Fytolan, Agrosan GN, Captan, RH-2161 and Octave. It should be noted that chemicals based on the same active compound are sometimes sold under different trade names by different manufacturers in different countries. Because of the larger market and use of agricultural seeds, most chemicals are accompanied with instructions and dosages for application for agricultural seeds only. Seed size and structure of the seed coat should be considered when determining the dose.

Some fungicides are only effective if they are in direct contact with the fungi; hence, fungi already present deep within the seed are likely to escape treatment (Christensen 1973; Gardner 1980). Systemic pesticides like triadimethol, ethirimol and metalaxyl are effective against deep-seated seed-borne fungal organisms (Mohanan and Sharma 1991). In Tasmania two calico bags each containing 50 g *p*-dichlorobenzene are added to each tin (approximately 12 l) of seed, one at two-thirds depth and the other at the top of the seed for fungal protection (Forestry Commission 1994).

The effectiveness of different fungicides to control different fungal species varies. In India, the relative efficacy of five commonly used fungicides, viz. Dithane M-45, Bavistin, Fytolan, Ceresan and Thyride (all 0.1% concentration), was tested on eight common storage fungi on seeds of three different tree species (Purohit et al. 1996). One of the fungicides, Fytolan had no effect on *Aspergillus niger*. The effect of Thyride depended on tree species.

4.11.2.2
Application of Fungicides

Most fungicides are applied as a dry powder mixed with the seeds. This method is mostly applicable to seeds with a relatively rough surface to which the powder will adhere. For larger quantities of seed the best method is to mix seed and powder by tumbling in a rotating drum. Where the seed surface is smooth, the fungicide may be applied by a dip and slurry method in which the seeds are dipped into an aqueous solution of the fungicide; sometimes a glue or binder may be added to improve the retention of the material. The dip and slurry method also assists in the absorption of the chemical (Agarwal and Sinclair 1997).

Fungicides can also be applied to pelleting material. During pelleting the seeds are tumbled with an adhesive material such as gum arabic, gelatin, methyl cellulose or the like, plus an inert filler such as gypsum talc, kaolin clay, limestone, peat or vermiculite. A fungicide powder may be mixed throughout the coating material or can be added in discrete layers or in the outermost part of the pellet (Mohanan and Sharma 1991; Agarwal and Sinclair 1997).

4.11.2.3
Biological Methods

There is little experience in the use of biological agents to control fungal development in tropical forest seed. Schaefer (1990b) reported that storing recalcitrant *Prunus africana* and *Podocarpus milanjianus* seeds in sawdust restricts fungal development, but it is not known whether the sawdust has any antifungal properties. In India, a *Eucalyptus* hybrid oil was found effective in controlling mould development in *Shorea robusta* seeds at high humidity. A minimum of 3 cm^3 of oil per 1,000 cm^3 of storage container was effective (Singh et al. 1979).

Biological control measures on fungi in agricultural seeds include the application of fungi antagonistic to pathogenic fungi (I. Knudsen 1997, unpublished report).

Seed Dormancy and Presowing Treatment 5

5.1 Introduction

Adequate moisture and appropriate temperature normally trigger germination; however, in order to avoid germination under conditions where seedling survival is likely to be low, many species have developed regulatory mechanisms to delay germination. The states in which viable seeds fail to germinate when provided with conditions normally favourable to germination (adequate moisture, appropriate temperature regime, a normal atmosphere and in some cases light) are collectively known as dormancy. 'Resistance' to germination can have different degrees, ranging from very slight, causing just a short delay of germination, to very strong/deep, in which seeds need a strong pretreatment to initiate germination. There is often a large variation between individual seeds in a seed lot with respect to dormancy (Bewley 1997). It can be discussed *how much* restriction should be imposed before the term dormancy applies. All dry seeds thus have some restriction to water absorption, and all seeds germinating from indehiscent fruits will experience some mechanical resistance to their expansion. However, the barrier must result in a significant delay in germination to justify the term 'dormancy' (physical and mechanical dormancy, respectively, in the two examples mentioned). Sometimes dormancy changes during the lifetime of the seed, usually as a response to external conditions. Hence, dormancy may be innate, develop, be broken and redevelop in seed.

There are several types of dormancy, and sometimes more than one type occurs in the same seed. In nature, dormancy is broken gradually or by a particular event. There is a close connection between dormancy type and type of dormancy-breaking event. For example, dormancy caused by a hard fruit or seed coat may be overcome by a gradual or an instant abrasion; darkness-induced dormancy is overcome by exposure to light. In seed handling the natural dormancy-breaking mechanism is applied or simulated during the process of pretreatment[1].

[1] 'Pretreatment' is a term used for conditions or processes applied to break dormancy prior to germination, while 'treatment' is used for application of pesticides for control of pest and diseases.

The type of pretreatment is thus closely related to the type of dormancy, and in seed handling the classification of dormancy is linked to the pretreatment procedure.

Dormancy can occasionally be advantageous, e.g. to prevent presowing germination of recalcitrant seed; however, for orthodox seed, where germination is easily prevented by drying, dormancy is mostly considered as an inconvenience or constraint in seed handling. Complex types of dormancy where seeds need a very specific pretreatment often result in poor and/or non-uniform germination. Seeds which have not been given an appropriate pretreatment to overcome dormancy may fail to germinate altogether, germination may be slow or germination of individual seeds in a seed lot may take place over a lengthy period. In seed testing, seed lots with low germination rate, where non-germinated seeds prove to be sound and viable in, for example, cutting or 2,3,5-triphenyltetrazolium chloride (TTZ) tests, are considered dormant.

Pretreatment thus has a dual purpose, viz. to ensure both that seeds will germinate and that germination is fast and uniform. Most dormancy types are relatively simple; complex combined dormancy types exist, and where they occur, they can be very difficult to overcome, but they are not common. Dormancies are products of different dispersal modes, regeneration strategies and taxonomic relation. For example, animal-dispersed fleshy fruits usually have dormancy caused by inhibitory substances, pioneers are often light-dependent and leguminous species often develop hard impermeable seed coats. Knowledge of species taxonomy, morphology, dispersal and regeneration biology thus often gives a clue to dormancy type. However, variation in dormancy between and within species, and variation within the same seed lots, implies a current challenge to refine and adopt the pretreatment method to maximise germination. Some dormancy types and pretreatment methods imply a risk of overtreatment in the sense that a seed lot exposed to a certain pretreatment method will result in some seeds dying from overtreatment; other seeds remain dormant because of undertreatment. Pretreatment methods thus often have to be adjusted to individual species and seed lots on the basis of experience and experiments. Pretreatment is often a matter of adjusting the strength or duration of already-known methods, rather than adopting new ones.

Pretreatment is a 'presowing' treatment, which is normally carried out just before sowing. Some types of physiological dormancy are overcome during germination rather than pretreatment. An example is where dormancy is overcome by light or fluctuating temperature.

Some pretreatment procedures are not directly related to seed dormancy, but are carried out in order to speed up the germination process or promote seedling establishment. Various hormones and nitrogenous compounds may help in breaking dormancy under certain conditions, and may simultaneously have a direct impact on germination. In priming, seeds are treated in a way to initiate germination without the process being carried as far as radicle protrusion. In pelleting, seeds are enclosed in a matrix to which may be added fertiliser, fungicide or microsymbiont inoculants. Both priming and pelleting are mainly used in connection with direct sowing.

5.2 Dormancy in a Regenerational Context

The ecological purpose of dormancy is to delay germination until the chances of seedling survival are high and/or to spread germination over a prolonged time. This makes sense for species regenerating in a seasonal climate and under variable conditions. Dormancy here serves to 'save' seed from waste germination efforts under conditions where seeds are exposed to conditions favourable to germination but where conditions for seedling survival and establishment may be poor or erratic (Bewley 1997). Some of these situations are 'predictable' or regular:

1. Species in cold seasonal climates, e.g. temperate conifers and deciduous broadleaved trees, mature during the temperate autumn. Germination conditions in terms of temperature and moisture are suitable for germination, but survival conditions for young seedlings are poor because of subsequent low winter temperatures. In these species germination is delayed by a temperature-related dormancy mechanism; seeds germinate during the spring period when the chances of survival of the offspring are much better (Bewley 1997).
2. Seasonal dry areas often have erratic rainfall. Light showers may be sufficient to trigger germination, but not to provide adequate moisture for the seedlings to establish themselves. By producing seeds with different degrees of dormancy, or dormancy which is gradually broken, the species saves part of the seed pool, so that some of the seeds are likely to germinate when conditions are favourable for seedling establishment (Mayer and Poljakoff-Mayber 1982).
3. Regeneration in fire-prone areas is greatly improved after fire. Many tree species from such areas, e.g. some legumes, pines, eucalypts and *Banksia*, have developed dormancy which is broken only by exposure to high temperature; some show response to smoke (Brown 1993; Brown and van Stead 1997; Dixon et al. 1995; Razanamandranto et al. 2005).
4. Seeds that happen to be buried under a thick layer of soil may be unable to reach the soil surface during germination. Such seeds may remain alive and dormant, and only germinate if they are uncovered. Light and temperature fluctuations are stimuli likely to trigger germination of the dormant seeds.

5. Seedlings of light-demanding pioneers cannot survive the shaded conditions under a forest canopy. They regenerate only after gap formation. Gap formation is associated with light and fluctuating diurnal temperatures. The two factors, sometimes interacting, sometimes replacing each other break dormancy and trigger germination of pioneer seeds.

Since dormancy is closely related to the regeneration environment, large variation in degree of dormancy can be observed between different provenances of species with wide ecological amplitude and geographical distribution. For example, highland and lowland provenances of the same species can show a significant difference in response to chilling pretreatment (Richards and Beardsell 1987; Close and Wilson 2002).

Some dormancy types relate to dispersal. In natural regeneration, germination must be postponed until the seeds have dispersed:

1. In dry fruits, maturation drying prevents germination. In fleshy, animal-dispersed fruits, seeds are surrounded by a juicy substance with high water content, which is actually sufficient for imbibition. However, the sugar content and chemical inhibitors prevent germination. Once the fruit flesh has been removed and possible remaining inhibitors washed out or diluted, germination can take place. This may happen during animal ingestion, or in non-ingested fruits by the action of bacteria and fungi. Seeds with an aril which does not surround the fruit or is not fleshy (*Afzelia, Sindora*) frequently have inhibitory substances in the aril.
2. Dispersal by ingestion requires strong protection against physical and chemical damage during passage of the seeds through the digestive track. Drupes have developed a strong protective stone. Berries and animal-dispersed legumes a hard seed coat. These coverings also often restrict water uptake and thus imbibition, but ingestion usually partly abrades the seed coats (Halevy 1974; Winer 1983). Seeds that are dispersed by a variety of animals, e.g. hard-seeded acacias, tend to have very strong seed-coat dormancy, but also large variation in this type of dormancy.

Seeds may develop dormancy if exposed to unfavourable germination conditions. This phenomenon, known as secondary or induced dormancy, occurs, for example, in light-sensitive seeds that are exposed to darkness for a prolonged period, for example the aforementioned examples of pioneer seed deposited under a canopy or covered with soil. Dormancy may be broken by an instant event like the aforementioned gap formation, ingestion or fire. In other cases, dormancy is broken gradually by the influence of external factors, e.g. sand abrading hard seed coats (K. Brown 1987, unpublished report), leaching of inhibitors by rainwater (Villiers 1972; Brown 1972) or natural decay of fleshy fruit substance (Mayer and Poljakoff-Mayber 1982).

Most species are exposed to some type of stress during dispersal and regeneration, and dormancy regulation is apparently so effective that most species have developed some type of dormancy. Complete absence of dormancy is found in some mangrove species, where seed maturation and germination are more or less continuous. Neither moist forest nor wind dispersal trigger dormancy development, and in fact wind-dispersed rain forest trees belong to the non-dormant category (e.g. dipterocarps). Many rain forest trees are animal-dispersed and have inhibitory compounds. Climax species of the humid tropical forests rarely have postdispersal seed dormancy; their seeds are adapted to rapid germination on the forest floor, where they often survive for long periods as dormant or suppressed seedlings, while awaiting improved light conditions for growth. Many wind-dispersed trees are from dry areas where they have developed some type of stress dormancy.

For most dispersal-related dormancy, the duration is short and overcome once the seeds have been deposited. Temperate species typically stay dormant over one winter period. Non-germinated seeds of pioneers in moist forest have a short viability because of predation and soil-living microorganisms. However, seeds with a strong innate protection (e.g. some Leguminosae, pines and eucalypts) may build up large soil seed banks of several years' accumulated seed production in areas with slow deterioration (e.g. cold or hot, dry areas), where predation is low, and where the frequency of dormancy-breaking events, such as fire or rainfall, is rare (Holmes et al. 1987; Cochard and Jakes 2005).

In a regenerational context it is important that dormancy breaking is an 'either/or' event. A theoretical maintenance of 'some dormancy' into the germination process would imply slow or delayed germination, which could leave the seed in a vulnerable stage. A 'threshold event' is an event that will overcome dormancy and have no aftereffect. For example, once a physical restriction to water absorption has been overcome, seed and plantlet can absorb freely. Once the threshold to overcome physiological dormancy has been exceeded, the germination metabolism will, as one of the essential events, produce hormones that counteract residual hormonal restriction to germination.

5.3 Physiology of Seed Dormancy

The reaction of any type of dormancy is in the embryo, but seed dormancy can be ascribed to different seed tissue. The location and type of dormancy can be detected experimentally by removing or treating various parts of the fruit or seed separately. For example, if dormant seeds germinate after removal of their seed coat, it can be concluded that the site of dormancy is the seed coat. If excised embryos do not germinate when treated with extract from the fruit, it may indicate that the fruit contains inhibitory substances (Thapliyal and Naithani 1996). These experiments have revealed that basically any part of the fruit or seed can be part of seed dormancy (Fig. 5.1).

The pericarp or seed coat may (1) form a mechanical barrier to the protrusion of the radicle or swelling of the embryo (mechanical dormancy), (2) be a physical barrier to water uptake and/or gaseous exchange (physical dormancy), (3) modify the light reaching the embryo (photodormancy), (4) contain inhibitory substances or (5) prevent escape of inhibitors from the embryo (Bewley and Black 1982, 1994; Ellis et al. 1985).

The pericarp and seed coat are exogenous structures. Their roles are more or less the same in different species. Fleshy pericarps occur in drupes and berries, fleshy testas in magnolias and podocarps. Hard structures with concurrent dormancy phenomena occur either in fruits or seeds. In general, the two structures interact such that fleshy or hard structures in fruits are associated with

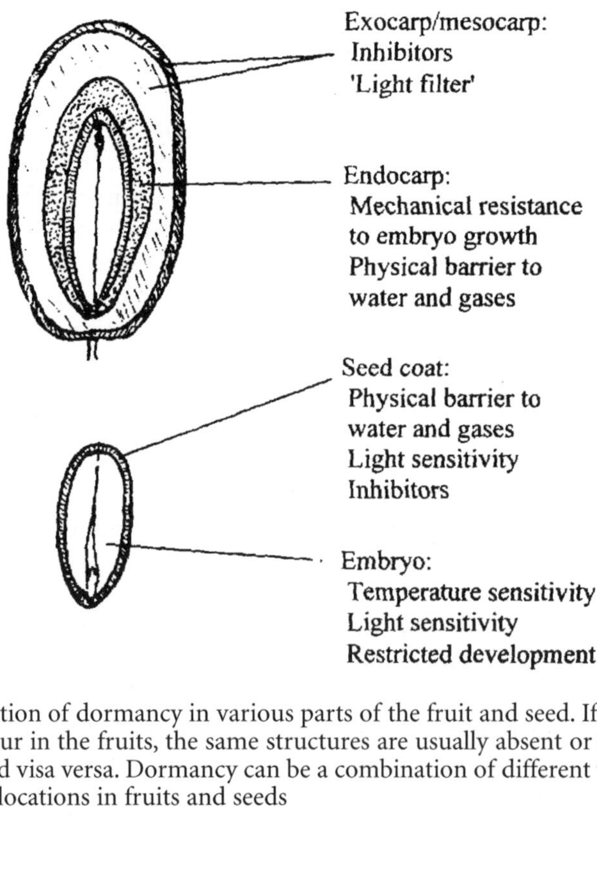

Fig. 5.1. Location of dormancy in various parts of the fruit and seed. If fleshy or hard structures occur in the fruits, the same structures are usually absent or less developed in the seed and visa versa. Dormancy can be a combination of different types and thus also different locations in fruits and seeds

thin and fragile testas. Seed coats are, however, often more specialised, and light sensitivity and moisture regulation are often associated with specialised structures in the seed coat.

The endosperm occasionally contains inhibitory substances but otherwise plays minor roles in dormancy (most seed have no endosperm at maturity), Dormancy caused by immature or underdeveloped embryos is obviously a pure embryo character, and thermodormancy can probably also be restricted to the embryo itself. Although both chemical inhibitors and light ultimately are sensed by the embryo, both inhibitors and photodormancy are normally associated with the outer coverings.

Dormancy in the embryo, i.e. the phenomenon that the embryo does not initiate metabolism and growth despite imbibition, temperature and other external conditions, is caused by various blocking mechanisms in the metabolic pathway. The plant hormone abscisic acid (ABA) plays a main role in these blocks, which may be simple or complicated. Temperature, light and chemical inhibitors are germination-controlling mechanisms that interact with ABA (Bewley 1997). Dormancy-breaking stimuli can therefore sometimes interact, such that one pretreatment compensates for another, and chemical compounds can have an indirect effect on breaking dormancy.

Dormancy frequently changes during development. Germination inhibitors are often located in the fruit and outer layers of the seed. Dormancy is overcome by removing the fruit structure, the seed coat and/or leaching out inhibitors. Stored seed often shows less response to such treatment because the inhibitors have moved from the seed coat into the embryo, thereby inducing embryo dormancy. Such induced or increased levels of dormancy have been shown, for instance, in *Corylus avellana* (Jarvis 1975, quoted in Richards and Beardsell 1987) and *Prunus africana* (Schaefer 1990a, b).

Physical, and sometimes also mechanical, dormancy is related to seed moisture content. Maturity stage and concurrent desiccation thus influence dormancy in the way that young seeds exhibit less dormancy than old ones. In most legumes, complete impermeability develops around a moisture content of 12–14%, but as drying progresses the cells of the seed coat become tightly packed. Part of the seed pool of many Leguminosae in humid areas never develops dormancy because the seeds do not dry out sufficiently (Fig. 5.2). Dormancy is here a phenomenon that develops if seeds happen to dry out either naturally or during processing. Duguma et al. (1988) found that all freshly collected seeds of *Leucaena leucocephala* germinated readily although the germination percentage was low (70%) for green seeds (moisture content 53%). In later maturity classes (light brown with a moisture content of 53% and dark brown with a moisture content of 40%) some seeds were dormant and at 4% there was pronounced physical dormancy (more than 80% impermeable).

Fig. 5.2. Germination of seeds of *Albizia gummifera* from fresh pods in a moist forest. The seeds do not dry out and hence do not develop physical dormancy under moist forest conditions; Kenya

5.4 Terminology and Classification of Dormancy

Several types of dormancy classifications exist. Traditionally physiologists, ecologists and seed handlers have used different classifications.

A plant-physiological approach classifies dormancy according to which part of the seed is responsible. Dormancy related to the embryo, e.g. immature development or chemical inhibitors located in the embryo, is referred to as **endogenous** or **embryo dormancy**. Analogously, mechanical resistance, physical impermeability, inhibitors or light sensitivity associated with the seed coat are called **exogenous** or **seed-coat (enhanced) dormancy.** 'Seed coat' is here used in the wide sense of any enclosing structure including, for example, endocarp or the entire pericarp. Water absorption and concurrent expansion of the embryo (imbibition) are purely physical processes, and restriction to either event is thus also physical. Processes in the embryo that lead up to germination are essentially life processes. Analogues, restricting factors that interfere directly with the onset of germination are physiological dormancy. Specific temperature regimes, lack of light and chemical inhibitors which 'block' the onset of germination thus represent various forms of **physiological dormancy**.

An ecological approach prefers classification based on development: (1) **innate** (or **primary**) **dormancy** is when dormancy is developed before dispersal (e.g. hard seed coat or chemical inhibitors), (2) **induced** (or **secondary**) **dormancy** is dormancy developing as a response to external environmental factors (e.g. drought or cold) and (3) **enforced dormancy** occurs when germination is constrained because of external conditions (Harper 1977). The last group does not, however, comply with the usual definition of dormancy since germination, according to that definition, does not take place *despite* favourable germination conditions, not *because* of unfavourable conditions. Seeds, which do not germinate because of external conditions, are more correctly referred to as '**quiescent**' (Villiers 1972).

Seed practitioners relate dormancy to the physiological nature and the method of pretreatment used to overcome them (Hartmann et al. 1997; Nikolaeva 1977). The system is simplified here to six main types: **mechanical dormancy** is used for structures which impede expansion of the embryo; **physical dormancy** refers to impermeability or serious delay in water absorption; **thermodormancy** here encompasses all types of temperature-related dormancies, whether high, low or fluctuating; **photodormancy** encompasses all light-related dormancy phenomena; **chemical dormancy** is used for all types of dormancy based on chemical inhibitors; **immature embryo dormancy** is used for the delayed germination caused by an undeveloped embryo at dispersal.

Where two or more dormancy types occur in the same seeds it is called '**double dormancy**' or '**combined dormancy**'. Double or combined dormancy is, for example, found in fleshy fruits with chemical inhibitors combined with, for example, a hard endocarp (**physical dormancy**), or immature embryos combined with other dormancy types. A summary of dormancy types according to the classification used in this book is shown in Table 5.1.

5.5 Dormancy Types and Pretreatment Methods

In the classification used in this book, dormancy is linked to a set of pretreatment methods used to overcome the particular type (Table 5.1). All dormancy types block some essential stages in the normal germination process. The simplest and probably most common or widespread form is a mechanical barrier to water absorption and embryo development found in physical and mechanical dormancy, respectively. These types are also pretreated mechanically as disintegration of the restricting morphological structures. The other dormancy types relate to some physiological pathways of the seed.

Pretreatment is a 'presowing-treatment' usually carried out in connection with sowing. There are cases where dormancy is overcome as a side effect of normal processing, e.g. removal of chemical inhibitors during depulping or aril removal, mechanical abrasion during extraction and embryo development during after-ripening. Hot-water treatment against insect or seed-borne pathogens may also occasionally serve as a pretreatment. However, deliberately

Table 5.1. Classification and characteristics of seed dormancy

Dormancy type	Characteristics	Examples of occurrence	Dormancy breaking stimulus: Natural	Dormancy breaking stimulus: Seed handling
Immature embryo[a]	Seeds are physiologically immature for germination	*Fraxinus excelcior*, *Gingko biloba*	Postdispersal development	After-ripening
Mechanical dormancy	Embryo development physically restricted owing to hard seed/fruit coat	*Pterocarpus*, some *Terminalia* spp., *Melia volkensii*	Gradual decomposition of hard structures, e.g. by termites	Mechanical cracking of restricting structure
Physical dormancy	Imbibition impeded because of impermeable seed coat or fruit	Mainly hard seed Leguminosae, plus some Myrtaceae and others	Abrasion by sand, high temperatures, temperature fluctuations, ingestion by animals, or other mechanical or chemical impact	Mechanical scarification (e.g. abrasion or burning), boiling water or acid pretreatment
Chemical dormancy	Fruit and seed contain chemical inhibitory compounds that prevent germination	Fleshy fruit such as berries, drupes and pomes, plus some dry seeds	Ingestion by frugivores, leaching by rain, gradual decomposition of fruit pulp	Removal of fruit pulp plus leaching with water
Photodormancy	Seeds fail to germinate unless exposed to appropriate light conditions/regime. Is operated by a biochemical phytochrome mechanism	Many temperate species, e.g. *Betula*. Humid tropical pioneer species, e.g. *Spathodea* and some eucalypts	Exposure to light conditions likely to promote seedling survival, viz. white light or light relatively rich in red light	Exposure to light, normally during germination, sometimes a distinct light–dark cycle of variable duration
Thermodormancy	Germination low without pretreatment with appropriate temperatures	Most temperate species, e.g. *Fagus*, *Quercus*, *Pinus*. Dry-zone tropical–subtropical pioneers, e.g. *Hakea*, *Pinus*, *Eucalyptus*, *Banksia*. Humid tropical pioneers	Exposure to low winter temperature. Exposure to grass, bush or forest fires. Diurnal fluctuating temperature in gaps	Stratification or chilling. High temperature, e.g. kiln or light burning. Fluctuating temperature

[a] Immature embryo here refers to the normal dispersal time. Many seeds undergo some changes after dispersal, including embryo development.

delaying dormancy breaking until just before sowing has various practical rationales. Firstly, maintaining seeds dormant during storage is a prevention against presowing germination. Secondly, some pretreatment methods hamper storability, e.g. by mechanical damage (e.g. scarification). Thirdly, some dormancy types may redevelop during storage and are only effective in close connection with germination (e.g. photodormancy). Fourthly, some dormancy types only react when seeds are physiologically active, i.e. imbibed, and also these are only effective in close connection with germination.

In the instance of low-temperature dormancy found in many temperate and some high-altitude tropical species, cold storage of moist seed is in itself a pretreatment. In some cases there is no practical distinction between pretreatment and germination conditions. Light and fluctuating temperature demand are strictly speaking dormancy phenomena in seeds where only initiation of germination is influenced by the two factors. In practice such dormancy is overcome by providing germination conditions suitable to overcome dormancy, rather than giving a special pretreatment.

Particular pretreatment methods are designed to overcome particular dormancy types. However, some methods may be effective for more than one type. Cold moist stratification may thus be effective on thermodormancy as well as on softening of the seed coat; soaking in water influences both physical and chemical dormancy (Boland et al. 1980). Light and temperature sometimes interfere and giving one pretreatment may compensate for lack of another.

In many species specific knowledge of seed dormancy is scarce. However, adoption of methods known to work for related species, or duplication or simulation of natural conditions believed to influence dormancy are often effective (Hartmann et al. 1997).

5.5.1 Mechanical Dormancy

Mechanical dormancy refers to the condition in which the embryo development is physically restricted owing to a hard enclosing structure. In the strict sense, mechanical dormancy does not include impermeability to water and gases, which is referred to as 'physical dormancy' (see later). In practice, most mechanically dormant seeds have some restriction to water uptake. The connection is rarer the other way around: only a minority of physically dormant seeds also exhibit mechanical dormancy.

Mechanically dormant seeds may imbibe water, but the radicle is unable to split or penetrate its enclosure, which is usually the fruit or part of the fruit. The term 'unable' is relative in the sense that most seeds germinating from indehiscent fruits will experience some physical barrier to embryo expansion, but almost all seeds will overcome the barrier after a certain time. Delay in germination after imbibition is natural in all germinating seeds. Mechanical resistance to embryo expansion may in some milder cases just delay germination. Mechanical dormancy can be found in hard-fruited samaras (*Pterocarpus*, *Terminalia* and *Heretiera*), drupes (*Melia volkensii*, *Canarium* sp.) and nuts (*Lithocarpus*). Mechanical dormancy has also been suggested in seeds of *Eucalyptus delegatensis* and *Eucalyptus pauciflora* (Bachelard 1967, quoted in Turnbull and Doran 1987).

Mechanical restriction to embryo development is overcome by softening or splitting open the fruit or seed. Splitting open may have the character of breaking hard structures or extracting the seeds from hard fruits.

Mechanical extraction or splitting of hard fruits always implies a high risk of seed damage. A hard pericarp is always associated with a soft and fragile seed coat. A very hard pericarp can be a strong physical barrier for germination; however, physical breaking of hard fruits by force implies a high risk of damaging seeds. Manual extraction or splitting fruits open will almost inevitably cause loss of seed. Random cracking with a hammer may be tempting but is not advisable.

The damage can be reduced by initially familiarising oneself with the fruit and seed and identifying the weaker and more sensitive part. Fruits naturally split open along the junctions of the carpels. These junctions are often visible on the fruit as grooves or depressions. The most sensitive part of the seed is the radicle of the embryo. The orientation of the seed in the fruit, or embryo, can be identified on a single seed or fruit. During splitting open or extraction, the potential impact on the seed radicle should receive special attention.

For example, fruits of *Pterocarpus* spp. are very hard, modified pods consisting of one carpel. Pods normally split open in two halves along one or two sutures. It is very difficult, however, to split the halves along these sutures without damaging the seed. The fruits can be cut transversely with a secateur and the one to three seeds can be picked out from their cavity. This implies a risk of cutting the seeds, but as minor damage to the cotyledons is not necessarily fatal, a cutting angle is adapted to avoid the radicle end. This can be done on intact fruits, where seed position and orientation can be estimated from the position of the petiole and the fruit apex (Fig. 5.3). These points of orientation are, however, lost where fruits have been dewinged. In these cases the best way is to clip the fruit with secateurs a few millimetres around and then split it open along the suture with the help of a strong knife.

Some pyrenes can be split open by using a knife, a chisel or other adapted tool to split the stones from the top. The tool is placed across the endocarp, where the fruit is to be split and applied with a few gentle blows with a hammer until the stone starts to crack (Fig. 5.4). A strong tool is inserted in the crack and the fruit split open by twisting. This method is used for, for instance,

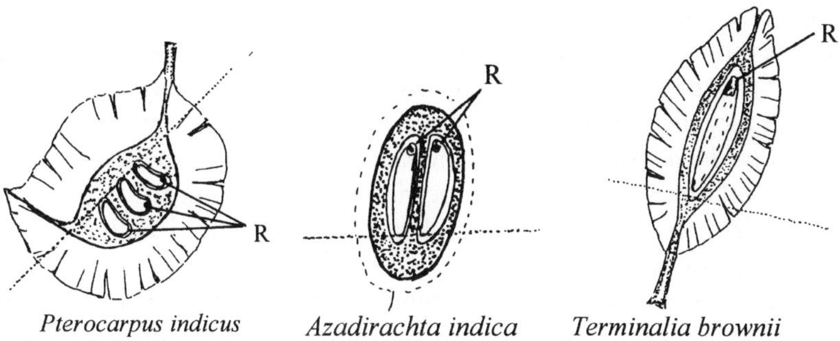

Fig. 5.3. Orientation of the embryo in some fruits with hard pericarp enclosure. *R* indicates the radicle and the *dotted line* the cutting line for extraction or mechanical pretreatment

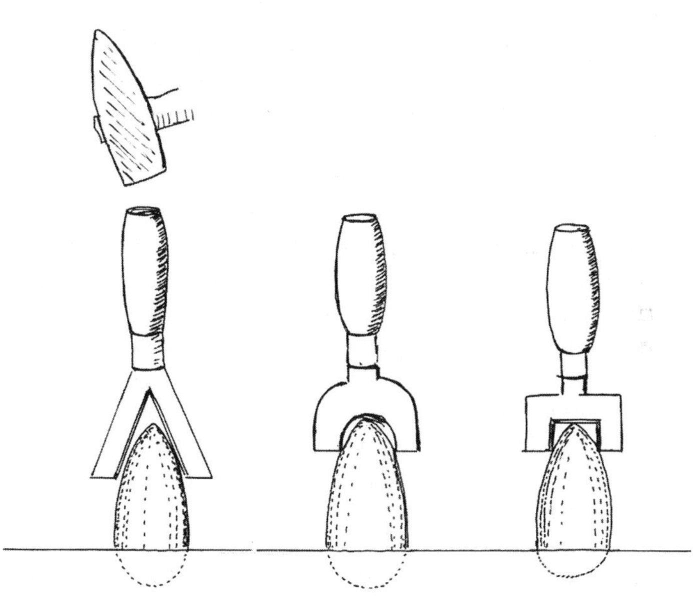

Fig. 5.4. Very hard endocarps can exert a physical restriction to embryo expansion. This type of dormancy can sometimes be overcome by mechanically splitting open the stone. To minimise damage to the seed, a special tool design is used, a split chisel, with a concave form adapted to the stone size

East African *Melia volkensii*, the seeds of which are enclosed in a very hard stony endocarp (Milimo 1986, quoted in Kamondo and Kalanganire 1996).

Softening the entire fruit or seed coat with the help of some dissolving method has the problem that if dormancy is purely mechanical, i.e. there is permeability to the seed, then the dissolvent must be physiologically harmless.

However, where mechanical dormancy is combined with physical dormancy (and this is often the case) various soaking agents can be used.

Acid pretreatment has been used successfully to improve germination of *Pterocarpus angolensis* (Groome et al. 1957, quoted in Willan 1985) and *Terminalia bellirica* (Bhardwaj and Chakraborty 1994). In the latter case, both total germination and germination speed were greatly improved by an optimal 12-min soaking in concentrated sulphuric acid compared with the control. Concentrated sulphuric acid (95–98% v/v) for 15–60 min greatly improved germination of *Terminalia superba* seeds, while any exposure to boiling water killed the seeds. It was suggested that the acid, having a higher viscosity than water, did not penetrate slits in the pericarp and hence did not come into physical contact with the embryo (Khasa 1992). Sodium hypochlorite (5.25% v/v) also greatly improved germination of this species, presumably because it penetrates into the slits of the pericarp and softens it from inside, yet is harmless to the embryo.

Moist stratification gradually helps softening seed coats or indehiscent fruit enclosures. The method is mostly used to break physiological dormancy in temperate and highland species. *Eucalyptus delegatensis* and *Eucalyptus pauciflora* are two species reported to exhibit mechanical dormancy together with physiological dormancy. Moist stratification tends to overcome both types (Boland et al. 1980).

Moist stratification has also been used for fruits of *Pterocarpus* spp. as an alternative to seed extraction. The fruit wings and outer softer coverings are removed during processing. After stratification the fruits are sown as an entity.

A combination of acid pretreatment and (warm) moist stratification can be used to shorten the period of stratification. The duration typically ranges between 3 and 5 weeks. This period can be shortened by initially treating the seed with acid, which scarifies the coat, but the treatment is stopped well before acid has penetrated the pericarp or seed coat. After careful washing, the seeds are exposed to moist stratification until dormancy has been overcome. This procedure both reduces the time required for moist stratification and reduces the risk of damage by prolonged acid treatment (Gordon and Rowe 1982).

Moist stratification at germination temperature will lead to germination, once possible mechanical restrictions have been overcome. Pretreatment is hence, in practice, just a long germination period. Stratification under physiological germination temperature is a better method to control germination, but many tropical seeds will survive only a short time under cold, imbibed conditions.

5.5.2 Physical Dormancy

Physical dormancy is caused by a hard and impermeable seed coat or fruit enclosure which prevents imbibition and sometimes also gaseous exchange.

The phenomenon is sometimes referred to as 'hard seed', because the seed coats remain hard and impenetrable during exposure to normal germination conditions. Physical dormancy is mostly known and described from Leguminosae, a family where the majority of the species exhibit this type of dormancy (Box 5.1), but it also occurs in, for example, some members of the families Myrtaceae (*Eucalyptus* and *Melaleuca*), Cupressaceae (*Juniperus procera*) and Pinaceae (*Pinus* spp.). Physical dormancy caused by the pericarp or part of the pericarp occurs in Rhamnaceae (*Ziziphus* spp.), Verbenaceae (*Tectona grandis*), Combretaceae (*Terminalia* spp.), Santalaceae (*Santalum* spp.), Ulmaceae (*Trema* spp.) and several others. Because most legume trees exhibit some

Box 5.1

The legume seed

The Leguminosae family exhibits some of the most advanced morphological structures of the seed coat to regulate physical dormancy. The seed coat (Fig. 5.5) consists of four distinct layers: (1) the outermost layer is the cuticle, which has a waxy and water-repellent character; (2) macrosclereids or palisade layer, which consists of long, narrow, tightly packed, vertical cells; (3) osteosclereids, which is a layer of more loosely packed cells; and (4) parenchyma layer, which is made up of a layer of little differentiated cells. Impermeability is caused by the cuticle and the palisade layer; scarification through the cuticle and halfway through the palisade layer is sufficient to overcome impermeability and seeds absorb water.

The thickness of the total seed coat as well as the relative thickness of the individual layers vary with species. Several *Cassia* species have relatively thick cuticles, while the palisade layer is relatively thin; in acacias, the cuticle is thin and the palisade layer thick. The seed coat is uniform except from a few special places: the hilar region is the region of funicle attachment, where also the micropyle and strophiole are located. As the seed loses water during maturation, the palisade cells of the seed coat become more tightly packed and the seed coat becomes more impermeable. During the latter part of drying, the strophiole functions as a valve, which allows the seed to lose water during dry conditions but prevents it from regaining moisture during humid conditions (Dell 1980; Chen and Fu 1984). The hilar region is the relatively 'weak' site where the seed is most likely to become permeable during pretreatment.

The cell structure of this region is slightly different with thin or no cuticle and the palisade cells are thinner and slightly modified. Hot-water pretreatment is believed to have a special effect on this region (Dell 1980). Another relatively weak area is the pleurogram, a horseshoe line found in the subfamily Mimosoideae to which, for example, *Acacia* and *Albizia* belong. The seed coat tends to crack along this line after, for example, heat treatment. Though the hilar region and the pleurogram are likely sites for initial water absorption, any part of the seed coat may be turned into the weaker site where water will ultimately penetrate (Werker 1980).

(*Continued*)

Box 5.1

The legume seed—Cont'd.

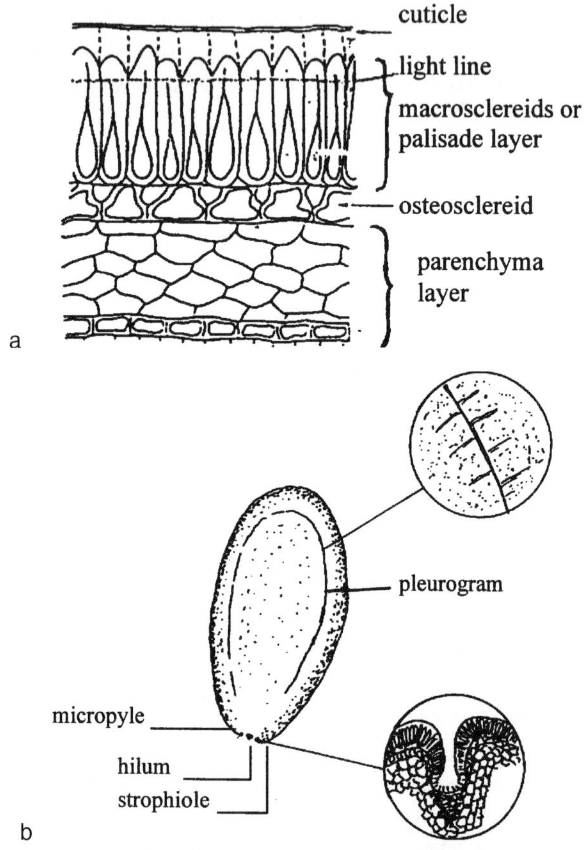

Fig. 5.5. Legume seed. **a** Cross-section of the legume seed coat. Seeds become permeable when the cuticle and the outer part of the palisade cells are penetrated. **b** Entire seed with demarcation of 'weak sites', sites most likely to become permeable during pretreatment. *Top insert*: Cracks along the pleurogram after hot-water pretreatment. *Bottom insert*: Cross-section of the seed coat in the strophiolar region

degree of physical dormancy, this dormancy type is by far the most common in tropical environments, particularly in arid zones. Physical dormancy is caused by highly specialised structures in the seed coats of Leguminosae. Impermeable endocarps are functionally analogous, and most of the same types of pretreatment methods apply. However, because of the anatomical differences between the seed coat and the pericarp, the character of the pretreatment sometimes differs.

Physical dormancy develops during and as a result of seed drying. Early mature seeds with high moisture content are thus typically less dormant than mature ones. Protective structures around the seed have two purposes: a dormancy function that is linked to seedling survival, and protection against digestion during dispersal by animals. The latter is apparently the stronger force, which can be concluded because ingestively dispersed seeds are far more hard-coated than, for example, wind-dispersed seed growing in the same environment.

Pretreatment of physically dormant seeds is basically a deviation of the same principle: to pierce the seed coat to an extent that will render it permeable to water so that imbibition can take place. Unless physical dormancy is combined with mechanical dormancy, penetration at one point is sufficient to ensure permeability. Because the impermeability in legumes is exerted by the outer layer of the coat and the palisade cells absorb water, a relatively superficial treatment may overcome dormancy in these seeds. A more homogeneous structure of the seed cover and a deeper impermeable layer of, for example, some endocarps may require a more drastic scarification.

Physical dormancy and thus pretreatment exhibit a wide variation between species, stage of maturity and degree of desiccation (Box 5.2). An individual seed lot thus also contains a wide variation from seeds absorbing readily in water to hard-coated seeds which need a very strong pretreatment to become permeable to water. Bulk pretreatment thus faces the problem that, when aiming at overcoming dormancy in the most resistant individuals in the seed lot, the seeds with relatively thin seed coats may be damaged by the pretreatment. We call this type of damage 'overtreatment'. Potential damage depends on the pretreatment type. Prolonged exposure to boiling water or acid can, for example, cause serious damage, while the risk of overtreatment is small in individual scarification. It is thus possible to achieve a very high germination rate after individual scarification, while bulk pretreatment will typically give a lower germination percentage.

In *Trichilia emetica*, a non-legume with physical dormancy, it has been found that the aril has a strong influence on dormancy. Removal of the aril was sufficient to break dormancy in the majority of seeds, while the remaining seeds needed an additional scarification (Masanga and Maghembe 1993). The importance of the aril in imposing physical dormancy is also known in, for example, *Afzelia xylocarpa* and *Sindora siamensis* (Pukittayacamee 1990). The aril physically blocks the weaker site around the strophiole and once removed this site is left exposed to water penetration. The aril also contains inhibitory substances which restrict germination (see later).

A large variation in seed-coat hardness exists between different species, between stages of maturity and within seed lots. Individual seeds in a seed lot vary from slightly to extremely dormant. Legume seeds achieve impermeability

Box 5.2

Comparing physical dormancy

Once physical dormancy has been broken (seed coat rendered permeable), imbibition takes place within a few hours. The response to a given treatment is thus that seeds become permeable, and the efficiency of the treatment can be measured as the percentage of seeds in seed lot that become permeable after the treatment. Comparing different methods for the same seed lot reveals their relative efficiency. Comparing different seed lots exposed to the same pretreatment reveals their relative seed-coat hardness.

Hot-water pretreatment is a 'standard' pretreatment that can be used for most legume species. Variation in physical dormancy between species, stages of maturity and individual seeds in a seed lot can be illustrated by a repeated pretreatment– imbibition experiment. In the two experiments illustrated in Fig. 5.6, the effect of several repeated standard pretreatments was measured on different seed lots. The seeds (5X100) were submerged in water at day 0. The number of imbibed seeds were counted and the imbibed seeds were removed after 1 day; these represent the non-dormant seed. The remaining seeds were pretreated with boiling water, left to cool and imbibe in the water, and the number of imbibed seeds was counted and the imbibed seeds were removed after 24 h. This pretreatment and imbibition procedure was repeated for 14 days. The cumulated average daily imbibition rate was plotted against time. The shape of the curve is a measure of relative seed-coat hardness.

Comparison of different species and seed maturity stages gives the following information:

1. Young/fresh seed of most species contains a high fraction of non-dormant seeds and the remaining seeds have relatively easily overcome dormancy.
2. Seeds of wind-dispersed species, e.g. *Acacia seyal*, *Acacia hockii*, *Acacia mellifera*, *Acacia polyacantha* and *Acacia reficiens*, have a significantly weaker seed coat than seeds of, for example, *Acacia nilotica* and *Acacia tortilis*, which are dispersed mainly by ingestion by large herbivores.
3. Within seed lot variation is very high. For example, 15% of *Acacia reficiens* seeds imbibed readily in cold water (non-dormant), while about 35% remained dormant even after 13 pretreatments with boiling water.

Because imbibition is a purely physical process and therefore independent of whether the seed is alive or dead, it is a more direct measure of physical dormancy than germination, but on the other hand only an expression of physical dormancy. Seeds with heat-sensitive embryos may be killed by boiling-water treatment and yet imbibe perfectly, apparently indicating that dormancy has been overcome. Repeated pretreatments gives a more effective distinction of relative seed dormancy than just prolonged soaking in water, used by, for example, Bebawi and Mohamed (1985). Prolonged soaking tends to cause little more additional imbibition than short-time soaking.

(*Continued*)

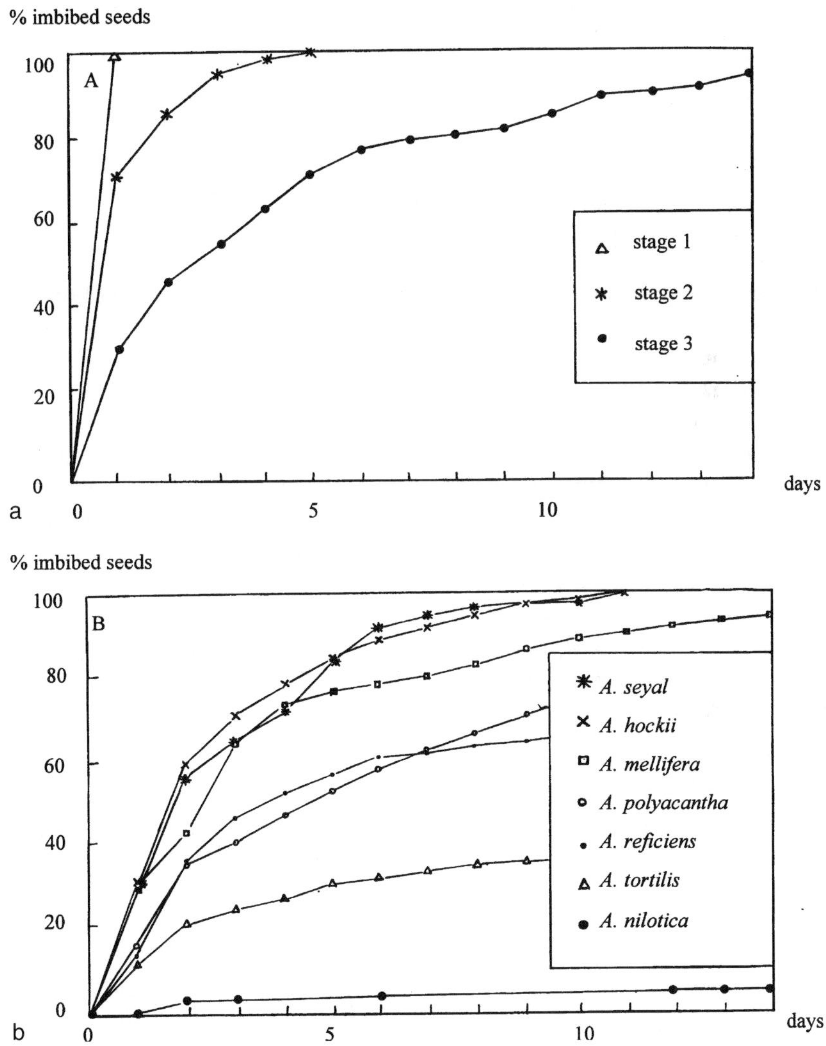

Fig. 5.6. Relative seed-coat hardness expressed by cumulative imbibition after soaking in cold water at day 0, and subsequent pretreatment with boiling water and soaking for days 1–13. **a** Three maturity stages of *Acacia mellifera*. **b** Seven *Acacia* spp. from eastern Kenya (moisture content 6–8%)

at around 10% moisture content, but the seeds can easily be dried to less than 5% moisture content. Palisade cells pack during desiccation and the drier the seed, the more impermeable it is. Fresh relatively immature seeds with high moisture content germinate readily without pretreatment, while very dry seeds need more severe pretreatment. Wind-dispersed seeds of humid zones, e.g. those of *Albizia* spp., exhibit little or no dormancy, while animal-dispersed seeds in dry areas are extremely hard coated. Comparing a selection of dry-zone acacias shows that seeds of animal-dispersed species are far more hard-coated than wind-dispersed seeds.

5.5.2.1
Mechanical Scarification

Scarification is an abrasion of the whole or the outer layers of the seed coat by piercing, nicking, chipping, filing or burning with the aid of a knife, needle, file, hot-wire burner, abrasion paper or the like (Fig. 5.7). If each seed is handled manually, the individual treatment is adjusted to seed-coat thickness and the treatment can avoid sensitive areas of the seed (the radicle end).

Virtually all seed can be made permeable, and the risk of overtreatment (damage) is small. It is thus often used as a reference method to which the effectiveness of other methods is compared.

Any site of the seed coat can be turned into a weak site where imbibition will start. In legume seeds, the cells of the palisade layer of the seed coat take up water, and the softening process spreads from the initial site of imbibition into the whole seed coat within few hours when the seed is submerged in water (Fig. 5.8).

Simultaneously the embryo imbibes and expands. In legume seeds, abrasion should penetrate at least through the cuticle and half way through the palisade layer (Fig. 5.5). Manual scarification is effective at any site of the seed coat, but the micropylar region should be avoided as it is the most sensitive site of the seed where the radicle is located (cf. discussion of mechanically dormant seed). Accidental damage to this region may damage the seed, while minor damage to the cotyledons is unlikely to affect germination (Cremer 1990).

Hot-wire burning as scarification has proven to be one of the most effective methods for manual pretreatment (Sandiford 1988). Compared with bulk methods, any manual scarification is labour-intensive, but with a few aids, e.g. sticking seeds to a tape to hold them in place, a person may be able to burn at least one seed per second. Large seed like those of *Afzelia* and *Sindora* are fairly thick coated and take longer to burn.

Another problem observed in connection with burning is fungal attack. Burning creates a small area of necrotic tissue around the burned site. This site is particularly vulnerable to fungi (Fig. 7.18). With optimum germination

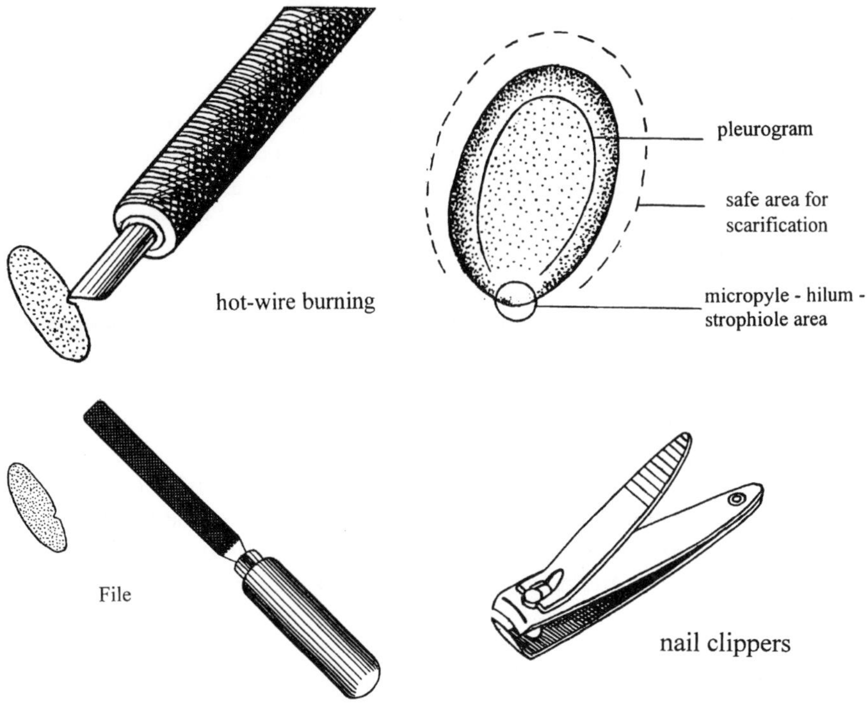

Fig. 5.7. Simple tools for mechanical scarification of hard-seeded legumes. Note that any point of the larger part of the seed surface may be scarified, while the micropylar region must be avoided. (P. Andersen)

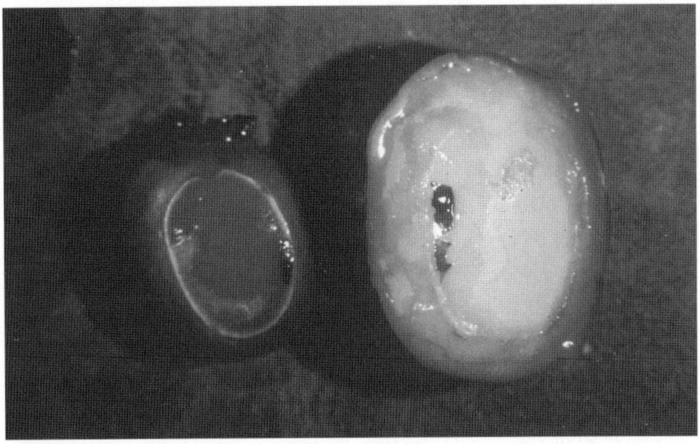

Fig. 5.8. Increase in size during imbibition of *Acacia tortilis* seeds. *Left*: Not imbibed; *right*: imbibed. Imbibition can start at any place on the seed coat and spread to the entire coat. Legume seeds imbibe 2–4 times their dry weight during imbibition

temperature, the seeds overcome the hazard, but under suboptimal conditions, the attack can be fatal.

Bulk scarification may be carried out by tumbling the seed in a cement mixer together with sand, gravel or any other sharp abrasive material. Smaller seed lots may be scarified by gently stirring them in a mortar with sharp sand. Abrasive material should obviously be of a size that makes it easy to separate it from the seed again (Chap. 3). The duration of treatment depends on seed type and should be determined by experience. In fast imbibing seeds such as Leguminosae, the efficiency in overcoming physical dormancy can easily be determined in an imbibition test: if the majority of a sample of seeds imbibe within a couple of hours, the pretreatment is sufficient; if only a few seeds imbibe, prolonged pretreatment is necessary. Mechanical bulk scarification may also be carried out with a so-called seed gun, the technical details of which are described in Fig. 5.9. During operation the seeds are filled into a central funnel with an outlet in a fast-rotating horizontal pipe. The seeds are slung against the wall of the enclosing concrete pipe and this causes their coats to crack (Poulsen and Stubsgaard 1995). The device has proven efficient for a number of species, but the number of damaged seeds can be fairly high.

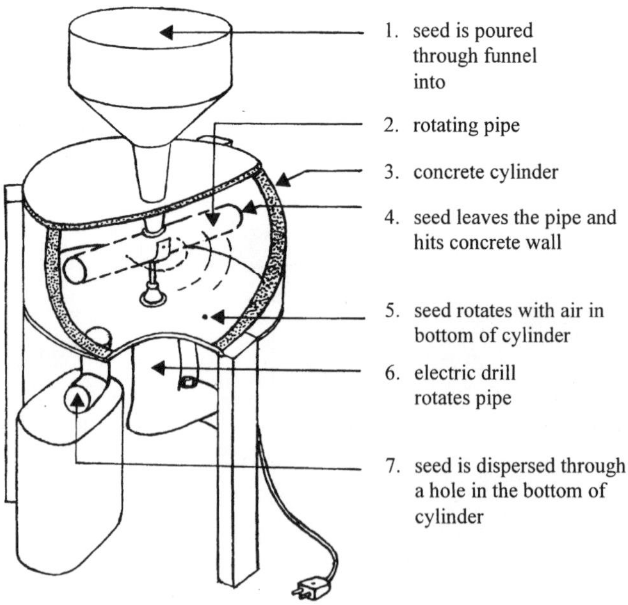

Fig. 5.9. Seed gun for bulk mechanical scarification of hard seeds. The speed of the revolving central pipe can be regulated and determines the centrifugal force and hence the treatment. For very hard seeds, the speed is increased; for relatively soft-coated seeds, the speed is decreased. (From Poulsen and Stubsgaard 1995)

5.5.2.2
Hot Water

Hot water overcomes physical dormancy in Leguminosae by affecting the strophiolar plug (Dell 1980), or by creating tension in the palisade cells which causes small cracks in the seed coat (Brant et al. 1971). The method is effective for most wind-dispersed seeds, but the treatment is often insufficient to overcome dormancy in some of the very hard coated animal-dispersed species. In the Sudan, the method was found inferior to both manual scarification and acid pretreatment of especially the very hard coated species *Acacia nilotica*, *Acacia nubica* and *Faidherbia albida* (Bebawi and Mohamed 1985).

The effect of hot water is greatest when the seeds are submerged in the hot water and are not heated together with the water. A quick dip is also better to avoid heat damage to the embryo. A common procedure is to pour the seeds into boiling water and then leave them to cool and imbibe in the water for 12–24 h. Keeping the seeds for a prolonged period at high temperature does not usually have an additional effect on overcoming dormancy. However, boiling-water pretreatment for 30–60 s with the seeds being left to cool in the water was the most effective method for non-leguminous *Juniperus procera* (Laurent and Chamshama 1987), and ATSC (1995) found that 2-min boiling was more effective than 1 min for some hard-coated Australian species; for some species boiling for up to 5 min is recommended.

Heat damage in connection with hot-water pretreatment is a current risk and must be balanced against dormancy breaking. Embryos are damaged by heat, but several factors influence possible damage during pretreatment:

1. Moist tissue is more sensitive and transfer of heat more effective for moist seed than for dry seed.
2. In thin-coated and non-dormant seed, heat transfer is more rapid.
3. The longer the seeds are exposed to hot water, the higher the risk of embryo heating and the greater the potential damage.
4. Tolerance to high temperature varies with species.
5. Strong heat damage causes loss of viability; less damage may cause loss of vigour or production of abnormal seedlings.

Most dry, thick-coated (animal-ingested) *Acacia* species tolerate a brief, less than 1 min., submersion in boiling water.

Cassia species are reportedly a group easily damaged by high temperature or prolonged exposure. For *Cassia siamea*, soaking in water at 95°C from 1 to 3 min caused rapid reduction of viability; it was 71% after 1 min, 47% after 2 min and 40% after 3 min. Soaking for 1–2 min in 85°C water or submersion at 85°C with subsequent cooling in the water for 12–36 h gave a germination percentage of

82–89. Longer soaking at 85°C decreased the germination percentage (Kobmoo and Hellum 1984). In *Cassia fistula*, a quick dip in boiling water killed 50% of the seeds, and 68% were killed after 5-min boiling (Babeley and Kandya 1988). High viability (75%) was maintained in *Cassia sieberiana* after 2-min boiling; longer boiling and any dry heat treatment rapidly reduced viability (Todd-Bockarie et al. 1993).

Boiling water was lethal to five out of 20 tested species in Ethiopia, viz. *Acacia seyal, Acacia tortilis, Acacia senegal, Cassia decapetala* and *Cassia spinosa* (Teketay 1996b). *Entada abyssinica* showed improved germination after a 5 s dip, but 15 s of boiling reduced viability to two thirds of that after 5 s – longer exposure was completely lethal. It should be noted that heat damage to *Acacia tortilis* seeds at 5 s as observed in the experiment mentioned is unusual; brief boiling is a common pretreatment method for that species elsewhere. Boiling-water damage has also been observed for *Paraserianthes falcataria* and *Albizia procera* (Sajeevukumar et al. 1995).

Tests of *Prosopis* species have shown different effects. No damage was observed for *Prosopis alba, Prosopis flexuoso, Prosopis chilensis* and *Prosopis tamarugo* after pretreatment by submerging their seeds in boiling water and leaving them to cool in the water (Lopez and Aviles 1988). Catalan and Macchiavelli (1991) found, however, seedling abnormalities greatly increased after high-temperature treatment (90–98°C) in *Prosopis alba* and *Prosopis flexuosa.*

The effect of hot water on overcoming physical dormancy and temperature sensitivity apparently varies both between and within species. Checking out the effect it should be recalled that the effect on dormancy is a physical phenomenon, which is revealed by the imbibition ability, while sensitivity to high temperature is a physiological phenomenon, which must be tested in a germination experiment.

5.5.2.3 Heating or Burning

Tension of the seed coat with consequent crack formation and permeability can be created by dry heat, e.g. brief exposure to oven-temperature heat or quick burning. As with hot water, the effect is mostly caused by temperature change and not by the temperature level. An effective method is to pour the hot dry seed into cold water. This will both enhance the cracking effect and reduce the risk of heat damage. Both temperature level and duration of exposure are crucial for the effect and possible damage.

Dry heat is often less effective than boiling water in overcoming physical dormancy, at least in legumes, but seeds may be easier to store after pretreatment provided they are cooled quickly without being left in water to imbibe. Dry heat is also frequently used in connection with dry extraction. The heat and subsequent cooling may serve as a dormancy break in some species.

A dry-heat pretreatment of *Acacia mangium* in Sabah is referred to in Table 5.2. As seen in Table 5.2, oven drying at 100°C for 10 min followed by immersion in cold water was found to be an effective pretreatment (Bowen and Eusebio 1981, quoted in Adjers and Srivastava 1993).

Grass burning over seedbeds is used for several species to break a type of physiological dormancy, where germination requires high-temperature exposure (see later). However, for pure physical dormancy, the method seems to be inferior to other pretreatment methods. Burning caused complete failure of germination in *Albizia procera* and *Paraserianthes* (former *Albizia*) *falcataria* in India (Sajeevukumar et al. 1995). In *Enterolobium cyclocarpum* and *Hymenaea courbaril*, burning enhanced germination but the effect was poorer and the damage greater than in other pretreatment methods (Brahmam 1996). The best results are reported from fire scorching of *Juniperus procera* seeds in Tanzania. However, although scorching improved germination from 0 to 50–60% (depending on fire intensity), the results were poorer than with hot water and acid pretreatment (Laurent and Chamshama 1987).

5.5.2.4 Acid Pretreatment

Strong acid causes some kind of wet combustion of the seed coat and works equally well in legumes and non-legumes (Fig. 5.10). The acid used for seed pretreatment is almost exclusively concentrated sulphuric acid (H_2SO_4), which is cheap and readily available in most places. Acid treatment is applicable only to species with thick and impermeable seed coats. The acid is highly lethal if it comes into contact with living parts such as the embryo. Sulphuric acid is very corrosive and dangerous to work with. Use of sulphuric acid requires the utmost observation of safety (Box 5.3). A 'politically correct' statement is that it should not be used at all because of these concerns; however, its efficiency in pretreatment is well documented and the remedy can hardly be ruled out.

Table 5.2. Effect of dry heat as pretreatment for breaking physical dormancy of *Acacia mangium* in Sabah. (Bowen and Eusebio 1981)

Dry-heat pretreatment		Effect on dormancy (%)	
Temperature	Duration	Imbibition	Germination
Ambient	Not indicated	3	3
100°C	5 min	80	67
100°C	10 min	Not indicated	83
100°C	15 min	95	80
100°C	60 min	95	50

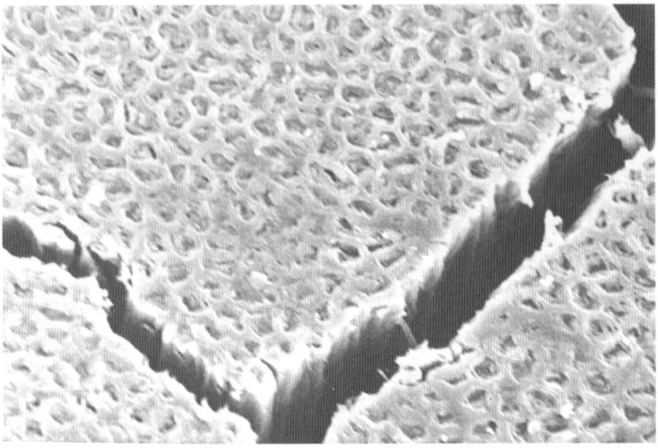

Fig. 5.10. Surface of seed coat treated with sulphuric acid. The acid attacks the cuticle and penetrates into the upper palisade cells. Acid is highly toxic to embryos but is harmless as long as it is in contact with the outer coverings

Box 5.3

Safety rules for sulphuric acid

Sulphuric acid(H_2SO_4) belongs to the group of strong acids and used in concentrated form (95%, 36 N) it implies high safety hazards during handling. Strict safety rules should therefore be observed:

1. Use acid only in a well-ventilated place as evaporated gases can cause serious irritation when inhaled. Avoid inhaling the gas when opening bottles.
2. Always use safety glasses, protective gloves (good-quality rubber gloves without perforations) and protective clothing (e.g. laboratory coat or apron).
3. Never pour water into undiluted acid; if the acid is to be diluted, carefully and slowly pour the acid into water.
4. Beware that even diluted acid can corrode skin, eyes and clothes. Protective clothing should be used throughout the operation, i.e. also during rinsing after pretreatment.
5. Store the acid locked up in a safe place when not in use. Make sure that the container used will not be corroded by the acid, that containers are not leaking, and that they are distinctly marked 'STRONG ACID'. This also applies for containers containing used acid.
6. Dispose of used and ineffective acid safely, i.e. heavily diluted with water.
7. Always have plenty of water, preferably a water tap, within easy reach during any handling of acid.
8. If acid is spilled on clothing or skin, rinse with plenty of water. If acid comes into contact with the eyes, rinse with plenty of water and contact a doctor immediately.

The practical application of acid pretreatment is as follows: A non-corrosive container should be used, e.g. a glass beaker for small lots under laboratory conditions (testing) or a thick plastic bucket or bowl for large quantities. Seed treated with acid should be dry (Willan 1985). Pretreatment should be at ambient temperature (15–25°C). The duration of treatment varies according to the following factors:

1. *Seed-coat thickness.* Depending on species, maturity, age, etc. Thick-coated seed coats or endocarps need longer treatment than thin-coated ones.
2. *Temperature.* Acid is more effective at higher temperature, and the treatment is thus shorter. In practice, most pretreatment takes place at ambient temperature.
3. *Strength of the acid.* Fresh acid is stronger than reused acid. Acid may be reused several times but its strength will gradually reduce, and the treatment time must be prolonged. The strength of the acid can be checked with a pH meter.
4. *Stirring.* Stirring during treatment reduces the duration of treatment compared with treatment with a still bath.
5. *Relative volume of the acid.* The larger the relatively volume of the acid as related to the volume of seed, the less the strength will reduce and the shorter the time required for pretreatment.

The duration of treatment can vary from a few minutes to several hours (Table 5.2). As with other pretreatment methods, too short is ineffective, too long causes overtreatment, which, in the case of acid treatment, causes death of the seed. Unfortunately, comparative studies exist only for different durations of pretreatment for different seed lots, while the conditions have not been systematically investigated. To avoid overtreatment by excessive soaking in acid, the duration of treatment must be adjusted.

Seeds should be rinsed carefully under running water for at least 10 min after pretreatment to remove leftover acid. The seeds must not imbibe acid; seeds that have already imbibed acid when they are removed from the acid bath can be discarded as they are no longer viable. It is possible to redry seeds pretreated with acid and keep them for at least 1–2 months.

The duration of acid pretreatment should aim at reaching a balance in which the seed coat (or pericarp) is sufficiently ruptured to permit the seed to imbibe, but without the acid itself reaching the embryo. Some guideline durations compiled from the literature are listed in Table 5.3. Considering the within seed lot variation in physical dormancy, the effect of acid pretreatment

Table 5.3. Duration of soaking in concentrated sulphuric acid to overcome seed-coat dormancy in some legume seeds

Species	Duration of acid pretreatment (germination %)	References
Acacia nilotica	>15 min	Rungu (1996)
Acacia tortilis	30 min–2h (100%)	Teketay (1996b)
Albizia lebbeck	40 min (85%)	Teketay (1996b)
Caesalpina spinosa	1–4 h (100%)	Teketay (1996b)
Cassia sieberiana	45 min (90–95%)	Todd-Bockarie et al. (1993)
Cassia fistula	45 min (75%)–90 min (84%)	Babeley and Kandya (1988)
Cassia siamea	15–45 min (98%)	Kobmoo and Hellum (1984)
Ceratonia siliqua	20 min (89%)	Martins-Loucau et al. (1996)
Cornus capitata	5 min (70–80%)	Airi et al. (2005)
Delonix regia	3–6 h	Sandiford (1988), Teketay (1996b)
Erythrina abyssinica	5–20 min	Laurent and Chamshama (1987)
Leucaena leucocephala	30 min (95%)	Duguma et al. (1988)
Prosopis alba	6–24 min (100%)	Lopez and Aviles (1988)
P. flexuosa	6–24 min (100%)	Lopez and Aviles (1988)
P. chilensis	6–24 min (95%)	Lopez and Aviles (1988)
P. tamarugo	6–24 min (95%)	Lopez and Aviles (1988)
P. juliflora	15–60 min (95–100%)	Teketay (1996b)
Senna bicaparis	60 min (95–100%)	Teketay (1996a)
S. didymobotrya	60 min (95–100%)	Teketay (1996a)
S. multiglandulosa	60 min (95–100%)	Teketay (1996a)
S. occidentalis	60 min (95–100%)	Teketay (1996a)
S. septemtrionalis	60 min (95–100%)	Teketay (1996a)

In all the experiments the seeds were carefully washed after pretreatment, allowed to imbibe in water and sown under optimal germination conditions. The numbers in *parentheses* indicate germination percentage after pretreatment

is quite remarkable – the duration of treatment must be significantly prolonged before any damage is observed.

In *Cassia siamea* 15–45-min treatment resulted in about 98% germination, while the amount of germination was lower for both shorter (1–10-min) and longer e (60-min) soaking (Kobmoo and Hellum 1984).

Similar results were found in experiments on the variation in the duration of acid treatment for several Ethiopian species (Teketay 1996b). In *Albizia lebbeck*, 40 min was effective, while both 20 and 60 min gave poorer germination; in *Caesalpina spinosa*, any duration of soaking within the tested pretreatment time from 1 to 4 h gave almost 100% germination. Eventually, in an acid pretreatment of *Hymenaea courbaril* and *Enterolobium cyclocarpum*, 15-min soaking was found suitable for both species, while a longer duration (20–25 min) gave slightly poorer results (Brahmam 1996).

Acid pretreatment is commonly used for Australian and African acacias and other legumes (Doran et al. 1983; Bebawi and Mohamed 1985). It must be considered one of the most effective pretreatments for hard seed, especially those with very hard coats, e.g. seeds of *Acacia nilotica, Faidherbia albida, Acacia bidwillii* and *Acacia stenophylla.* For thin-coated species, other treatments are available and preferred. Acid treatment also greatly improved germination of non-leguminous *Juniperus procera* (from 0 to more than 70%) (Laurent and Chamshama 1987).

Pericarp pretreatment by sulphuric acid requires a long time. Vasista and Soni (1988) investigated the effect of up to 60-min soaking of drupes of *Trema politoria* and found that germination increased proportionally with duration of soaking. However, for *Terminalia bellirica,* Bhardwaj and Chakraborty (1994) found that 10–12-min dipping in concentrated sulphuric acid was the most suitable pretreatment, which almost doubled the percentage of germination as compared with the untreated control. In *Ziziphus mucronata,* 20-min soaking in concentrated acid was the most efficient to overcome dormancy, but at the same time it killed several seeds (Hassen et al. 2005).

Acid pretreatment is probably the most effective method of bulk treatment for very hard coated seeds. It is widely applicable and effective for both legumes and non-legumes. A side effect is that it efficiently eliminates fungal spores (Nan et al. 1998). In addition to the risk of overtreatment mentioned already, a major drawback is safety risk. Pretreatment with sulphuric acid should be carried out with utmost care, since the chemical can cause serious injuries if it accidentally comes in contact with skin or eyes. A number of safety rules are summarised in Box 5.3. It should also be noted that acid causes corrosion of a lot of materials, such as fabric and metal, while glass and most plastics are resistant.

5.5.2.5 Other Chemicals

A number of alternative chemicals have been tested for breaking physical dormancy, none of which, however, have given results comparable to those of conventional pretreatment methods. Among 66 methods compared, none of them came close to sulphuric acid and manual scarification (Todd-Bockarie et al. 1993). Hydrogen peroxide (H_2O_2) has been shown to have a promoting effect on germination. Because the liquid has low viscosity and no harmful effect, it may be applicable to overcome seed physical dormancy. Chien and Lin (1994) suggested that the observed improvement of germination of *Cinnamomum camphorum* from 0 to 11% before treatment to 51–58% after treatment with 15% H_2O_2 could probably be ascribed entirely to the chemical helping to release physical dormancy.

5.5.2.6
Biological Methods

Ingestion by large animals (wild or domestic) scarifies the seeds and helps overcome physical dormancy. Thus, seeds of *Acacia* species extracted from goat faeces are often less dormant than non-ingested dry seed (Ahmed 1986). Experience has shown that many seeds are digested; those passing through the digestive track are the most hard-coated ones and sometimes a small fraction of the total lot ingested. In the comparative study of Todd-Bockarie et al. (1993), ingestion of *Cassia sieberiana* by sheep gave a poor result. On the small scale, ingestion may be applicable as it is cheap and simple. For example, in rehabilitation of degraded land with legumes, goats can be 'seed collectors', pretreat seed and do 'direct sowing' when left on the rehabilitation site.

5.5.2.7
Selection of Pretreatment Method

Several comparative studies on the relative effectiveness of a range of pretreatment methods on one or several species have been carried out (Teketay 1996a, b; Masamba 1994; Bebawi and Mohamed 1985; Khasa 1992). Apart from scarification of each individual seed, which universally seems to be the most effective method, no single pretreatment method is equally effective for all species. Since relative dormancy also varies within species, preliminary trials are often necessary to find the best method. In most instances, time (duration of treatment), safety risk (primarily acid treatment), available equipment and their relative costs are factors to be balanced against the physiological advantage; if seeds are abundant, a less effective pretreatment may be most efficient; if seeds are rare and expensive, a more efficient method is chosen. It should be recalled that overcoming physical dormancy is not necessarily the same as germination; see the remarks earlier on overtreatment.

5.5.3
Chemical Dormancy (Inhibitors)

Chemical germination inhibitors are prevalent in fleshy fruits or fleshy arilate seed, which naturally mature and are dispersed surrounded by a very watery structure. Inhibitors also occur in relatively dry arils on, for example, legumes (Fig. 5.11). Germination inhibitors are prevalent in animal-dispersed seed: germination is prevented as long as the seeds are not dispersed, and the inhibitors are removed by the dispersing animal (Traveset and Verdu 2002;

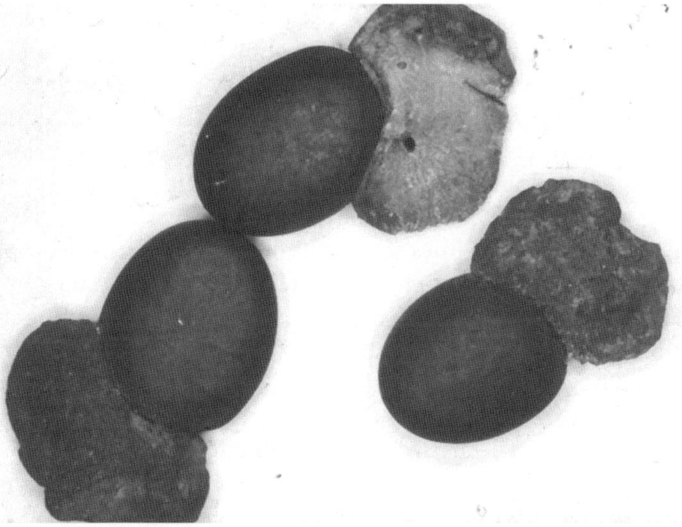

Fig. 5.11. Arilate seed of *Sindora sinensis.* The aril contains inhibitors which delay germination

Cipollini and Levey 1997). Inhibitors interfere with metabolic processes that initiate germination in imbibed seed. Germination inhibitors contained in fruit pulp make up a diverse group including lipids, glycoalkaloids, coumarin, ABA, hydrogen cyanide and ammonia (Cipollini and Levey 1997). Dormancy imposed by inhibitors is overcome by removing the pulp with the inhibitors. In addition to the aforementioned dispersal process, this happens in nature during decomposition of fruits and by rainwater leaching. In seed handling, germination inhibitors in fleshy parts of the fruits or seed will usually be removed during processing (depulping and removal of arils).

Germination inhibitors have been demonstrated in different fruit and seed parts. In *Dobera glabra,* an East African species with soft fruits which are dispersed by hornbills, inhibitors were demonstrated both in the outer green coreaceous exocarp and in an inner red soft mesocarp: removal of the exocarp increased germination from 8 to 57%; removal of the mesocarp further increased germination to 70% (Schaefer 1990b). In *Prunus africana,* very low germination was achieved when whole fruits were sown, while 75–90% of the seed germinated after extraction/depulping. Seeds of *Gmelina arborea* completely failed to germinate without extraction. Extraction followed by thorough washing in running water, to leach out inhibitors, enhanced germination to 50–90% depending on the preceding fermentation/softening procedure (Ogunnica and Kadeba 1993).

Depulping with consequent removal of inhibitors is usually routinely carried out as a part of seed processing. Delayed depulping, however, appears to affect the ease of removing inhibitory substances, as the inhibitors apparently tend to move into the fruit after some time. In the above example of *Prunus africana*, 20 days' delayed depulping reduced germination by 50% (Schaefer 1990b). Species with thin and relatively dry pulp can be stored as dry fruits, and in some species this is sometimes done deliberately if the seed coat is fragile and easily damaged during depulping, or if depulping is difficult, for example if the pulp is sticky. Some *Vitex* species have the combination of sticky pulp and fragile seed coat, and the fruits are sometimes sown without depulping, or depulping is carried out as a pretreatment just before sowing. Both methods show poor germination compared with that of fresh depulped seed. Depulping used as a germination pretreatment is carried out as washing and rinsing in running water. Several days is sometimes needed for thorough leaching of inhibitors.

Arils are sometimes firmly attached to the seed and difficult to remove during normal processing. The arils have the same effect as pulp and can seriously delay germination. In *Afzelia* and *Sindora* species, the aril also forms a strong physical barrier against imbibition (physical dormancy).

While inhibitors present in fruit structures are readily removed by extraction, those located in non-removable structures, e.g. remaining pericarps, seed coats, endosperms or embryos, must either be removed or be inactivated/overcome by special seed treatments. Water-soluble inhibitors are often effectively removed by leaching. Seeds are either subjected to running water or soaked in several changes of water.

The treatment may work both by physically removing the inhibitors with the discharged soaking water and by a gradual decomposition. Once the inhibitors have been adequately diluted, the seeds are capable of germinating. In teak (*Tectona grandis*), several alternate cycles of soaking and drying seem to gradually reduce chemical dormancy simultaneously with breaking physical dormancy (see later). Also stratification, primarily designed to overcome thermodormancy, may reduce the levels of inhibitors.

Dormancy in legumes has generally been ascribed only to the impermeable seed coat, but Sajeevukumar et al. (1995) also found indications of the presence of water-soluble inhibitors in the seed coat of *Albizia procera* and *Paraserianthes* (former *Albizia*) *falcataria*. The presence of an inhibitor has also been shown in seed coats of *Albizia odoratissima* (Kannan et al. 1996). Soaking for 24 h in running water after scarification is therefore recommended as a standard pretreatment for these species.

5.5.4 Photodormancy

Light plays a crucial role for seedling survival, especially so for light-demanding pioneer species, and photodormant seeds have adapted a system of impeding germination in the dark and in deep shade, e.g. under a forest canopy.

Germination of photosensitive seed is regulated by a phytochrome system, which in a simplified presentation works as follows.

Phytochrome appears in two forms, P_r and P_{fr} (with the subscripts meaning 'red' and 'far red'), which can be reversibly converted to either form by radiation at different wavelengths (Fig. 5.12). Germination is determined by the amount of P_{fr} relative to the total amount of phytochrome. Phytochrome in the P_r form inhibits germination, whereas P_{fr} allows germination to proceed.

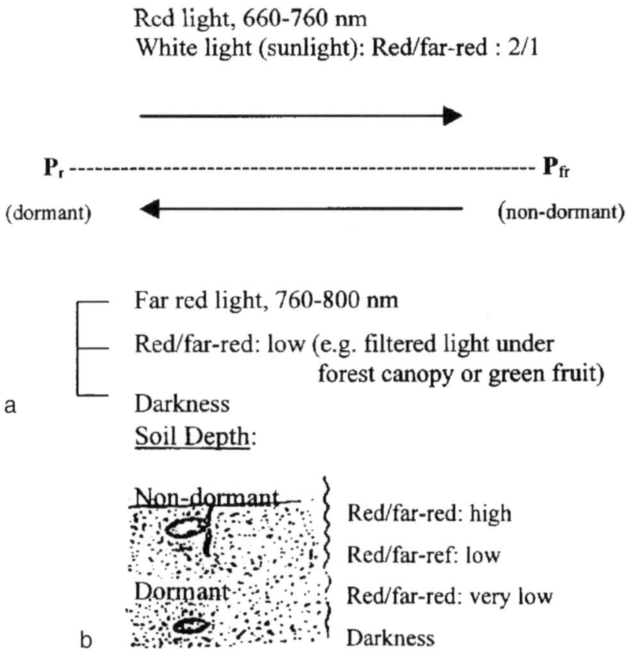

Fig. 5.12. **a** The principle of the conversion of phytochrome P_r to P_{fr} and phytochrome P_{fr} to P_r, respectively, under the influence of different light types. Red light and white light (high red to far red ratio) may convert P_r to P_{fr} and thus break dormancy (*top arrow*). Far-red light or light with a low red to far red ratio (e.g. filtered light) will convert P_{fr} to P_r and thus induce dormancy in seeds with a phytochrome dormancy system. In complete darkness, P_{fr} may revert to P_r and the seed consequently becomes dormant. The three conditions are indicated by the *lower arrow*, going from P_{fr} to P_r. Notice that subscripts *r* and *fr* refer both to a stage of the phytochrome and to the wavelength of the light that transforms the phytochrome. **b** An example of the conversion of phytochrome at different soil depths. Since red light has a lower penetration into the soil than far-red light, the relative amount of light of the two wavelengths changes. At the upper soil levels, the light will be rich in red light and there will be no dormancy. At some depth in the soil, very little red light will penetrate, and dormancy may be induced; the same will happen at greater depth, where no light penetrates. (Redrawn from Mayer and Poljakoff-Mayber 1982)

Dormant seeds have a large quantity of P_r; in non-dormant seed the phytochrome is mainly in the P_{fr} form. Dormancy in a photodormant seed may be broken by exposure to light with a high red to far red ratio, e.g. white light. Conversely, non-dormant seed may turn dormant (induced or secondary dormancy) if exposed to illumination with light relatively rich in the farred wavelength. The latter occurs, for example, where light is filtered through a dense canopy (Mayer and Poljakoff-Mayber 1982; Richards and Beardsell 1987) or where seeds are enclosed in a chlorophyll-rich (green) fruit or seed coat (Cresswell and Grime 1981). Eventually, seeds exposed to dark conditions (e.g. buried or dark storage) gradually develop dormancy because P_{fr} is converted to P_r.

The phytochrome dormancy mechanism seems to some degree to be influenced by temperature. High or fluctuating temperatures appear to overcome photodormancy in some instances. In nature, light and temperature are obviously interrelated.

Although photodormancy has been most frequently documented from herbal species, it also occurs among some tree pioneers. The phytochrome dormancy system has been documented, for example, in *Cecropia obtusifolia*, *Piper auritum* and four Latin American *Ficus* species (Vasquez-Yanes 1982; Vasquez-Yanes et al. 1996). *Cecropia obtusifolia* showed the strongest dependence and had very low germination under dark or canopy-shaded (far red light) conditions. None of the *Ficus* species germinated in the dark, but there was a great difference with regard to far-red illumination (which simulates a forest canopy): two of the species had largely the same germination rate as in white or red light; only *Ficus insipida* showed significantly reduced germination under far-red conditions. Also, germination of many *Eucalyptus* spp. is believed to be determined by light (Boland et al. 1980).

The condition of light requirement in pioneers is the simplest sort of photodormancy. In some species, seeds require a specific duration of light–dark cycles for germination to proceed. Under tropical conditions, a cycle of 12 h of light and 12 h of darkness is prevalent. Under temperate conditions, longer exposure to light is sometimes required, and corresponds to the longer daylight hours during the temperate spring and summer.

Most photodormant seeds require only a brief illumination after imbibition to break dormancy – the requirement is, however, different in different species (Casal and Sanches 1998). The duration of illumination depends on the balance between P_r and P_{fr}, the rate of conversion and the rate of spontaneous dark reversion. In practice, photodormancy is usually not overcome by pretreatment, but by germinating seeds under appropriate light conditions that will break the dormancy. Light may thus be regarded more as a germination-stimulating factor than as a dormancy-breaking factor, the definition is a 'matter of semantics' (Baskin and Baskin 2004).

5.5.5 Thermodormancy

The term 'thermodormancy' is here used in its widest sense to cover all types of dormancy in which temperature plays a role in the development or release from dormancy. Seeds with thermodormancy require exposure to a temperature regime which is often different from that required for the actual germination process. In dormant seeds of eucalypts, pines, acacias and others, benefiting from fires for their germination and regeneration, the cause can be both mechanical and physiological. In terms of physical dormancy, it is just one of several methods to break dormancy – in other cases it plays a physiological role. The distinction is, however, not always clear.

Low-temperature thermodormancy is experienced in most temperate species, e.g. *Fagus*, *Quercus*, *Pinus*, *Abies* and some highland tropical species of pines and eucalypts. Seeds of such species need exposure to cold, moist pretreatment for a period to break dormancy. Any cold and moist condition is called chilling. Prechilling applies specifically to the conditions when applied to breaking dormancy. Prechilling was previously undertaken in practice by placing the seeds in alternate layers with a moist medium in a cold environment, e.g. an outdoor pit exposed to ambient low winter temperatures. The common term 'stratification' or 'cold stratification' originates from this practice of layering (Box 5.4). Warm stratification is analogously used for any type of warm, moist pretreatment (Bonner et al. 1994). Warm stratification is used in connection with after-ripening, for overcoming dormancy caused by an underdeveloped embryo and for softening hard pericarps or seed coats (mechanical dormancy).

High-altitude (alpine) eucalypts, e.g. *Eucalyptus delegatensis*, *Eucalyptus pauciflora* and *Eucalyptus glaucescens*, require cold moist pretreatment to overcome dormancy. Stratification at 3–5°C for 4–8 weeks is recommended pretreatment for these species (Boland et al. 1980; Turnbull and Doran 1987; Close and Wilson 2002). In other eucalypt species, e.g. *Eucalyptus camaldulensis*, *Eucalyptus tereticornis* and *Eucalyptus nitens*, it has been shown that stratification may substitute for light requirement, another example of linkage of photodormany and thermodormancy. For *Terminalia chebula*, a highland Indian species, Bhardwaj and Chakraborty (1994) found improved germination after cold moist stratification in cowdung. Stratification for 5 weeks was considered optimum and less than 3 weeks was insufficient to overcome dormancy.

In order to break thermodormancy by cold moist treatment, seeds must be imbibed; hence, general cold storage of dry seed does not substitute for stratification since the seeds only respond when moist. Since imbibed seeds respire (albeit at a low rate at low temperature), good aeration must be provided during pretreatment. The necessary period of pretreatment varies, but as long as

Box 5.4

Stratification pit

1. The stratification pit (Fig. 5.13) should be dug where the temperature is relatively low, e.g. a shaded site on a north-facing slope in the northern hemisphere, south-facing in the southern hemisphere, with good drainage.
2. The size of the pit is adapted to the volume of seed. A trench with a depth of 60–80 cm is convenient – the length is then adapted as needed. The sides of the pit may be supported with a frame to protect the sides from falling in. As protection against rodents, the sides and the bottom may be lined with wire mesh.
3. To ensure good drainage, the bottom of the trench is covered with a layer of sand or gravel.
4. The seeds to be stratified should be mixed with moist sand 4 times their weight, or filled into the pit in alternate layers of seeds and sand in the above proportion. The pit is filled to 15 cm from the top. The top 15 cm is filled with pure sand.
5. The pits are covered with a wire-mesh cover as protection against rodents.

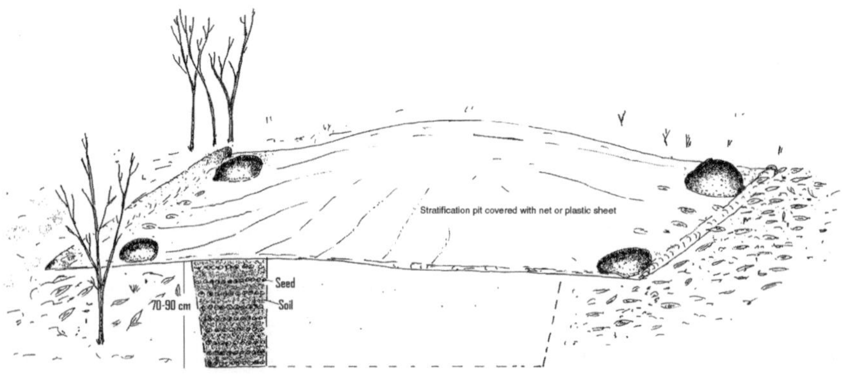

Fig. 5.13. Outdoor stratification pit as used in temperate regions but also applicable to some tropical highland species

the temperature is kept too low for germination to proceed, there is little risk of damage by overtreatment. However, most seeds requiring cold moist treatment also germinate at fairly low temperatures, so in practice it is difficult to avoid initial germination once dormancy has been overcome. In temperate regions, where thermodormancy is very common, stratification takes place during the late winter months up to the normal sowing time in early spring. Because the temperature increases during that period, seeds may germinate once dormancy has broken. The onset of splitting of the seed coat and radicle protrusion is an indication of terminated dormancy. Seeds may be transferred directly from stratification to the seedbed before the radicles have elongated. Immediate sowing is necessary to avoid mechanical damage to the radicle during handling (Aldhous 1972). In comparison with other dormancy types, thermodormancy requires a fairly long pretreatment period; therefore, appropriate scheduling according to time of sowing is important. A practical method of stratification used in the temperate region but also applicable to tropical highlands is described by Aldhous (1972), here slightly modified:

Where cool rooms are available, cold stratification is preferably carried out indoors at 1–5°C. The seeds are soaked in water for 24–48 h, then mixed with a moisture-retaining medium, e.g. moist sand, vermiculite, peat or a mixture. Occasionally seeds are prechilled 'naked', i.e. without mixing with a moisture-retaining medium, but that procedure makes control of moisture and temperature within the seed lot more difficult during treatment. Willan (1985) recommends use of a medium for long-term prechilling and any warm moist pretreatment, while 'naked' prechilling is suitable for species needing only a few weeks' cold pretreatment.

Indoor prechilling may take place in various types of containers. The main requirements are sufficient drainage and ventilation during the process. Boxes, cans, drums, trays or woven bags all make up suitable containers, although bags are obviously less applicable where seeds are mixed with sand. Polythene bags (100 μm) are suitable since they retain moisture, yet allow some ventilation. Where polythene bags are used, they should be only loosely closed, opened regularly and the seed should be stirred to avoid heating and ensure ventilation. Prechilling of naked seeds in trays in cool rooms is now the most common method for several temperate species (Fig. 5.14). Seeds are regularly moistened during the period. Moisture content during prechilling is crucial. Too low a moisture content slows down or stops the dormancy-breaking process; too high a moisture content may cause deterioration. During the latter part of the prechilling period, too high a moisture content may induce germination. Measuring moisture content (Chapter 7.7) of samples during the treatment period helps in adjusting the moisture content and hence in controlling germination during the pretreatment process.

Thermodormancy can in some instances be partly or fully overcome by chemical pretreatment (Sect. 5.5.3).

The benefit of high-temperature exposure is probably often a simple physical dormancy phenomenon, but may also be linked to the physiological initiation of the germination process. Burning a cover of grass over a seedbed is a widely used presowing treatment for eucalypts and legmes, for example. In the Philippines, seeds of *Aleurites moluccana* are pretreated by burning a 3 cm-thick layer of imperata grass (*Imperata cylindrica*) covering the seedbed. After burning, the seedbed is immediately sprinkled with water. Seeds are fully

Fig. 5.14. 'Naked prechilling' is a cold treatment where imbibed seeds are exposed to a period of low temperature in cold rooms without a moist medium

or partly covered by soil during the burning (Seeber and Agpaoa 1976). The same method is used in India for *Enterolobium cyclocarpum* and *Hymenaea courbaril*, for example (Brahmam 1996). However, although it promoted germination, it was noticed that many seeds were damaged by burning of too high intensity (thick layer of grass).

5.5.6 Underdeveloped Embryo

Seeds with underdeveloped embryos at the time of dispersal are unable to germinate under normal germination conditions and thus comply with the term 'dormancy'. The phenomenon is sometimes classified as morphological dormancy, referring to the immature morphological stage of the embryo. In principle, immature embryo development is the same as the condition of immature embryos in early collected seeds, and the pretreatment method is similar. But as a dormancy phenomenon it refers normally to species where immaturity of the embryo is prevailing at the time of seed dispersal. Species which normally disperse seeds with immature embryos are various palms (Arecaceae), *Gingko biloba* and several *Fraxinus* species. The stage of embryo development at the time of dispersal differs between different species with this type of dormancy. In *Gingko biloba*, even fertilisation may take place after dispersal; in *Ilex opaca* and some palms, the embryo consists of a core of undifferentiated cells, while in *Fraxinus* the embryo is fully differentiated but small. *Pinus* spp. from northern latitudes and high elevations are also reported to have morphological dormancy (Bonner et al. 1994).

For germination to proceed the embryo must grow to full size, which is promoted by a period of warm moist treatment; it is in practice an after-ripening/precuring similar to that used for early collected seeds (Sect. 3.4). Dormancy caused by immature embryos is often combined with other dormancy types, e.g. thermodormancy in *Fraxinus* spp.

5.5.7 Combined Dormancy

Where two or more types of dormancy are present in the same species, dormancy must be broken either by successive methods that work on different dormancy types or by methods with multiple effects. The latter is usually applied in the combination of mechanical and physical dormancy. Where the two types occur together, any method aimed at breaking physical dormancy will also work on mechanical dormancy. In practice it is often difficult to distinguish between the two. Since some pretreatment methods work on different types of dormancy, the nature of dual or combined dormancy is not always evident. For example, *Terminalia tomentosa* evidently has a pronounced physical dormancy with almost no germination unless mechanically scarified (Negi and Todaria 1995). However, since seeds extracted completely from the fruit showed greatly improved germination compared with those that were only scarified, it is likely that there is a second dormancy type, which could be mechanical or caused by inhibitors in the fruit.

Fraxinus spp. are commonly known to possess two types of dormancy, viz. underdeveloped embryo and thermodormancy, a combination also found in, for example, *Euscaphis japonica*. The former is broken by warm moist stratification, the latter by subsequent cold moist stratification (Piotto and Piccini 1998, Yang et al. 2005). Teak (*Tectona grandis*) is one of several species where physical dormancy is combined with chemical inhibitors in the fruit. In addition, the fruits often need a period of after-ripening which must be carried out before the seeds respond to other pretreatment procedures (Bedell 1989). A recommended pretreatment of teak fruits is alternate soaking and drying, plus sometimes sun baking. The duration of each treatment and the number of cycles vary; Keiding (1993) and Willan (1985) list variations of the procedure:

1. Soaking four times and drying three times, each of 30–35 min for scarified seed.
2. Five to ten cycles of soaking for 1 day and 3–5 days' drying and sun baking.
3. Alternate 24-h soaking and 24-h drying for 2 weeks.

Prolonged soaking in running water for one to several days also serves both to leach inhibitors and to soften fruit or the seed coat. This method is also applicable to teak (Keiding 1993). In India, Yadav (1992) found prolonged soaking a suitable alternative to the alternate treatment.

Chemical inhibitors in combination with physical dormancy have also been suggested for two *Albizia* species (Sect. 5.5.3).

5.6 Accelerating Germination

Fast-germinating species with no or successfully broken dormancy germinating under favourable conditions may complete germination to a stage of two unfolded leaves in a matter of 4–6 days. In most species, the rate is significantly longer. Whether seeds are sown in the nursery or directly in the field, an accelerated germination process is desirable since its helps to shorten the total establishment period and to overcome the most vulnerable period of seedling establishment. Under direct sowing it is desirable for germination to progress as fast as possible because germinating seeds are exposed to field stress and they do not have the competitive advantage which planted seedlings have. The germination process can be accelerated by different means during the three main germination phases:

1. *Stage 1: imbibition rate.* The higher the water pressure, the faster the imbibition. And the faster the imbibition, the faster the seed enters into the next phase.
2. *Stage 2: lag phase.* The second phase of germination is mobilisation of the metabolic system. The seeds use primarily their internal resources, but experience shows that various hormones can help shorten the germination process.
3. *Stage 3: growth phase.* The third stage of germination is the growth expansion of the embryo, which continues into the seedling growth phase, where seeds become increasingly reliant on their own absorption.

5.6.1 Soaking in Water

Water plays a role in most dormancy types. It helps in breaking mechanical dormancy by gradually reducing mechanical resistance to embryo expansion, softening seed coats for overcoming physical dormancy, leaching out or diluting chemical inhibitors in fruits and seeds and interfering with all types of physiological dormancy since overcoming these types of dormancies requires that the seeds are imbibed.

The direct effect of soaking on breaking dormancy is, however, weak and prolonged soaking implies a risk of anoxia, i.e. seeds die because of lack of oxygen. Soaking for more than 12 h normally requires aeration and if seeds are soaked for more than 1–2 days, water should be changed. Softening of hard structures often requires several days' soaking, e.g. 6 days was recommended for overcoming physical dormancy in teak (*Tectona grandis*) in India (Yadav 1992).

For non-dormant seed, soaking serves to ensure fast imbibition and thus entrance into the second phase of germination. Species with thin and fragile seed coats become more sensitive to mechanical damage after imbibition.

5.6.2 Growth Regulators

Chemical compounds that interact with physiological mechanisms, e.g. as hormones, have an effect on germination, including physiological dormancy. Application of various compounds can in some cases partly or fully substitute for temperature or light pretreatment, or for leaching of germination inhibitors. Some compounds stimulate individual metabolic processes during germination without being directly linked to dormancy. The same type of hormone is, for example, often involved in both dormancy release and germination processes. The effect of germination stimulants is often most evident under suboptimal germination temperatures. Total germination percentage, germination speed and seedling vigour may be promoted by application of germination stimulants. The main hormone group responsible for suppression of germination (dormancy hormone) is ABA; the main hormone group that stimulates germination and growth is gibberellic acid (GA) (Thomas 1992). The strength of dormancy is often determined by a balance between the two groups of growth regulators. GA play a central role in the early germination processes by activating enzyme production and mobilising storage reserves. GAs (usually GA_3) has been shown to help overcome thermodormancy (e.g. induced dormancy caused by high temperatures), photodormancy (e.g. inducing dark germination in light-sensitive seeds) and chemical dormancy (overcoming the effect of inhibitors) (Bewley and Black 1982; Villiers 1972). Murthy and Reddy (1989) used a concentration of 200 ppm for stimulating germination in *Ziziphus mauritiana*. These seeds were apparently not dormant, but GA_3 had a particularly positive effect on shoot development and vigour.

Cytokinins, a group of common natural plant hormones, are essential for cell division. The interaction between cytokinin and another plant hormone, auxin, is well established in plant propagation: a high auxin to cytokinin ratio favours root development; a high cytokinin to auxin ratio favours shoot development. Application of cytokinins or their synthetic equivalent, benzyl adenine (BA), can promote germination but because of their specific effect on shoot development, both germination and seedling development may be abnormal. Seeds treated with cytokinins sometimes germinate with the shoot before the radicle (Bewley and Black 1994). Applied cytokinins (kinetin) have been reported to overcome high-temperature dormancy in lettuce seed (Smith et al. 1968, quoted in Hartmann et al. 1997). In *Ziziphus mauritiana*, BA (100 ppm) stimulated both total germination and vigour of seeds, but the effect was generally lower than for the other compounds tested (Murthy and Reddy 1989).

Various nitrogenous compounds such as potassium nitrate (KNO_3) and thiourea promote the germination process (Vleeshouwers et al. 1995). KNO_3 is frequently used as germination stimulant both in connection with testing (ISTA 2001) and in operational plant propagation. KNO_3 had a strong effect on both germination percentage and vigour on *Acacia nilotica* seeds pretreated with acid (Palani et al. 1995). At 1% concentration, germination increased from 37 (control) to 79%, and at 2% concentration it increased to 85%. In *Casuarina equisetifolia*, germination increased from 46% in the control to 65% after soaking in 1.5% KNO_3 for 36 h. Both higher and lower concentration, and shorter duration of soaking showed a lower germination in that experiment (Maideen et al. 1990). In seed testing, 0.2% is the recommended concentration (ISTA 1996). Thiourea has a stimulating effect on breaking dormancy, possibly by deactivating the effect of inhibitors, e.g. ABA (Hartmann and Kester 1983). It has proved effective in overcoming photodormancy in a number of light-sensitive seeds (Mayer and Poljakoff-Mayber 1982; Sasaki and Asakawa 1974). In temperate *Quercus*, *Larix* and *Picea* species it has been used instead of stratification (Deubner 1932; Johnson 1946, quoted in Mayer and Poljakoff-Mayber 1982). The beneficial effect of smoke on germination of some fire-prone species may at least to some extent be ascribed to the nitrogenous compounds in the smoke (Razanamandranto et al. 2005).

Comparative studies on the effect of different germination stimulants have been carried out on seeds of *Ziziphus mauritiana* (Murthy and Reddy 1989), *Casuarina equisetifolia* (Maideen et al. 1990) and *Acacia nilotica* (Palani et al. 1995). In *Ziziphus mauritiana*, thiourea proved the most effective germination stimulant; 24-h soaking in a 1% solution increased the total germination percentage from 41 (control) to 78% at 30°C (optimal germination temperature). In seeds germinated at suboptimal temperatures, thiourea alleviated the detrimental effects in terms of both total germination and vigour. KNO_3 was in this study less effective than GA_3, thiourea and BA for all germination parameters except root length (Murthy and Reddy 1989).

Neither hormones nor other germination compounds are much used in practical seed propagation, but as more knowledge is generated, they may become suitable alternatives to overcome complicated physiological dormancy constraints.

5.6.3 Priming and Fluid Drilling

Priming is a method to promote rapid and uniform germination of seeds, by controlling imbibition to an extent where germination is initiated, but is insufficient to cause radicle emergence. The priming process carries germination further than pure imbibition, viz. as close as possible to phase three, the radicle expansion phase, in the germination process (Fig. 5.15). During

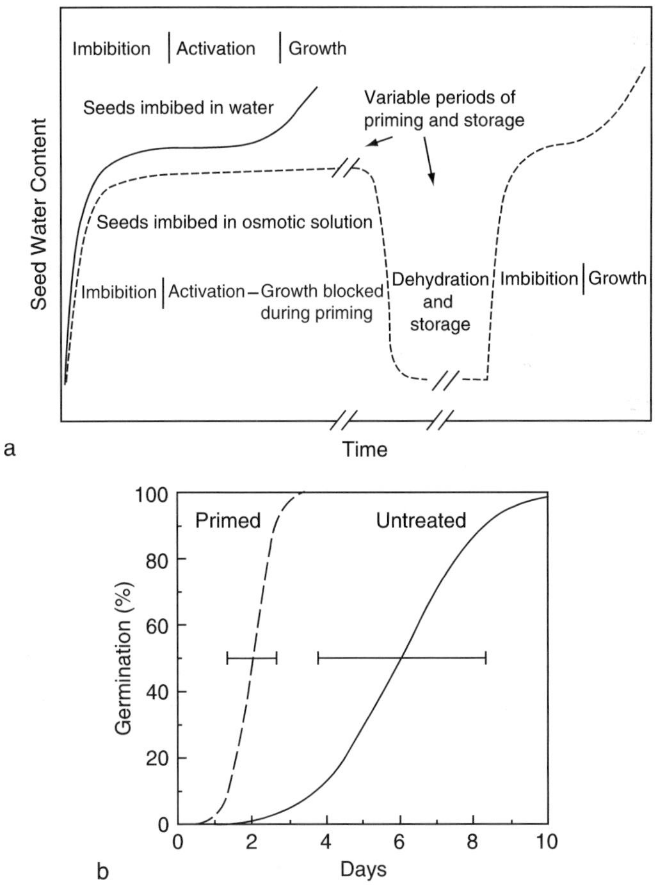

Fig. 5.15. Priming enhances germination. **a** Seeds imbibe and carry out initial germination processes, but germination is blocked before the phase of radicle protrusion and growth. **b** Primed seeds germinate fast and uniformly compared with untreated seeds. (From Bradford and Bewley 2002)

priming the variation in the initial imbibition rate is overcome: all seeds tend to reach a stage where they are ready to germinate once they are provided with optimum germination conditions (Maude 1996; Bradford 2004). Because the germination process is in the second 'lag' phase without protrusion of the sensitive radicle, primed seeds can be dehydrated, stored and handled at least for some time without damage to the seeds. Priming is not a dormancy-breaking treatment; possible dormancy must be broken by an appropriate pretreatment prior to priming. Under normal nursery practice for forest trees, priming is not much used, but the method is becoming increasingly important in connection with direct sowing in dry areas. Because the germination process has already started before sowing, germination and seedling establishment is fast and primed seeds thus have a competitive advantage under field conditions.

The simplest form of priming is a moist warm stratification for a duration that will carry germination up to the stage of radicle protrusion. This is used, for example, for various pines after cold moist stratification (prechilling) (Doody and O'Reilly 2005). In osmopriming, seeds are soaked in a priming fluid with high osmotic pressure to control the absorption rate. Usually poly ethylene glycol (PEG) is used as the priming fluid. The conditions and the duration of priming vary with species. Hartmann et al. (1997) indicated a range of conditions of osmotic potential (i.e. PEG concentration) from –5 to –15 bar (i.e. from –0.5 to –1.5 MPa), temperature from 10 to 25°C,and a duration of treatment from 1 to 15 days. A common priming condition is 15°C for 5–10 days. Stirring or bubbling is essential during priming of large quantities in containers, both to ensure uniform treatment and to ensure proper aeration while the seeds are being soaked (Fig. 5.16). Small quantities may be primed on filter paper irrigated with PEG (Maude 1996). Once priming has been completed, the seed lot is washed, dried superficially and coated with a film, e.g. sodium alginate. The priming fluid may be reused. The drying rate and the coating depend mainly on the time of priming in relation to the sowing date. Seeds to be sown immediately are only slightly dried, seeds to be sown later may need slightly more drying, e.g. by warm air, and protection against fungi. Fertiliser, pesticide or inoculant may be added as an integrated part of the coating process (Sect. 5.7). Fungicides are also occasionally added to the priming fluid (Maude 1996).

In fluid drilling, the germination process is allowed to proceed until radicle emergence. Germination takes place in aerated water, and once the radicle has emerged, the seed is mixed with a viscous gel to protect the radicle from mechanical injuries and desiccation. Sodium alginate, hydrolysed starch–polyacrylonitrile, gour gum or synthethic clay may be used as the gel (Hartman et al. 1997).

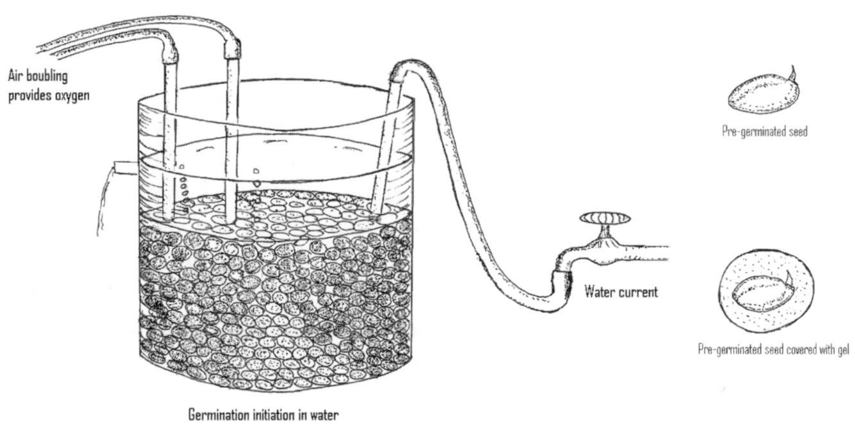

Fig. 5.16. Principle of fluid drilling. Seeds are germinated in flowing and aerated water to prevent anoxia. The germination process is stopped once the radicles have protruded. The germinated seeds are then covered with a viscous gel to protect them from mechanical damage and desiccation until sowing

5.7 Seed Coating and Pelleting

Coating and pelleting denote the practices of covering seeds with an inert substance during processing or as a presowing treatment. In coating, seeds are covered with the substance with or without an adhesive applied to the seed coat. The coating does not significantly increase seed size or weight. In pelleting, the functional substrate is mixed with an adhesive before application. Pelleted seeds thus achieve a larger, heavier and more uniform size which facilitates some types of handling, e.g. machine sowing. The coating material in both types of treatment yields in itself some protection to the seed. Special coating material may add particular protection, e.g. alginate as an antidesiccant and lime at low pH. In addition, various substances which promote germination and early seedling development may be added to the coating or pelleting materiel. Functional substances include fertilisers, growth regulators, fungicides or insecticides, rodent and bird repellents, and microsymbionts (mycorrhiza, rhizobia, frankiae) (Brockwell 1962). It is usually not possible to apply all these types to the seeds at the same time. For example, fertilisers are antagonistic to *Rhizobium* and fungicides cannot be applied together with mychorrhiza inoculants. In pelleting, the major purpose is to increase the seed size, so the major component of pelleting material is a filler, e.g. kaolin clay, vermiculite, gypsum or peat (Brockwell 1962).

Coating and pelleting are rarely economically feasible when seeds are raised in nurseries where moisture and other planting conditions are easily controlled, and fertiliser, microsymbionts, etc. can be applied directly to the seedbed or nursery soil. The methods are thus mostly applicable to direct seeding. Application of protection or essential substrates may be necessary or highly advantageous in direct sowing. Where seeds are small and machine sowing is applicable, an increased and more uniform size created by pelleting eases uniform sowing. For very small seeds, direct sowing without pelleting is difficult.

Technically, coating and pelleting are carried out by rolling or tumbling seeds with the covering substrate until sufficient substrate adheres to the seed surface. Coating and pelleting differ basically in the thickness of the covering layer:

- If seed-coats are relatively rough, and only a small amount of coating material is necessary, the coating can be applied by wetting the seed surface before tumbling in the dry substrate. Seeds pretreated with acid tend to get a rough surface which promotes adhesion. Inoculants and pesticides may be applied this way.
- Where seed coats are smooth and larger quantities of coating material are necessary, seeds can be rolled with an adhesive/binder prior to tumbling with dry coating powder. As adhesive 40% (w/v) gum arabic, 1.5% (w/v) methyl cellulose or vegetable or paraffin oils are usually used. The seeds are rolled in the sticker until evenly coated, then transferred to the dusty substrate.
- Larger amounts of coating material can adhere to the seed if seeds are tumbled or rolled in a wet slurry and then dried. However, there is a risk of losing the material during mechanical handling.
- Film-coating is an advanced method where seeds are covered with a polymer covering, which is applied by spraying as seeds fall through a specialised machine. This method is especially developed for application of fungicides and pesticides by a method that will reduce exposure to waste or material rubbed off by workers (Bradford 2004).
- A large and thick layer of substrate can be applied by mixing the dry substrate with a binder material. The binder is, for example, gum arabic or methyl cellulose. Seeds are rolled until evenly covered. As much substrate as required can be applied by extending the time of rolling. When sufficient substrate has been applied and the seeds have reached a reasonable size, the seeds are rolled in powdered rock phosphate, calcium carbonate or the like to avoid them agglutinating. The pelleting protects possible inoculants applied with the substrate and the seeds are easy to handle.

Coating and pelleting can be carried out before storage, but as pretreatment, for example for dormancy release, cannot be performed after the application of surface material, it is usually carried out just before sowing. Pelleting may in some instances delay germination since the pelleting material needs to dissolve before imbibition, and the pellet may act as a physical barrier to water and oxygen absorption.

Sowing, Germination and Seedling Establishment 6

6.1 Introduction

All seed handling activities up to the moment of germination basically concentrate on avoiding germination. The seed, from the moment it left the mother tree, has been a passive object to what was done to it. It left the mother tree under the impact of gravity, wind or a frugivore. It was extracted and transported by external forces and has been kept inactive until it eventually lands at a place suitable for germination. During all these processes the seed has switched off or turned down its metabolism to a minimum for survival or to what desiccation would allow; possible metabolism has only served to keep the seed alive.

Germination demarcates a drastic transition. From being dependent on food sources from the mother plant, it will now establish an independent plant capable of absorbing nutrients and growing on its own. From being a more or less quiescent or dormant object, it will now switch on the whole metabolic apparatus with its multitude of interlinked physiological and biochemical mechanisms. Germination is a growth process and as such it is irreversible – once started it must go on.

As any other transition, germination is a sensitive phase. All newly started processes are imperfect. The seed or fruit coat, which has been the protective covering, is ruptured. The embryo faces an environment where water, light and temperature stress is common and changing, and where it is exposed to new types of pathogens. New defence mechanisms must be established and until that happens, seedlings are vulnerable. Germination conditions have two integrated purposes in this connection:

1. To provide an environment with as little stress as possible
2. To speed up the germination process so that the seedlings pass through the most vulnerable stage as fast as possible

Stress reduction means to adapt germination conditions as close as possible to the physiological optimal for the species. Although this cannot always be accomplished under nursery or field propagation condition, it should always be the target for the tree planter.

Three key factors determine germination success:

1. The seed's physiological quality (germination capacity and vigour). The physiological quality is to a large extent determined by development aspects, e.g. maturity and ageing.
2. The stage of dormancy (release from dormancy is overcome by an appropriate pretreatment). Dormancy is a physiological 'stage' or 'condition' which is independent of physiological quality.
3. The germination environment, such as water, temperature, substrate, light and freedom from pathogens.

The physiological process of germination is quite similar for all plant species, but different species exhibit a wide variance with regards to specific germination conditions and tolerance ranges. Temperate and high-altitude species often germinate at a few degrees centigrade, while most tropical lowland species require temperatures of between 20 and 28°C for germination to proceed. Pioneer species typically have a much wider tolerance level than climax forest species. Water absorption capacity is typically also related to natural growth habit – dry-zone species being able to imbibe water at lower water pressure than moist-zone species. The third key germination regulator, light, is rarely a critical factor during the actual germination process but quickly becomes a key regulator for plant growth. Light in connection with germination is a dormancy factor, viz. phytochrome-determined photodormancy (Chapter 5.5.4), but it is conveniently overcome by providing seeds with appropriate conditions that will break dormancy in connection with germination rather than giving the seeds special pretreatment.

The transition from the seed using its own storage reserves to assimilation is not sharp, and in practice root absorption starts quickly after radicle protrusion. The substrate and conditions related to the substrate, e.g. pH, salinity, nutrients and drainage, quickly become important. Although some resistance against pest and diseases builds up in seedlings after germination, pest and disease management continues to be crucial for plant propagation. Seedlings growing at nursery densities also make up a potential contamination and pathogen reservoir which, under appropriate conditions, can affect a large number of the nursery plants. Pest management includes, for example, nursery soil sterilisation and inoculation with soil microsymbionts such as mychorrhiza, rhizobia and/or frankia. Optimal conditions should normally be maintained until the seedlings are well established. After that, stress is gradually applied to harden the plants in preparation for the field environment.

Because of the sensitivity during germination, as trees are often poor competitors against weeds during establishment, and because of the better land-use efficiency in keeping plants at close nursery spacing when they are small, it is customary to sow seeds and raise plants under nursery conditions. Under such protected conditions plants can be given optimal conditions. Before they are planted out they are gradually adapted to the harsher field conditions by exposing them to some stress. The nursery phase also allows a good timing so that seedlings of plantable size are ready at the time of the year when seedlings have the best chances of surviving, i.e. under seasonal tropical conditions at the beginning of the rainy season. Despite the obvious advantages of nursery-raised plants, the cost involved is significant. The attempt to avoid the nursery phase by direct sowing is becoming increasingly attractive especially in countries where labour is expensive.

6.2 The Physiological Events of Germination

Seed germination starts with imbibition, and germination is considered concluded by radicle protrusion and embryo enlargement (ISTA 1996, 1999, 2006). The intermediate phase consists of a number of internal physiological events, which include mobilisation of storage resources, repair and turnover of cellular components and start up of embryo growth processes. This 'normal' sequence of germination is most pronounced in orthodox seed where there is usually a distinct period of inactive metabolism during dispersal (and storage), and a distinct imbibition phase where seed weight increases drastically.

The transition between maturation and germination is often more or less continuous in recalcitrant seed. Even if most recalcitrant seeds do undergo some kind of maturation drying, their moisture content is always so high that they maintain, at any time, a certain level of metabolic activity. Since desiccation itself leads to irreversible damage, there is no drought-imposed quiescence. Some recalcitrant species maintain a certain period of seed integrity during dispersal and during relatively dry or cold conditions, but physiologically, as the germination process is defined as beginning with imbibition and metabolism, germination is always more or less a continuation of maturation (Berjak and Pammenter 1996).

Continuity is even more pronounced in viviparous seed. These seeds germinate while still attached to the mother plant. Vivipary is common in some mangrove genera of the Rhizophoraceae family and is a common feature in high-arctic grasses. True vivipary is not common in forest trees, but precocious germination, i.e. seed germination while the seeds are still attached to the mother trees, is sometimes observed in dipterocarps and other humid-forest species under very moist conditions.

Orthodox seeds are physiologically adapted to maturation drying and such drying usually happens during maturation on the tree. However, under very moist conditions there is very low maturation drying and even orthodox seed can here experience a more or less continuous process from maturation to germination (Fig. 6.1).

6.2.1 Imbibition

Imbibition is absorption of moisture by the seed in connection with germination. For orthodox seed, imbibition is the distinct first event of germination, which is the precondition for the onset of the subsequent metabolic processes. Metabolism is the precondition for life processes, of which radicle protrusion and embryo enlargement are the observable evidences. Because of the above

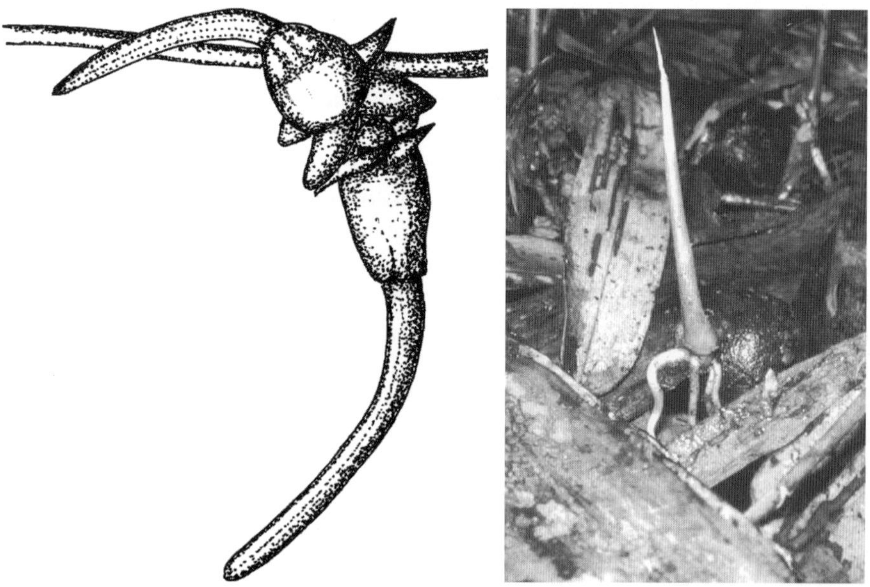

Fig. 6.1. Vivipary and precocious germination. Vivipary is where there is no true seed stage, and the dispersal units are seedlings or propagules. Vivipary is common in, for example, mangrove species. Precocious germination occurs under high-humidity conditions where species do not dry out but germinate immediately after maturation, here a bamboo in Vietnam

mentioned absence of maturation drying and the constant high moisture content in recalcitrant seeds, the term imbibition is more difficult to use in the context of these seeds.

Although imbibition in orthodox seed is a precondition for starting up the metabolic processes, which ultimately leads to completion of the germination process, imbibition is a purely physical process, which occurs whether the seed is able to complete germination or not. Imbibition occurs whether seeds are dormant or non-dormant (except physical dormancy), viable or non-viable (Bewley and Black 1994; Mayer and Poljakoff-Mayber 1982). Imbibition is thus not necessarily linked to and does not necessarily lead to germination. Orthodox seeds may thus tolerate repeated events of drying and wetting as long as the physiological events of germination are not started, e.g. because of dormancy or inappropriate temperature.

Seeds in soil seed banks are often fully imbibed but germination can be hindered by, for example, photodormancy or thermodormancy or germination may not take place because the temperature is not suitable. In physically dormant seed (hard seed coat), there is a close link between imbibition and germination in the sense that impermeability of the seed coat prevents germination. However, hard-coated seeds are not necessarily alive, and impermeability of seed coats can be one of several reasons for failed germination.

Imbibition can start if there is sufficient moisture. The absorption mechanism is basically the inverse of the principle of seed drying and moisture principles (Chap. 3). Whether imbibition will take place and the rate of imbibition depend on the water potential of the seed and the soil. Water potential (in physiological literature designated by the Greek letter ψ) is an expression of the energy status of water. Water will tend to flow from a place of high water potential to a place of low potential, and the larger the difference, the higher the flow rate. In common terms it implies that water will flow from a moist medium to a dry one, thus from moist soil to a dry seed. The higher the water potential of the soil, i.e. the damper the soil, the faster the seed will imbibe. And the drier the seed, the faster it will imbibe. Measuring imbibition as a function of time typically shows a pattern as shown in Fig. 6.2.

Dry tissue tends to form some physical barrier against water; but once the tissue is slightly moist, water movement increases. Legume seed coats are normally impermeable to water (physical dormancy). When the seed coat is scarified, imbibition starts from the small localised site of scarification but gradually spreads to the rest of the seed coat.

After some time of imbibition, the water pressure difference between the inside and the outside of the seed gets smaller and consequently the imbibition rate declines (last part of the imbibition curve). In the soil the imbibition rate is normally lower than the rates depicted in Fig. 6.2 because the water potential is lower in soil than in pure water, and as water moves into the seed, the

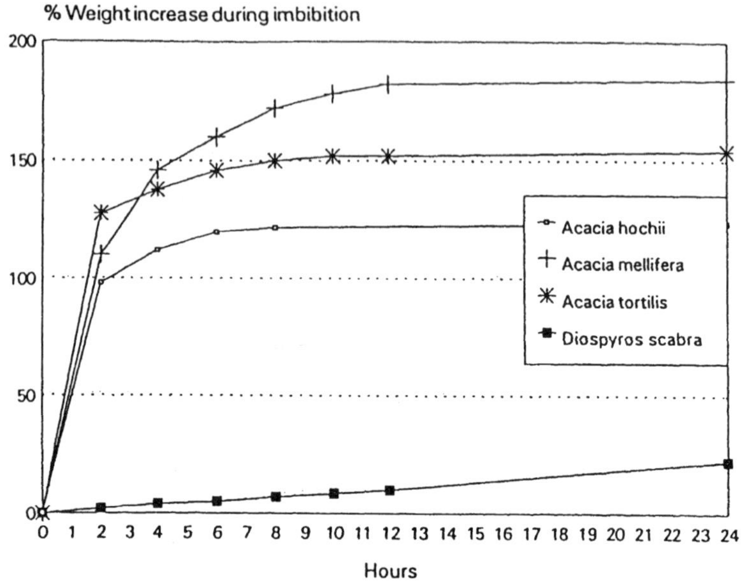

Fig. 6.2. Imbibition rate of *Acacia tortilis, Acacia mellifera, Acacia hockii* and *Diospyros scabra*, measured as the average weight increase during the imbibition process as a percentage of the initial weight. Seeds of the three acacia species were scarified prior to imbibition

water potential declines. Water from the soil in the vicinity will move and replace that taken up by the seed. The rate of water movement in the soil depends on soil structure and moisture content.

The rate of imbibition also depends on the size, morphology and internal structure of the seed as well as on temperature. Many dry-zone species show a very fast imbibition rate if adequate moisture is available. In dry-zone legumes, for example, seeds are fully imbibed within a few hours once physical dormancy has been broken (Fig. 6.2). Small seeds, seeds that produce mucilage, and seeds with relatively smooth coats tend to be the most efficient in absorbing water owing to their greater contact with soil and their larger surface-area to volume ratio (Bewley and Black 1994). The imbibition rate also tends to increase with temperature (Bewley and Black 1994).

6.2.2
Start of Metabolism – 'Lag Phase'

Water absorption with concurrent increase in weight proceeds until the seeds have imbibed as much water as is possible. Then follows a period of no or very little increase until the seeds again start to gain weight as a result of embryo

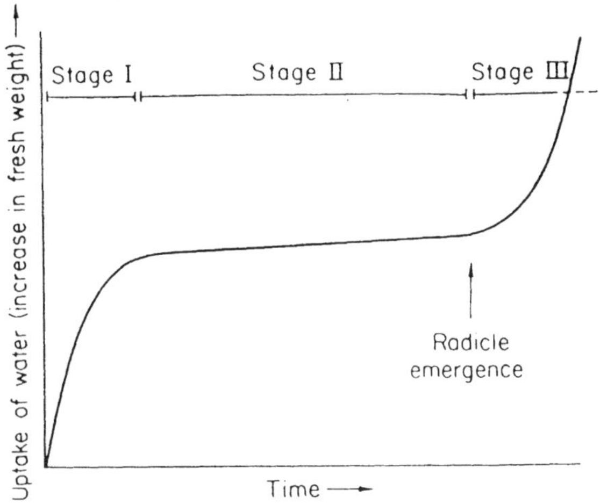

Fig. 6.3. Model of water uptake during the three phases of germination. *Stage I*: imbibition, the seed tissue is rehydrated. *Stage II*: lag phase, cell repair and start up of metabolic system. *Stage III*: cell elongation and mitosis, the growth phase, which continues as the plant grows. (From Bewley and Black 1994)

enlargement and growth (Fig. 6.3). This period is called the 'lag phase' because of absence of visible changes, but is in fact a phase of crucial activity, where the whole metabolic mechanism is switched on. Both dormant and non-dormant seeds become metabolically active as can be verified by, for example, dehydrogenase activity, the enzyme forming the basis of the tetrazolium viability test (Chap. 7). However, in dormant seeds there are some types of blocking in the biochemical system so that metabolism does not lead to germination (Chap. 5).

Onset of the germination mechanism involves a range of physiological events, e.g. mobilisation of stored food reserves (protein, starch and fat), activation of metabolic enzymes, and repair and turnover of cell components. As metabolic processes require oxygen, excess moisture with concurrent low oxygen around the seed may easily inhibit processes necessary for germination and the seed may experience delayed germination or in extreme situations it may rot owing to anoxia.

6.2.3 Embryo Differentiation and Growth

Embryo differentiation varies between seeds of different species. Some conifers have a very small, rudimentary and little-differentiated embryo. Germination in these species is often delayed for a long time while the embryo expands. It is a matter of definition whether it is called immaturity, dormancy or delayed germination, but the fact is that the seed requires a long period of moist (imbibed) and physiological appropriate conditions until radicle protrusion occurs. In many other species, e.g. legumes, the embryo is fully differentiated and basically just needs to be unfolded to make a plant (Fig. 6.4).

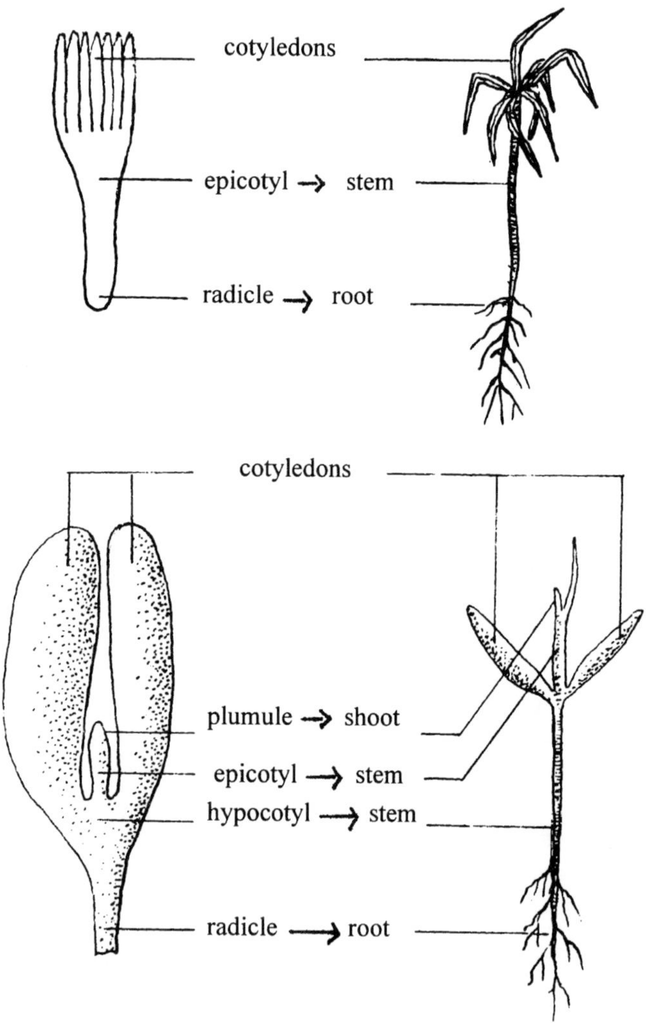

Fig. 6.4. Embryo differentiation and development during germination. The seed embryo contains all essential structures recognised in the seedling, i.e. root, stem, shoot and leaf primordia. Gymnosperms usually have a whorl of cotyledons, angiosperms have one (monocotyledons, e.g. palms, bamboo and rattan) or two (dicotyledons – most forest trees)

Seeds initially live on their stored reserves but as soon as they are exposed to light they will start photoassimilation. Water absorption happens, after the initial rehydration of seed tissue, almost exclusively through the root. Once absorption starts, the embryo goes into the third phase of germination with rapid mitosis and cell elongation which continues into the active growth phase of the seedlings.

The first visible event of germination is usually elongation and protrusion of the radicle with later appearance of epicotyl, hypocotyl and cotyledons. Radicle emergence is essentially a growth manifestation, and physiologically seed germination is considered completed on emergence of the radicle. Germination is usually considered completed when the seed is fully developed with all essential parts (Sect. 7.8).

6.2.4
Germination Types

The aforementioned sequence of embryo development starting with radicle protrusion and followed by elongation of the part of the embryonic axis that develops into the main stem is the 'normal' type. In a few species elongation takes place in the reverse order (Ng 1991).

The seedling stem is divided into the hypocotyl, which is the section below the cotyledons, and the epicotyl, which is the section above the cotyledons (Fig. 6.4).

If the hypocotyl does not expand or expands only slightly, the cotyledons (and hence the seed) remain below the ground during germination and do not become photosynthetic (cryptocotylar). This type is called hypogeal germination (Fig. 6.5).

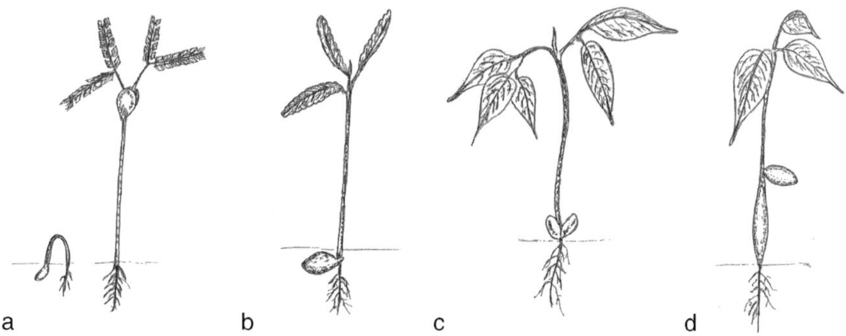

Fig. 6.5. Germination types. **a** Epigeal germination (*Albizia gummifera*). **b** Hypogeal germination (*Antiaris toxicaria*). **c** Semihypogeal (*Pithecellobium* spp.) **d** Durian type (*Durian* spp.)

If the hypocotyl expands, the cotyledons are pushed above the ground, often together with the seed coat and possible remaining endosperm. This is called epigeal germination (Fig. 6.5). The first appearance of an epigeal germination is often the 'loop' of the elongated hypocotyl above the ground. As the hypocotyl straightens, the seed is lifted up. The cotyledons then normally separate from each other and become the first photosynthetic leaves (phanerocotylar). Seedling cotyledons in angiosperms are sometimes named paracotyledons to distinguish them from the embryonic cotyledons (Vogel 1980). Paracotyledons of angiosperms are normally morphologically different from subsequent leaves as they do not expand, have no veins and retain a fleshy structure. As the epicotyl expands and normal leaves appear, the paracotyledons often quickly lose their importance and wither. In gymnosperms, which also have epigeal germination, cotyledons resemble subsequent leaves and are normally retained for a longer time after germination.

Intermediate types occur. Semihypogeal is a type where the hypocotyl remains small but cotyledons emerge, sometimes because of elongation of the cotyledonary stalks. In the durian type, the hypocotyl elongates but the cotyledons are non-emergent and hence do not become photosynthetic (epigeal, cryptocotylar *sensu* Vogel 1980). The latter occurs, apart from in durian types, also in viviparous mangrove seedlings and some dipterocarps. Detailed descriptions of the germination system, classification and morphology of seedlings have been made by Burger (1972), Vogel (1980) and Ng (1991) among others. Masanga (1998) included germination type in a description of a number of East African tree seedlings.

Epigeal germination is by far the most common in woody plants. All gymnosperms, and the major families of angiosperms have epigeal germination. Germination types do not, however, truly reflect the taxonomic system. Few families have exclusively hypogeal germination, but the germination type occurs in many families with prevailing or partly epigeal germination. Germination types for some forest species according to the traditional classification are shown in Table 6.1.

6.2.5 Seedling Establishment

Seedlings are adapted to juvenile life, which is very different from that of adults. The environment and the regeneration strategy have a strong influence on seedling ecology and tolerance. Variation includes, for example:

1. *Light adaptation.* Sensitivity to bright light prevails in some close forest species, which usually germinate and grow during the first years in deep shade. Light stress causes withering in these species even where

Table 6.1. Germination types according to families or genera of forest tree species

Epigeal germination	Hypogeal germination
All gymnosperms	Lauraceae
Myrtaceae	Most Moraceae (*Antiaris, Artocarpus*)
Apocynaceae (*Alstonia, Dyera*)	Anacardiaceae (*Mangifera, Swintonia*)
Bignoniaceae	Most Fagaceae (e.g. *Quercus*)
Casuarinaceae	Some Leguminosae (e.g. *Milletia, Erythrina, Pithecellobium* (semi-hypogeal)
Dipterocarpaceae (*Dipterocarpus* spp. are durian type)	Some Meliaceae (*Swietenia, Aglaia, Trichilia, Xylocarpus*)
Euphorbiaceae	*Prunus*
Boraginaceae (e.g. *Cordia*)	*Gonostylus*
Most Leguminosae	
Terminalia	
Some Meliaceae (*Azadirachta, Chukrasia, Toona*)	
Rhamnaceae (*Ziziphus, Maesopsis*)	
Sterculia	
Some Moraceae (e.g. *Ficus* spp.)	
Gmelina (except semihypogeal *G. elliptica*)	
Eleocarpus	
Durio (durian type)	

they have adequate water supply. Light requirement changes with age and most seedlings tolerate full sunlight after a weaning period.

2. *Shoot–root balance.* Plants establish a certain balance between shoot and root. In nurseries, roots are usually pruned to avoid them anchoring themselves to the ground. Root development is thus restricted but will continue after planting out. In dry environments, water supply is critical and dry-zone species often start development with a very deep growing taproot, while there is little height growth. The phenomenon is genetic but is also strongly influenced by the environment. If water is scarce, roots can continue down several metres before any significant height growth appears (Fig. 6.6). Species with strong taproot development are often sensitive to root pruning and must be planted out at small height with a large root volume. Pioneer species in a humid environment, in contrast, often develop large shoots ('top heavy'), which is presumably an adaptation to give them a competitive advantage over weeds.
3. *Stress tolerance.* Most seedlings are adapted to withstand some stress, which frequently occurs in their environment. Many seedlings will tolerate some water stress and just cease growing. Humid-forest species are usually sensitive to water stress but will survive very long periods of

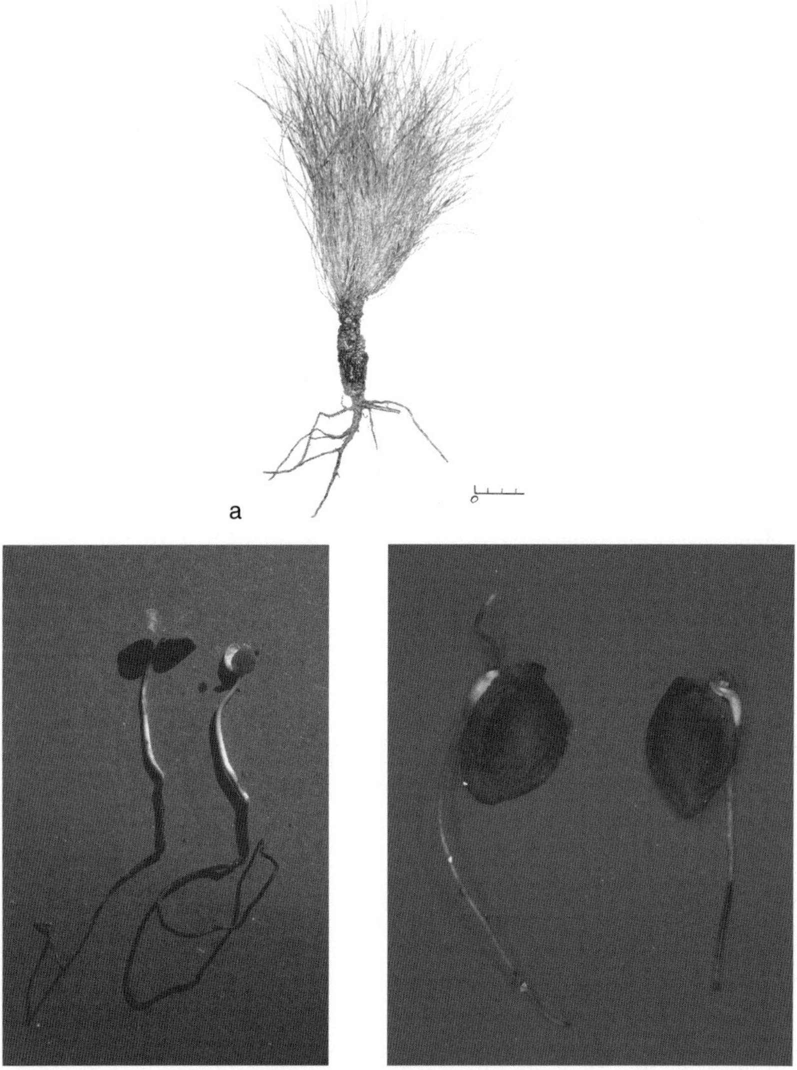

a

b c

Fig. 6.6. Early seedling development. In most germinating seeds **a** (*Khaya senegalensis*) the radicle will grow to 2–4 times the seed length before shoot elongation; seedlings often maintain a shoot–root balance of 50-50%. Dry-zone species **b** *Diospyros* spp.) develop a very deep root, sometimes more than 1 m, before height growth. Seedlings regenerating in fire-prone areas often have a strong shoot protection. *Pinus merkusii* **c** develops a so-called grass stage in which the plant has a short stem and a very dense needle coverage

time in deep shade (Whitmore 1984). Species growing in dry and/or cold areas often have special morphological protection of the shoot. In fire-prone areas, some species have developed advanced shoot protection. In some pine species, shoot elongation is suppressed for one to several years while the seedling develops a thick carrotlike root and a dense cover of needles. This so-called grass stage (Fig. 6.6) is common in, for example, *Pinus merkusii* and *Pinus roxburghii* (Sirikul 1990; Turakka et al. 1982; Koskela et al. 1995). Stress tolerance is usually much improved by mychorrhiza symbiosis. Selection of particular mychorrhiza species can sometimes enhance stress tolerance.

6.3 Raising Plants from Seed

Most trees are raised in nurseries and planted out when they have grown to a certain appropriate size. A number of crucial factors can be manipulated in the nursery, but these are difficult to control in the field. A main rationale of nursery establishment is that plants are established under close-to-ideal conditions and thus given a good start; exposure to stress is delayed until they have a better chance to overcome it. However, nursery practice must aim at field conditions as the plants will eventually grow in the field. The target nursery plant is thus one that will have a good chance of survival in the field, i.e. a healthy, robust and good-sized plant. It is important that nursery and field conditions are interlinked. Nurseries must adapt to the field, as field conditions are difficult to change.

6.3.1 Sowing Time

The sowing time is adapted to produce plants of plantable size at the best time of outplanting. In seasonal climates, the planting time is definite and often short. Generally planting is done at the beginning of the rainy season when the soil is moist but other vegetation still small. If field watering is possible, expediting planting a few weeks before the rain can give plants a very good competitive advantage over weeds. In highland areas, temperature rather than rain can be critical. Planting will always be spread over a period because of labour utilisation. Raising plants should fit to this practicality, so that late plants are not oversized (Kijar 1990). Sowing at regular intervals is a practical method to spread seedling size over a prolonged planting period. It should here be kept in mind that a difference in sowing time of, for example, 1 week does not necessarily mean the same difference in seedling size towards the end of the growing season. If growth conditions improve, e.g. higher temperature, later-sown seed may almost catch up with the early-sown seed (Napier and Robbins 1989).

Orthodox seed can be sown any time, and hence the sowing time can be scheduled entirely according to the planting season, i.e. 'counting back' a nursery season. For example, the planting time in Malawi is from mid-October to mid-December. A good planting size for most species is about 60 cm. Experience shows that *Gmelina arborea* takes about 4 months from germination to reach this height. The sowing time is thus from mid-June to mid-August. The duration of raising nursery plants varies very much. Fast-growing pioneers like *Gmelina, Paraserianthes, Senna* and eucalypts take 3–5 months from germination to good seedling size. Species like teak are typically kept in the nursery for about 1 year before planting out and some conifers need 2 years' nursery care to reach a plantable size.

Recalcitrant seeds gives fewer time options as they must usually be sown immediately or shortly after collection, and this frequently does not fit conveniently with the planting time. Most recalcitrant species bear fruit at the beginning of the rainy season, i.e. the time of planting, and though being a good natural regeneration strategy and fitting to direct sowing, it is ill adapted to the nursery calendar because it implies that seedlings must be kept for about 1 year before outplanting. This is inconvenient for two reasons: (1) seedlings may grow too tall and their height growth be difficult to control; (2) seedlings must be kept in the nursery over a dry season, where there may be water shortage or there is a risk that they will be simply forgotten because it is a season with little other nursery activity. Pruning, light and water stress are management methods to control growth and thus avoid seedlings growing too large.

Soil temperature is a crucial factor in seasonal climates in the marginal tropics and in highlands. Low soil temperature impedes germination for some species that require a certain threshold temperature. In many other species, the germination and growth rate is very low, and low temperature can both result in seedling abnormalities and make seedlings more susceptible to fungal attack, e.g. damping-off diseases (Hartmann et al. 1997). In the Himalayan region, it is recommended to schedule sowing according to a minimum springtime soil temperature of 10°C. (Negi and Todaria 1993).

More practical considerations may interfere with the biological planning. Species grow at different rates, and nursery space, equipment and labour can be crucial bottlenecks during nursery operations, which also have indirect influences on schedules.

6.3.2
Germination and Growth Medium

Usually small seeds are sown in seedbeds and transplanted into pots or planting tubes; large seeds are sown directly[1] into the pots. This is because tiny seeds are generally more sensitive to, for example, too deep sowing or washing away by rain, and it is easier to protect the sensitive stage and give a more gentle treatment in a seedbed.

The key properties of germination as well as growth substrate are to create a good balance between water and air. A loose but fine structure ensures a good contact between seed and soil so that water can be supplied continuously, yet provides adequate aeration for respiration by the roots. At the same time, the soil structure should allow easy penetration by the roots. Both too loose and too compact soil may influence germination and establishment negatively. Generally, small seeds should have a finer and more compact medium than larger seeds. The soil should be free from clods and the surface should have a texture that will not form a crust (Hartmann et al. 1997). Crusting may hinder aeration and be a physical barrier to penetration by the emerging seedling, the latter especially for small-seeded species.

The growth medium is the product of base substrate and the preparation and management. Although species can be found growing on almost any soil type and distribution and growth is strongly correlated with soil type, species are amazingly uniform in their preference for germination and seedling substrate. Most species prefer a medium loam texture, not too sandy and not too fine. A growth medium can be adjusted by mixing various components into the prevailing soil type:

1. Sand is course-grained minerals that improves drainage and aeration but does not hold water well. River sand is normally free of toxic salts and thus better than seashore sand.
2. Clay, loam or other fine particles have the opposite effect to sand. They have a high water-holding capacity and thus reduce drainage and aeration.
3. Organic materials like peat or other materials with high organic content improve the water-retention capacity. In Southeast Asia, coconut husk is the most appreciated potting medium in planting stock production. The husk has good water-holding capacity and friability, and it is readily available at low cost (Kijkar and Pong-anant 1990). Various types of compost have the same effect, but with the drawback that they often contain pathogens.

[1] Not to be confused with 'direct sowing' which refers to sowing in the field without a nursery phase (Sect. 6.5).

4. 'Forest soil' is a mixture of natural mineral soil and organic debris. It is often used as a potting medium, because it contains organic material and nutrients as well as mychorrhiza and other beneficial soil symbionts. A disadvantage of forest soil is that it may carry pathogens. Forest soil can be mixed with other components to modify its structure, e.g. to increase or reduce water-holding capacity by increasing or reducing the amount of water-retention material (organic material and fine particles).

In addition to water retention and aeration, transplanting properties should be considered. When plants are transplanted, their roots can easily be damaged. Loose material is thus preferred for both seedbeds and bare-root plants. The need for water retention often necessitates a certain amount of fine material. However, the more efficient the watering system is, the less important are the water retention properties of the soil. Automatic watering systems which apply water at regular events thus allow the use of course and loose material, which in turn eases planting.

The physical properties of the soil/germination medium are also influenced by management. Aeration of any soil can, for instance, be greatly improved by loosening treatment. Stamping has the opposite effect.

Both water and air are necessary for plants, but excess water tends to replace air and, in fine particle soil, cause clogging of soil particles. The best seedbed is prepared under slightly damp, but not wet conditions. Once the seedbed has been worked, any physical compaction such as that caused by walking should be avoided (Seeber 1976).

Pathogens often accumulate in nursery soil and necessitate pathogen management operations. The medium may be treated with pesticides or fumigated, or germination soil may be renewed.

6.3.3 Temperature and Light

Temperature plays an important role in seed germination but as temperatures are fairly high and constant in lowland tropics, the practical impact is mostly relevant in seasonal highland climates. Germination temperatures which are much above or below the optimal conditions for the species can result in both poor germination and abnormalities of seedlings. In addition to adjusting the sowing time, the microclimate can be modified. Shading reduces daytime temperature and sometimes increases nighttime temperature i.e. reduces fluctuations. This is often wanted because it reduces water stress and because most seedlings prefer some degree of shade. However, temperature fluctuations can have a direct influence on germination, e.g. by promoting germination in some pioneer species. Greenhouse germination is applicable if consideration to other facets of plant production necessitates sowing during the cold season.

Light is generally managed together with temperature, as shading will reduce both light and temperature during daytime. Special conditions apply to seeds with photodormancy which only germinate in light with a high red to far red ratio, e.g. direct sunlight (Chap. 5).

The change from the dormant to the non-dormant stage of light-sensitive seeds occurs only when the seeds are imbibed. Hence, seeds that are sown deep in the soil may remain photodormant, or in extreme cases even develop photodormancy because of the relative enrichment of far-red light at greater depth. Germination of seed under the shade of a green canopy may also give insufficient light stimulus for sensitive seeds, as the light is 'filtered' (Fig. 5.12). In practice, light stimulus to overcome dormancy is provided during germination, simply by germinating light-sensitive seeds in light, i.e. only slightly covered.

6.3.4
Water and Air

The balance of water and air is achieved by applying an appropriate growth substrate, making good soil preparation and applying a balanced watering procedure. Excess water tends to replace the soil air and cause compactness, which in turn restricts respiration. Further, excess water promotes development of fungal diseases like 'damping off'. Germinating seed and young seedlings do not need much water, but water must always be available. Careful adjustment of application must be observed and excess water drained off. A seedbed raised slightly above the ground and with a bottom of coarse-grained material helps drain off surplus water (Fig. 6.7). Water should be applied frequently and the seedbed sheltered to reduce desiccation.

Damage caused by excess water can nearly always be ascribed to lack of aeration. Crusting of the soil surface may also restrict gas exchange. Measures to improve aeration by improved soil structure and drainage were described earlier.

6.3.5
pH

Soil acidity has a very important influence on plant growth and competition in the field, e.g. because it influences nutrient availability. As germinating seeds rely on their own nutrient resources and the initial water absorption is primarily

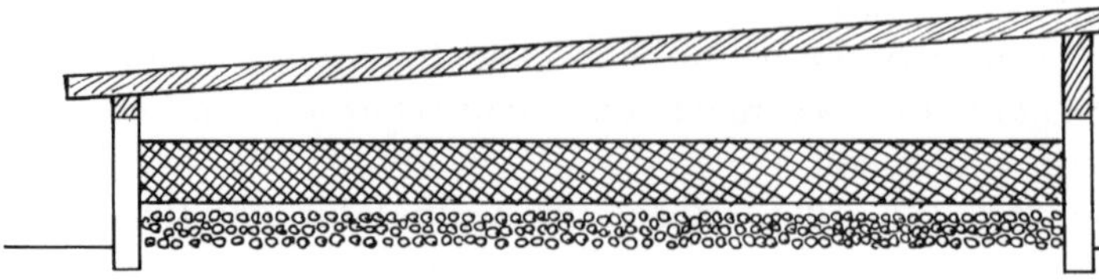

Fig. 6.7. Cross-section of seedbed. Seedbeds should have a good drainage system, e.g. a layer of coarse-grained material under the sowing medium. The beds are usually shaded to avoid drying out and overheating of sensitive germinants. (P. Andersen)

for dissolution and transport, many seeds are quite pH-indifferent during germination. pH becomes increasingly important after germination. Some species show a strong pH effect on germination. Lacey and Line (1994) found, for instance, detrimental effects of alkaline conditions in *Eucalyptus regnans*. The effect was observed both on the total number of germinating seeds and on seedling survival. In nature, very alkaline conditions are often experienced after burning because the pH of ash is high. Very acidic conditions (pH 2–3.5) were shown to inhibit germination in *Cunninghamia lanceolata* (Fan et al. 2005).

6.3.6 Sowing Depth

The practice of covering seed with soil when sowing has three purposes: (1) to maximise soil contact and thus water absorption, (2) to prevent predation and temperature stress and (3) to stabilise seed position, i.e. avoid seeds flowing away during watering or rain showers (Napier and Robbins 1989).

If surface stress factors can be overcome, e.g. by an improved water system, seed can easily be sown on the surface.

After germination seeds readily anchor themselves into the soil and start absorption themselves. Germinating seeds are for a certain period dependent on the nutrient reserves of the seed and remain so until they become self-assimilating. Since small seeds store less material than large seeds, the emerging seedling of a small seed is only capable of growing through a shallow layer of soil. Hartmann et al. (1997) state as a rule of thumb that seeds should be sown at a depth that approximates 3–4 times their diameter. This holds for small to medium-sized seeds; large seeds (more than 1.5–2 cm diameter) need only a sowing depth of twice their diameter. For any seeds, too deep sowing delays the emergence, and where seeds are sown very deep, emergence may fail altogether. Seeds that need light for germination should obviously only be covered with a shallow layer of soil, but in practice all light-sensitive seeds are relatively small, and are sown shallowly because of their size.

Practical nursery sowing is either in seedbeds or directly in 'poly-tubes'. Small seeds are broadcasted on top of the seedbed, then possibly covered with a thin layer of soil either by raking the upper half centimetre or covering the seeds with a thin layer of soil or coarse sand (Napier and Robbins 1989). Larger seeds are normally sown in drills or directly into poly-tubes, and then covered with soil.

6.3.7 Orientation

Seeds with an asymmetrical shape tend to deposit themselves during dispersal in a position which is favourable to germination (Box 6.1). Seeds which are sown by broadcasting will tend to find the same position. Large seeds which are sown individually may require some attention to avoid unfavourable positions. In some seeds, appendices like wings which help positioning seeds during natural dispersal are removed during processing. Roots always penetrate from the micropylar end, and roots always grow down, but a position with the root facing down is not necessarily the best.

In a study by Mahgoub (1996) on germination in relation to sowing position, he found that the germination depends on germination type (hypogeal or epigeal), seed size and seed shape. Most seeds showed the highest germination percentage when sown in a horizontal position.

In *Derris indica*, Swaminathan et al. (1993) found that vertical sowing with the micropyle (radicle end) down was superior to any other positioning. This position was also recommended by Flores (1992) for *Dipteryx panamensis.* Very small variations were observed between different sowing positions in dipterocarps in a study by Otsamo et al. (1996) – dipterocarps always position themselves with the radicle upwards, the same direction as the wings. On the basis of these and other studies it is recommended to orient seeds with the radicle end (micropyle) down when seeds are sown individually. Where the micropyle is difficult to identify, oblong and flat seeds should be placed in a horizontal position; kidney-shaped seeds should be oriented with the grooves upwards.

6.3.8 Fungal Problems, 'Damping-Off' Disease

'Damping off' is a collective name for a number of fungal diseases attacking germinating seeds and young seedlings. Pathogens causing damping off may be seed-borne or soil-borne. The latter is especially the case when the same nursery soil is used for two successive seed lots where the soil was not sterilised in-between. The fungi attack non-lignified (soft) parts of the plants and

Box 6.1

Gravitropism – finding the way down

Some seeds tend to position themselves with the root down, when they land. Many flat seeds tend to land with the radicle in a horizontal position, and some winged seeds, e.g. dipterocarps, tend to land with the root end upwards. However, no matter how the landing position is designed or how the seeds happen to be deposited or planted, the root will always find its way down into the soil (Fig. 6.8). This is because specialised cells of the root (statocytes) possess gravity sensors, so-called statoliths, which orient themselves according to gravity (Taiz and Zeiger 1991). The phenomenon is called gravitropism (or geotropism – the latter is a broader concept, which includes any three-dimensional directional orientation). Gravitropism also works in shoots. Shoots mostly orient themselves according to light (phototropism), but shoots growing in the dark or in diffuse (non-directional) light will grow against gravity. Gravitropism by statoliths is also found in other organisms which orient themselves in a three-dimensional space, e.g. water-living creatures.

Gravitropism and phototropism are the principal directional forces determining plant growth (Correll and Kiss 2002). Gravitropism is the principal mode of orientation in young roots, but it is soon balanced by other forces. Secondary roots will thus primarily grow horizontally and according to water and nutrients; shoots and leaves will mainly be influenced by light. These factors can occasionally inverse the gravity orientation. For example, roots subjected to waterlogging will start to grow upwards, and in some mangrove plants a special type of root, a pneumatic root, always grows vertically against gravity.

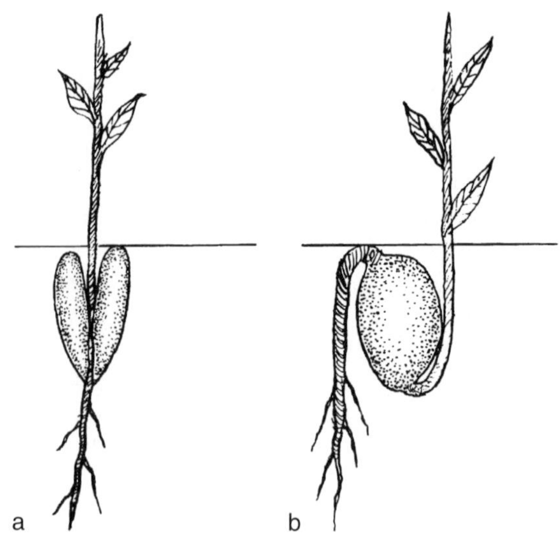

Fig. 6.8. Gravitropism/geotropism in germination. The radicle always penetrates the seed coat at the radicle end – also when this end is facing up, but the root will immediately bend and grow under influence of the gravity

the diseases thus mostly affect plants during the first stages. There are two principal types:

1. Preemergence damping-off causes seeds and sprouts to rot before the plant has broken through the soil surface.
2. Postemergence damping-off causes rotting of the stem at soil-surface level, causing the seedlings to fall over and die.

Damping off usually starts in spots or patches and spreads from these localised areas from one plant to another through direct contact or by soil or water movement (Fig. 6.9a). Under particular conditions, the infections may spread rapidly to the entire seedbed (Cremer 1990). Conditions causing rapid spread of the disease are, for example, excessively wet soil, alkaline soil, poor light conditions, too high or too low temperature and crowded seedbeds (FRIM 1987).

Damping-off diseases are preferably dealt with by preventive measures rather than treatment. Preventive measures are, for example, preventing dispersal and reducing multiplication. Some routine methods apply:

1. General nursery hygiene, e.g. cleaning tools, boots and gloves between handling different seed lots and plants (Fig. 6.9b). Equipment may be disinfected with, for example, 2% bleach solution (Cremer 1990).
2. Optimise conditions for plant growth. Generally, healthy and vigorous seedlings are less prone to attack than poor ones. Optimal germination and growth conditions help plants overcome attack by mobilising the inert resistance system and make plants pass the vulnerable stages before attacks are damaging.

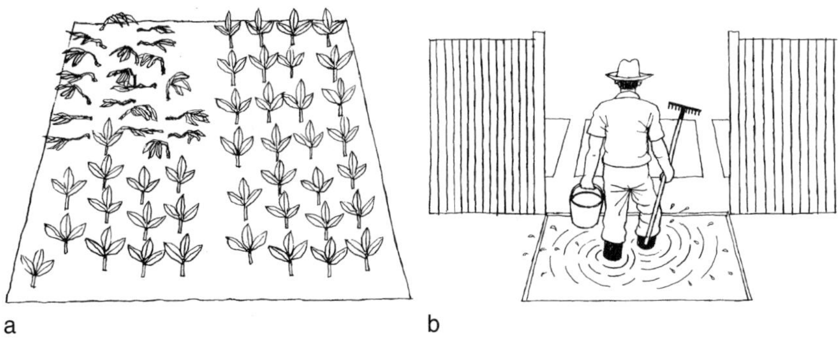

Fig. 6.9. **a** Nursery seedbed showing a patch of seedlings dying from postemergence damping off. **b** Foot bath containing disinfectant at nursery entrance to control fungal spreading via footwear. (P. Andersen)

3. Suppress spread and development of fungi from plant to plant or via soil by adequate spacing, a substrate with good aeration, proper moisture and light management plus adequate ventilation of seedbeds and transplanting beds.

Where conditions are optimal for both plants and fungi, the balance may shift in favour of the plants by providing conditions more tolerable to plants than to fungi. For example, light and drought are more stressful to fungi than to plants. Temperature preference and tolerance differ between different species of plants and fungi and in practice it is difficult to manipulate temperatures in seedbeds. The problem with temperature manipulation as disease management is that there are many damping-off fungi with different development.

Soil-borne fungi may be eliminated by fumigation (Box 6.2). Severe cases of damping-off diseases may be controlled by fungicides applied to plants

Box 6.2

Fumigation

Fumigation is the process of soil sterilisation by gas treatment. The objective of fumigation is to prevent infection of plants by soil-living organisms, e.g. damping-off diseases, insects and nematodes. In addition, fumigation kills most weed seeds. The most commonly used chemical for fumigation is methyl bromide, a volatile, odourless gas applied by injection from pressurised containers into the soil covered by a plastic sheet.

Hartmann et al. (1997) suggest an application rate of 333 ml or 0.6 kg methyl bromide per cubic metre of soil to be treated. The plastic cover is left over the soil for 48 h and the soil can only be used as a sowing or planting medium after some days' aeration.

A major drawback of methyl bromide fumigation is that it is toxic to humans and animals and must be applied only by trained staff. It has been banned by several countries because of its alleged detrimental effect on the earth's ozone layer, and its use is predicted to be phased out (Hartmann et al. 1997). An alternative to fumigation is heat treatment. Kiln heating to, for example, 80°C for 15–30 min will kill most soil-living organisms.

Soil sterilisation has some drawbacks: In addition to pests and pathogens also beneficial organisms are killed. Microsymbionts, pest-controlling organisms and earthworms are, for example, beneficial organisms.

Because beneficial organisms are killed together with the target organisms, invading pathogens may spread more easily when their natural predators have been eliminated.

after germination. FRIM (1987) suggests application of Captan or Zineb as follows. Captan 50% wettable powder: 0.06% active suspension (12 g powder in 10 l water) is applied through a fine rose at a rate of 5 l suspension per square metre of seedbed or surface area of plant tubes. Zineb 65% wettable powder: 0.13% active suspension (10 g powder in 5 l water) is applied through a fine rose at a rate of 3 l of suspension per square metre of seedbed or surface area of plant tubes. The application is repeated 14 days later. Also Thiram and Bordeaux mixture may be used, although Captan and Zineb are preferred as they are not phytotoxic (FRIM 1987). Captan, Zineb and Thiram can be applied as seed dressing before sowing. Spraying with Dithane M-45 or Blitox (25 g powder mixed with 5 l water) has been used for damping-off control in Nepal (Napier and Robbins 1989). Kommedahl and Windels (1986) mention Busan and Metalaxyl having an effect on damping-off fungi in maize, but their effect on the disease in forest tree seedlings is not known.

6.4 Seedlings in the Nursery

6.4.1 Light and Shade

Seedbeds and polythene tubes are shaded during germination and the early seedling stage. Shades or shelters protect seeds and young plants from direct sunlight, large temperature fluctuations, desiccation, heavy rain, and, in some areas, frost and hail (Napier and Robbins 1989). The shade is raised 30– 60 cm above the seedbeds, and usually 2 m above polypot tubes to allow a convenient working height (Fig. 6.10). Maintenance of shades over young plants depends on species. In Malawi, shades are removed completely from pine seedbeds at high altitudes a few days after germination, while the shade is maintained for some time for other species (FRIM 1987). The density of the shade must be adjusted according to the species. Too dense shade may result in etiolation (thin weak seedlings) of light-demanding species. Too little shade may provide inadequate protection from the aforementioned factors. Shade is gradually reduced as the seedlings grow, except for very shade demanding species like *Khaya*, dipterocarps and others, which are normally planted under a shelter of pioneer trees in the field. Where shade consists of, for example, grass mat frames, a gradually increased exposure may be achieved by removing the frames initially a few hours a day, and increasing the duration of full exposure (Napier and Robbins 1989).

a

b

Fig. 6.10. a Plastic tubes ('polypots') are used for transplanting seedlings from seedbeds or seeds are sown directly in the tubes. **b** Shelter construction over young seedlings. As the plants grow, the shelter is usually removed in order to 'harden' the plants for planting. Thailand. Photo: A.B. Larsen.

6.4.2 Moisture

Water requirements will differ according to species and weather conditions. Too little moisture causes reduced growth or, in the worst case, wilting; too much water causes problems in root respiration and often promotes fungal diseases. Young germinants are especially sensitive and must be watered frequently. As the seedlings grow, their water demand increases, and watering should be increased accordingly. However, established seedlings also tend to achieve a certain tolerance to desiccation. The frequency of watering can be reduced from, for example, several times a day to only once or twice. It is important that seedlings are thoroughly wetted through the full root system. If water reaches the upper layer only, root development may be superficial. Waterlogging is less likely when the seedlings have become established because they continuously consume water. Generally it is advisable to wet thoroughly at intervals and allow a certain degree of drying out in order to facilitate aeration, rather than adding water very frequently to keep the soil permanently wet. Moisture regulation is relatively easy during the dry season as long as water is available. Overwatering by heavy downpours during the rainy season cannot be avoided. It is therefore important that excess water can be drained off easily, whether the seedlings are kept in transplanting beds or polyethylene tubes. If seedlings tend to grow too fast (cf. Sect. 6.3.1), reduced watering can be used to control their growth (Napier and Robbins 1989). Towards the end of the nursery period, watering should always be reduced as part of 'hardening' to adapt them to field conditions.

6.4.3 Fertilisers

The need for and the type of fertiliser application depend on the nutrient content of the soil, the size of the seedlings and the length of time they will spend in the nursery. Where forest top soil is used in germination beds or as potting soil, application may be unnecessary. Where planting soil is relatively poor in nutrients, application of a granular NPK or other fertiliser will be beneficial. Also, fertilisers may be necessary where seedlings are held in the nursery for long periods where large seedlings are required, e.g. for ornamental/amenity planting or for grafting. The composition and strength of NPK fertilisers are indicated by a set of three numbers, for example 12:24:12 meaning that the fertiliser contains 12% nitrogen (N), 24% phosphorus (P) and 12% potassium (K). A fertiliser relatively rich in phosphorus is usually recommended, both because phosphorus is the limiting factor in many soil types and because it encourages root development and stimulates the development of nitrogen-fixing bacteria in Leguminosae. Conversely, nitrogen encourages leaf and shoot development and discourages nitrogen-fixing bacteria; neither is desirable in the nursery.

Fertilisers for small seedlings are usually applied in liquid form with a watering can. For container plants, a few granules may be applied to each potted plant. It is important that granules do not remain on the leaves since this may damage the leaves. Seedlings should be thoroughly watered after application of granular fertilisers to dissolve the granules and ensure root contact. It should be noted that excess fertiliser may reduce mychorrhiza and rhizobia development.

6.4.4 Pruning

Potted plants are pruned for a number of purposes:

1. To reduce overgrowth in the nursery
2. To facilitate the physical planting process
3. To promote side root development for potted seedlings

The ideal of potted seedlings is that their root system fills up most of the pot volume, but roots must not grow outside the pot and anchor the seedlings into the soil below. Bare-root seedlings may be root-pruned in order to facilitate planting since a large spreading root makes practical planting difficult. Another reason is that deep roots are easily damaged when removing the seedlings from the nursery. Bare-root seedlings may be root-pruned mechanically by under-cutting the whole seedbed.

Potted seedlings may be root-pruned by lifting the individual pots and cutting roots that have grown out of the polythene tube with a knife (Fig. 6.11). The time and number of prunings vary with species and conditions. Potted seedlings are inspected and root-pruned when roots start to grow through the pots. Frequent root pruning (every 7–14 days depending on the growth rate) is better than delayed pruning, which shocks the plant. More frequent pruning is usually necessary by the end of the nursery season when the plants have grown large. The last pruning is usually scheduled 2–3 weeks before outplanting. If the pruning has involved cutting of many roots, seedlings must be kept under shade and watered thoroughly for the first few days after pruning to help them recover from the shock (Hoskin 1983). If planting cannot be undertaken as scheduled and seedlings again tend to grow out of the pots, a new pruning and recovery period must be allowed.

An alternative to root pruning, vertical root development can sometimes be controlled in potted plants by moving the pots regularly. This method, known as root wrenching, will stress the roots and prevent them from growing into the nursery bed. Since the roots are not cut, it is often less stressing to the whole plant than pruning. Root wrenching rather than pruning may also be applied during the period just before outplanting to minimise the need for a recovery period.

Nursery practice should ensure that there is a reasonable ratio between root and shoot. Overgrown, top-heavy seedlings occasionally need top pruning to reduce evaporation. In Malawi, it has been recommended to cut back seedlings of eucalypts and *Gmelina* exceeding 18–24 cm (depending on planting tube size) to 2-cm stumps. The seedlings will recover by setting new shoots (FRIM 1987). Where the sowing time is determined by seed viability rather than by the

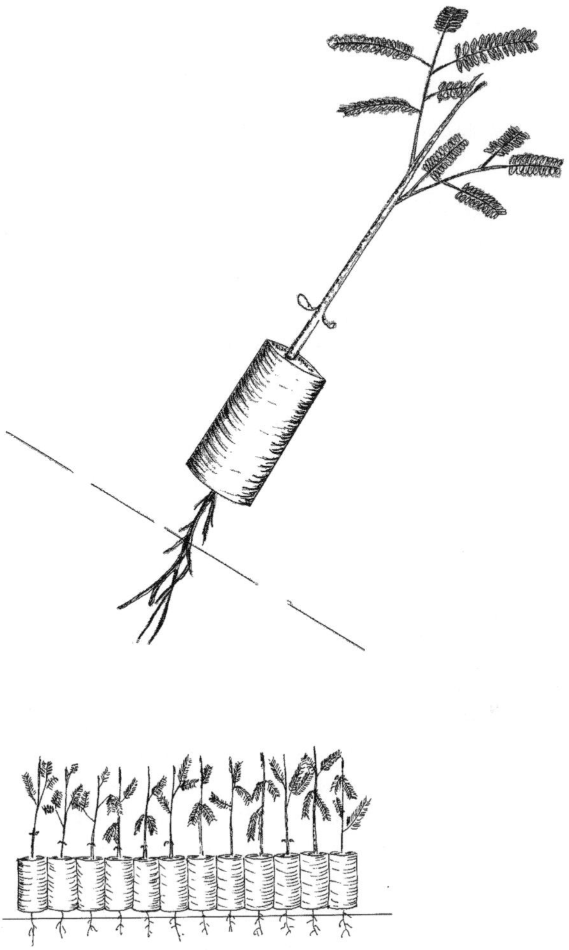

Fig. 6.11. Root pruning of a container plant to promote side root development within the pot and to avoid the plant anchoring itself to the nursery soil

planting time, e.g. as for *Azadirachta indica*, top pruning to produce stumps may be necessary to ensure survival during a prolonged dry season (Lauridsen and Souvannavong 1993). It should be noted that the natural shoot-to-root ratio differs greatly between species. Many dry-zone species will have poor height growth until deep roots have formed.

The ability to recover after pruning differs between species. Pruning of smaller roots is tolerated by most species, but species that form deep taproots with few superficial side roots are often sensitive. Several dry-zone Leguminosae, e.g. *Faidherbia albida*, are quite sensitive to pruning, especially if the root has overgrown. Top pruning is tolerated by most dry-zone species, but it may lead to undesired branching or forking of the stem. It is generally only applicable to species where multiple stems are acceptable.

6.4.5 Hardening or Conditioning

A few weeks before the seedlings are transplanted into the field they must be hardened to adapt them to the harsher field conditions. Watering is reduced and fertilisation stopped. Shelters are usually removed to expose the seedlings to full sunlight, although some shelter may be maintained for species that are to be planted under shelter trees, e.g. some rain forest trees. Hardening should normally be initiated some days after root pruning so that the seedlings will have some days to recover from the pruning shock (Hoskin 1983).

6.5 Direct Seeding

Trees are generally slow starters compared with, for example, herbal plants and grass and therefore often suffer high competition mortality during natural regeneration. Hence, trees are usually raised under protected nursery conditions and kept there for a period until they have grown to a size where they have a better competition over weeds when planted out. A second reason to raise plants in a nursery is land-utilisation efficiency. Seedlings take up a very small space when small compared with when they have achieved their adult size, and except from agroforestry practices where trees and crops are combined to make the best use of land, young plantations have very low area efficiency. Area efficiency is a major concern in commercial plantations, where the aim is to maximise production from any land unit resource at any time.

Better survival rate, less maintenance, better initial growth and better area efficiency usually make nursery raising profitable compared with direct seeding, despite its high direct cost. Nursery raising has thus become the 'normal' for tree propagation. However, the cost–benefit balance between nursery and direct seeding sometimes favours the latter, e.g. where labour costs are high and where terrain conditions make use of labour-saving farm machinery applicable (Table 6.2).

Direct sowing (or seeding) is applicable under a limited set of conditions, where seeds of woody plants can germinate and the seedlings can establish themselves fast in situ and in competition with other plants, and where land-use efficiency is less important. Such conditions prevail, for example, in some

Table 6.2. Comparison between cost and efficiency of conventional forest establishment by planting and direct sowing

Operation	Conventional establishment by nursery seedlings	Direct seeding
Seed cost	High seed efficiency, use of improved seed applicable	Low seed germination rate – improved seed relatively costly
Sowing cost	Low in nursery	High – depending on terrain
Seed predation	Low owing to effective nursery control	High – especially in broadcasting methods
Plant establishment	Requires entire nursery operation	Not available
Planting out	Labour-demanding, especially in difficult terrain	Not available
Site preparation	Thorough preparation	Thorough preparation
Maintenance/weeding	Good competition with weed	Poor competition with weeds
Survival rate	Expected high	Low owing to germination and establishment mortality
Labour cost	High	Low

new or degraded afforestation sites like alang-alang grassland, mine spoils, mud areas or new volcanic areas (Bird and Lawrence 1993; Cremer 1990; DPI 1994; Coffey and Horlock 1998). The efficiency depends both on field conditions and on the plants. Denuded and degraded areas are also stress areas for direct sowing because of a harsh microenvironment, e.g. with high fluctuations in temperature and water availability (Rao and Singh 1985; Uniyal and Nautiyal 1999). Germinants of species used for direct sowing must thus be able to cope with a high level of field stress, e.g. germinate at relatively low water regime (Uniyal and Nautiyal 1999).

The colonisation/establishment rate can be improved by site preparation and management, and by supporting the establishment and growth ability of the plants, e.g. by:

1. *Reduction of competition from other vegetation.* In open sites, land is usually cleared prior to sowing, e.g. by burning and/or mechanical soil preparation in the same way as for planting. Herbicides may in some cases be applicable, e.g. if burning is difficult to control and mechanical clearing cannot be undertaken owing to safety or terrain constraints. Direct sowing is occasionally used for agroforestry practices for hedgerow establishment (alley cropping). Weeding is here undertaken as part of the normal farming practice (Holt 1999).

2. *Species selection.* Species should have fast germination and establishment i.e. be 'aggressive' pioneers. In humid climates weeds are the main limiting factor, and the faster trees cover and shade other plants, the higher are the chances for survival.
3. *Ensuring a fast germination rate.* Hard seed must be pretreated and some species can be primed to make them 'just ready' to germinate when sown. Application of fertiliser and/or inoculation with microsymbionts as a 'start package' can enhance the establishment rate. Application can be done by, for example, pelleting (Vargas-Maciel 2003).
4. *Appropriate sowing time.* Sowing should be done at the beginning of the wet season where moisture is sufficient for seed to imbibe, germinate, establish a firm root grip and establish seedlings that can survive a subsequent dry season. Too late sowing may fail to give the plants the necessary competition against weeds, and the time may be too short for their development, so they may be killed during the subsequent dry season (Cremer 1990). Sowing into the rainy season, especially in high-rainfall areas and on sloping terrain, implies a risk that heavy showers/rainstorms will completely wash away seeds and new germinants (Ezell 2004).
5. *Reducing seed predation.* Exposed seeds sown by broadcasting are prone to predation, e.g. by birds and rodents. Covering seeds significantly reduces the rate of predation for pine seeds in Sweden (Nilson and Hjalten 2002). In Vietnam, pine seeds broadcasted from aeroplanes were treated with pesticides.

Direct sowing basically has three forms:

1. *Broadcast sowing of small seeds on cleared land.* In both China and Vietnam, broadcast sowing of vast highland deforested grassland sites has been done from aeroplanes (aerial sowing) (Wang and Xu 1985; CAF 1981). The narrow-sense economy may be doubtful but if the activity is carried out as an aviation practice or training exercise, the cost is hidden in other core budgets. Alternatively, smaller and more easily accessible areas may be sown by manual broadcasting (Silc and Winston 1979). Seed broadcasting has the advantage of a large area coverage in a relatively short time and as such is efficient for remote areas[2] and difficult terrain.

[2] 'Remote' is often used as a geographical distance from cities or capitals and therefore sometimes ignores the fact that people live there. Socioeconomic implications of, for example, aerial sowing are obviously essential before launching such an activity.

2. *Precision sowing.* The seeds are placed in the soil and covered with soil using various types of sowing equipment. This method is common in Australia when reforesting barren land (Bird and Lawrence 1993; DPI 1994; Coffey and Horlock 1998). Precision sowing of hedgerow and alley cropping species (e.g. *Sesbania sesban*) is used in farm forestry and agroforestry (Owour et al. 2001).
3. *Sowing individual seeds of usually larger-seeded species.* Single-seed sowing may be on cleared land or under other woody vegetation, e.g. climax species under pioneers. Oak and beech trees in temperate regions are sometimes established by direct sowing using a sowing stick. The survival rate is higher because germination sites are selected and as seeds are covered they are protected against predation and other adverse conditions (Coffey and Horlock 1998; Greening Australia 2004).

Direct sowing inevitably implies higher mortality than planting of good-sized seedlings. In Australia, the survival rate of eucalypts was only 0.1%, that of acacias about 5% and that of most others about 1% (DPI 1994). Its use is thus limited to situations where such mortality can be tolerated, i.e. cost balanced with 'traditional' planting. This is first and foremost large-scale afforestation in remote and difficult terrain where planting costs are very high. High planting costs also apply to some agroforestry methods, e.g. alley cropping, which typically uses high densities of relatively small trees (Owuor et al. 2001). High labour costs generally favour direct sowing, as it is far less labour-intensive.

Species, which are difficult to raise under nursery conditions can have a higher survival chance by direct sowing. There are two main categories:

1. Species with recalcitrant seed, most of which are shade-tolerant (or shade-demanding) when young, are often difficult to raise and keep in the nursery. They are often pregerminated when collected and suffer during transplanting. On the other hand, they survive under some shade and can thus cope with some competition from weeds in the field. Mangrove plants such as *Rhizophora* and *Bruguiera* have little competition from other plants in the field and are best established by direct sowing/planting of the viviparous seed. Some species are very sensitive to root damage during transplanting and are for that reason preferably established by direct seeding.

2. Dry-zone species form deep-growing roots before they grow in height. Root pruning is usually applied in nurseries to avoid the plants anchoring themselves to the nursery. However, in dry-zone species root pruning implies a severe stress. For that reason, direct sowing has been seen as a suitable alternative method for afforestation in the Sahelian region (Eden Foundation 1992).

In both of these cases physiological complications of nursery propagation could point towards the direct sowing alternative.

Land rehabilitation or reforestation activities with a strong biodiversity element may use direct sowing as a suitable method as multiple species are far easier to handle as seeds than as seedlings (Holt 1999). Less common, albeit with increasing importance, is the application for establishment of an understorey of a climax forest species under a canopy of pioneers or a part of forest conversion (Ammer et al. 2002). Seedlings must be sown individually and the method requires a relatively open understorey and minimum weed competition.

A major drawback of direct sowing is the high-quantity seed use because of the excess mortality rate. This makes the method most applicable to species with cheap seed and discourages use of improved-quality (seed orchard) seed.

6.6 Microsymbiont Management

Microsymbionts encompass three main types of soil-living organism that form symbiosis with plant roots, viz. mychrorrhiza, rhizobia and frankia (Fig. 6.12). The symbiosis may be obligate or facultative for a wide range of plants which thrive poorly without the symbiosis. Mychorrhiza have a number of roles but are especially important for phosphorous absorption. The mychorrhiza 'sheet' on the roots of outplanted seedlings often forms an effective protection against environmental stress. Rhizobia and frankia are two distinct groups of soil bacteria that form nitrogen-fixing root nodules on host plants. Rhizobia are closely linked to species of the Leguminosae family; Frankia form root nodules on so-called actinorhizal plants, taxonomically a very diverse group with *Alnus* and *Casuarina* as the most important forest trees.

Some types of microsymbionts are very host specific in the sense that a particular species or strain will only form symbiosis with one or few related species of host plants. Others are broad-ranged and form symbiosis with many types of host species, both herbs and trees (Somasagaran and Hoben 1994).

Microsymbionts are applied to plants by inoculation. The inoculant may consist of a concentrated culture of the microsymbiont, e.g. bacterial culture, fungal spores or vegetative fungal mycelium. In other cases it consists of crushed,

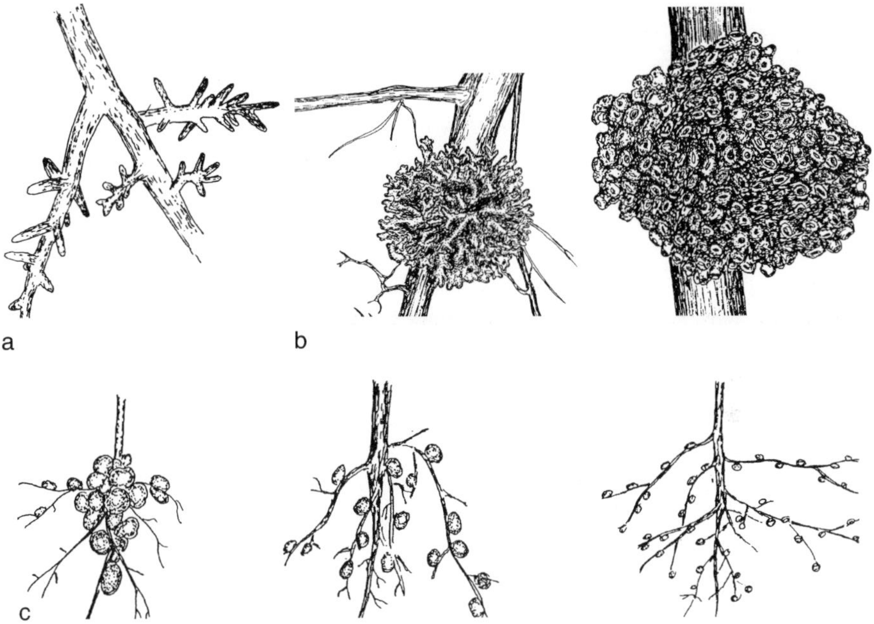

Fig. 6.12. Microsymbionts. **a** Mychorrhiza are fungal symbionts that cover the plant roots and help in nutrient absorption (particularly phosphorous) and protect the roots against stress. **b** Frankia is a special group of actinorhizal bacteria that form nitrogen-fixing root nodules with trees, e.g. *Casuarina, Alnus* and *Hipophae.* Frankia nodules are large and often coral-like **c** Rhizobia are nitrogen-fixing bacteria that form symbiosis with a lot of plants from the plant family Leguminosae. Rhizobium nodules are small and spherical. Inoculation of roots with microsymbionts can be done by applying soil or crushed roots/nodules to nursery soil, or the symbiont can be applied as a laboratory concentrate, e.g. as pellets or granules

infected[3] plant roots or infected soil. Because of host specificity and the difference in microsymbiont efficiency (there is sometimes a range between the symbiosis from parasitic, in which the microsymbiont mainly extract nutrients from the plant, to true mutualistic, in which both host and infecting organism benefit), there is increasing interest in handling microsymbionts as an integrated part of seed procurement and supply. Inoculants can be supplied together with seed as a separate bag, 'tablet' or granule, or as seed cover in a pelleted seed. If seeds are pelleted, they are automatically inoculated during sowing. This method has a number of drawbacks, which were described in Chap. 5. If seeds are not pelleted, the inoculant is applied shortly after germination. Most roots need to be a few weeks old before they can be infected by microsymbionts (Somasagaran and Hoben 1994; Molina and Trappe 1984; Marx 1980).

[3] Infected means the presence of a microorganism in the plant or soil. The same term is used in a negative context with disease-causing pathogens.

Seed Testing

7

7.1 Introduction

Seed testing is a quality test pertaining to the seed's physiological quality, i.e. its life processes. Testing does not in itself improve quality – a tested seed lot is not necessarily better than an untested one, but it helps distinguishing good and bad and can thus be a guide towards better quality.

A status or assessment of seed quality is of relevance for different reasons in seed handling, e.g.:

1. Investigation of seed behaviour in applied seed research
2. As a guideline during seed handling
3. As documentation for seed quality during seed trade and dispatch

Whether the 'normal' seed behaviour is investigated in a designed research programme or appears as a result of more informal accumulation of experience, the compilation of data gradually establishes a reference frame against which succeeding results are related. For example, when moisture content is used as an indirect quality parameter it is based on the experience that moisture has an impact on seed storage and hence quality.

Research and trade documentation set the same high standard for exactness: research because it establishes the objective reference frame; trade documentation because it has legal economic implications. A number of simpler tests are carried out during seed handling. Simple tests, some of which may rather be referred to as examinations or checks, are quicker and less exact, yet are correlated with more elaborated standard tests.

The ultimate figure of interest for seed users, whether seed lots are traded or not, is how many live plants can be produced from a given quantity of seed. Seed weight, purity and germination/viability are all parameters in this calculation which ultimately points towards seed demand (Karrfalt 2001). Moisture content has primarily relevance for storability and longevity and is thus an indirect indication of the validity of the test results in some near future.

Simple guideline tests are not used for documentation, partly because they are not formalised, partly because they are implicitly used to suggest a further process, which makes the result of the test invalid.

Research and trade documentation set the same criteria for objectivity and replicability, viz.:

1. The tests must be unbiased and objective, i.e. independent of the person doing the testing.
2. The tests must be replicable in the sense that a subsequent test of the same seed lot must give the same result within the range of statistical error.

The test result gives a status which is representative for a given seed lot at a particular time. There is thus no guarantee that the test will be valid for the seed lot if the seed lot undergoes progressive change during the testing period. How long it is valid depends on the possible changes from the time of testing onwards. Moisture content may change because of absorption or desorption of water either from the external environment or from internal respiration. Seed weight changes very little over time and purity does not change unless the seed lot has been subjected to cleaning or contamination. Viability declines for all seed lots over time as they deteriorate, but the rate is very low for most orthodox species under good storage conditions. It makes sense to apply expensive and time-consuming testing only when the results have a certain timely validity. Most seed lots are tested once, viz. after final processing just before or during early storage. Seeds that are stored for a long time may be tested at intervals from harvest until the seeds leave storage to be dispatched or sown in the nursery, the interval depending on the decline in viability (Hor 1993).

Standards for seed testing of all species, including agricultural and horticulture seed, have been laid down by the International Seed Testing Association (ISTA)[1]. Standards are revised and updated every 3 years from the first issue formulated in 1931.

Seed research is not necessarily dependent on/bound by ISTA or AOSA standards but may apply its own method, as long as it lives up to the aforementioned standards of objectivity and replicability. As seed research covers all aspects of seed physiology and biochemistry, it is also not limited to the few parameters included in standard testing.

[1] Several American countries follow the rules of the Association of Official Seed Analysts (AOSA). The two sets of rules differ only in minor aspects.

Seed testing standards, procedures and laboratories were originally designed for agricultural and horticultural seed. Forest seed differs in the context of testing:

1. Quantities are generally small. Test quantities, in particular repeated tests from seed storage, can for some species take up a significant part of the stored lot.
2. Many species are large-seeded and recalcitrant and contain various dispersal appendices.
3. Many species have complex dormancy.

7.2 Timing Seed Testing

Simple testing of purity and moisture content may be carried out during processing to guide how far cleaning should continue, and the necessity for further drying (Hor 1993). A standard seed test is usually carried out after final processing and prior to storage. Seed weight and purity will normally not change much during storage and the moisture content should vary very little, if seeds are stored dry and hermetically sealed. The main interest is thus the decline of viability during storage. A second test would be relevant when viability is anticipated to have declined significantly from the first test. During long-term storage, regular testing is relevant to verify the quality at any time and, for very old and deteriorated seed lots, to determine when the seeds should no longer be stored. Karrfalt (2001) suggested testing of orthodox seed should take place with 3–5-year intervals during long-term storage. However, the need for testing can be reduced using knowledge of the pattern of seed deterioration.

If a seed lot has a theoretical longevity and the rate of seed deterioration is predictable (Chap. 6), then a single test would suffice to calculate seed quality at any future time. Unfortunately ageing parameters have only been established for very few species. However, from the viability equations it is shown that the pattern of viability decline is likely to follow a straight line when plotted on probit graph paper. Further, it is known that seed lots of the same species exposed to the same type of storage conditions are likely to show the same pattern of decline in viability, i.e. the slope of the viability curve is the same (Ellis and Roberts 1980). Hence, if some previous records are available for the same species and storage conditions, a viability graph can be constructed. After the result of one germination test has been plotted on probit paper, viability can be predicted at any time, using the slope of comparable seed lots. In reality, seed lots differ and, for example, the stage of maturity at collection, the moisture content and the genotype influence progress in deterioration (Fig. 7.1). A few tests during long-term storage are thus advisable in order to adjust viability curves.

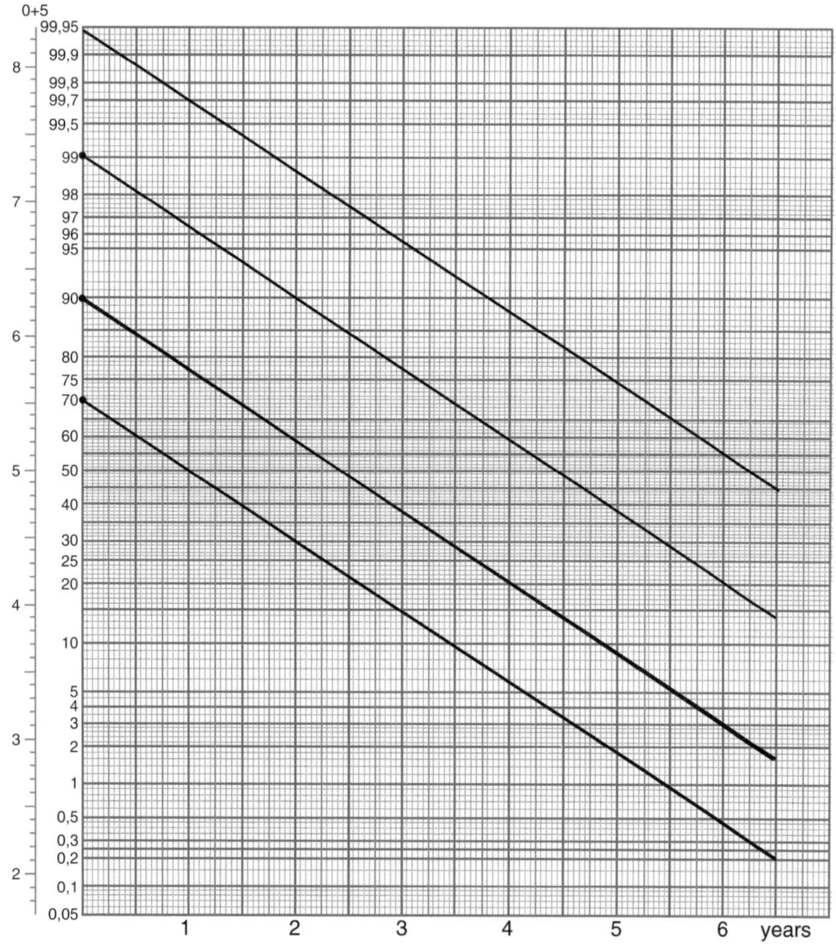

Fig. 7.1. Probit viability. A reference seed lot with initial germination percentage of 90% shows a viability of 70% after 1.5 years and 38% after 3 years. Considering that the relative deterioration follows the same rate, viability curves for seed lots with 100, 99 and 70 initial germination, respectively, can be drawn as parallel lines, and their predicted viability read at any time on the scale. It is noticed that a seed lot with an initial viability of 100% takes 6.25 years before half of its initial viability is lost (50% viability); in a seed lot with initially 70% viability, the viability is reduced to half (35%) in 1.75 years. Predictability is valid within a relatively narrow range of similar storage conditions and genetic variation (species, provenances, etc.)

The predictable pattern of viability decline is under the precondition that the storage condition is unchanged and is thus not applicable under ambient conditions where temperature and humidity vary and possible fungal infection interferes with normal deterioration events. Seed health may change during storage and where this is likely to occur, e.g. under ambient conditions, supplementary viability and health checks are relevant. Vigour tests may be relevant for seed lots which show high germination capacity under test conditions, but where some deterioration is suspected, e.g. after prolonged storage.

The latest test result represents the latest and thus the most valid status of seed quality and replaces former tests. However, the development in seed quality contains important information for the seed user. For example, a seed lot where viability has declined during storage from, say, 65 to 60% indicates that the seed lot had an initial low viability, which may be due to, for example, a large fraction of empty seed, which is not an ageing factor. A seed lot where quality has declined from, say, 95 to 60% indicates a progressed deteriorated seed lot. Progression of seed deterioration could be further documented by a vigour test, but the latest viability compared with the initial viability already contains relevant information on viability history.

7.3 Standard Seed Testing

Trade documentation has adopted international rules on seed testing in order to be able to compare documented quality parameters from different laboratories. Most European, Asian and African countries follow the ISTA rules, while the Americas follow the AOSA rules – in practice the two sets of rules are very similar. Test rules contain a standard set of parameters and prescription on how to measure them. Standard parameters are seed weight, purity, moisture content and germination. Tests of many other types of quality parameters may be performed when or as required, e.g. vigour test and phytosanitary tests.

The ISTA rules[2] (ISTA 1996, 1999, 2006) contain, in addition to standard seed testing procedures, specific guidelines for a number of species. Specific guidelines on testing tropical and subtropical forest species occur in the *ISTA Tropical and Sub-tropical Tree and Shrub Seed Handbook* (Poulsen et al. 1998). In addition, the ISTA has published a number of more elaborate handbooks on individual seed testing procedures (ISTA 1986, 1991, 1995, 1996, 1999, 2006; Poulsen et al. 1998; Kruse 2004). Official testing guidelines from the ISTA and the AOSA make up the basics for national rules and guidelines on

[2] The ISTA publishes an updated set of rules every 3 years. The rule set of 1996, 1999 and 2006 has been used in this compilation.

seed testing. Adopting national guidelines must necessarily take into consideration, what type of laboratories and equipment are available. Standard seed testing requires the capacity to control all parameters. Many laboratories are capable of carrying out tests of seed weight, purity, moisture content and short viability. Standard germination tests require a relatively high investment in equipment for germination chambers with control of temperature, light and moisture (Box 7.1).

Box 7.1

Laboratory hygiene

Cleanliness or hygiene is important in any laboratory work in order to prevent contamination of samples and make test results reliable. Dust, spiders' webs, soil, insect, etc. usually appear in laboratories; design and laboratory routines should help prevent these factors from becoming a nuisance and contamination. Hygiene becomes much easier if the laboratory is only used for seed testing, and there is no open connection to other functional rooms, e.g. via open or partly covered windows and doors. Interior building material should be made of easily cleanable material. Ceramic tiles and glass are some of the easiest materials to clean, and they are resistant to most chemicals and detergents; wood is less suitable as cracks and points of decay tend to collect dirt and fungi. Tiles are conveniently used for walls, floor and possibly tables. Glass is often used for cabinet covers. Drawers and cabinets are necessary in laboratories to store materials and samples not in use. Some simple laboratory routines help maintain a high standard of hygiene:

1. Keep files, labels and registration forms in an orderly manner during testing and avoid leaving such papers lying around in the laboratory.
2. Prepare all equipment before the start of each test, e.g. scales, glassware, desiccator, drying oven, and germination trays and cabinet.
3. Keep clean and dirty material distinctly separate.
4. Have separate boxes or containers for different types of waste material, e.g. organic (plant, soil, seed), paper (waste forms and labels), glass (broken items), chemical, and other material. Place the containers close to each other, mark them distinctly with labels and make sure they are used only for the type of material for which they are meant.
5. Have a special washing place for cleaning glasses, trays, tools, etc. Put washed items on a rack to dry.
6. Put washed, clean glasses, trays, etc. back into the right cabinet or drawer once they are dry: avoid using the drying rack as a storage place.
7. Avoid keeping things other than those necessary for laboratory use in the laboratory. Keep literature, instruction books and files in a separate cabinet.

Dispose of samples after the end of testing, and put aside containers and equipment to be washed.

Seed testing is carried out on a *sample*, which is a small representative part of the seed lot (Sect. 7.4). A *seed lot* is, according to the ISTA (1986), defined as a 'stated portion of the consignment assumed to be reasonably uniform'. 'Reasonably uniform' means that there is a relatively small within-lot variation, and that the test results represent a definite genetic identity and physiological history. What this more exactly implies is ultimately the judgement of the seed handler. Normally seeds of the same provenance and the same seed source, collected on approximately the same date, are bulked before processing and hence typically make up a seed lot. It is impractical to keep too many seed lots separate, and if a 'reasonable' level of uniformity exists, seed lots from different collections may be bulked into one larger lot. Provenances should, however, always be kept separate. It may also be reasonable to split up a large collection into smaller seed lots. This is typically the case for very large quantities of seeds (ISTA rules set an upper limit for the size of seed lots, typically 1,000 kg, or 5,000 kg for very large seeded species), or if different parts of the seed lot are exposed to different conditions likely to influence uniformity, e.g. during processing or storage.

A standard test has a certain *design* which describes how a test is carried out. It contains a number of *replicates*, which are similar tests carried out on the same number of seeds from the sample. Five replicates mean that five identical tests are carried out with an equal number of seeds. Replications allow calculation of statistical parameters such as mean and variance, and minimise the risk of an erroneous result.

7.4 Sampling

A sample must comply with the basic rule of being *representative* in any aspect of the whole seed lot to be tested. A sample should thus have the same average seed size, purity, moisture content and viability as the whole seed lot. Only if that is the case can the result be considered valid for the whole seed lot. Or expressed negatively: if a sample is *not* representative, then the quality of the seed lot cannot be concluded from the test results and the whole exercise is wasted. There are statistical methods to test whether sampling is representative: if sampling is ideal, the results of two individual samples should give the same result with regard to all tested aspects within the magnitude of statistical error. Thorough theoretical background and practical guidelines on sampling are found in the *ISTA Handbook on Seed Sampling* (ISTA 1986; Kruse 2004).

In a homogenous seed lot any sample is representative. In the real world a seed lot is never fully homogeneous. For example, seeds stored in bags or containers tend to stratify themselves according to gravity and any other physical features during handling (ISTA 1986). Stratification is a higher risk for seeds with a high variation in morphological characters. In a seed lot of pines containing seeds with and without wings, the seeds will typically stratify themselves with the winged seeds on top and the dewinged seeds at the bottom. Further, the environment in the immediate vicinity of the seed, which influences seed characters, sometimes differs according to where in the container a seed is located. In open bags the environment at the upper, lower or outer positions is significantly different from that at the centre of the seed lot. A sample taken from the top of a container may typically contain seeds which are on average smaller, lighter or drier than or have different viability from the average seed. Also impurities tend to be stratified by the impact of mechanical handling. If a seed lot contains a lot of inert matter, purity of seeds taken from the top and the bottom of containers or bags may vary significantly (Peterson 1987). Seed-borne pathogens or infected seeds are frequently not evenly distributed, since pathogens tend to multiply and infect neighbouring seeds where the environment is conducive to their development (Morrison 1999). Seeds stored in cold stores may form condensed water on the surface when removed, and it is thus advisable to allow them to reach ambient temperature before the container is opened.

Seed size, maturity and infestation are but some of the characters which vary in a seed lot. There are three ways to compensate for variation:

1. To homogenise seed lots before sampling. This is done by thorough mixing.
2. To compile samples from several subsamples representing possible variation across the seed lot.
3. To increase the sample size.

Sample size is a critical factor. The larger the sample, the greater the likelihood that it will contain seeds with different characters and thus represent greater variation. But in practice, individual tests rarely comprise more than four to eight replications of each 25–100 seeds.

In practice, all three methods are used: seed lots are thoroughly mixed, samples are taken from different positions and samples are significantly larger than needed for the tests.

7.4.1 Drawing Samples

In principle, there are two ways of drawing test samples: (1) by subsequent divisions after mixing (Fig. 7.2) and (2) by triers taking out samples from different parts of the seed lot and then mixing them into a larger sample (Fig. 7.3).

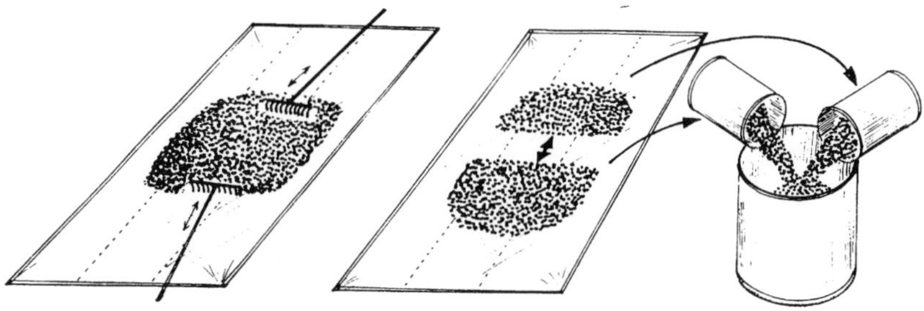

Fig. 7.2. Manual mixing of a seed lot prior to sampling. (P. Andersen)

7.4.1.1
Mixing and Division

The method of mixing largely depends on the quantity. Small seed lots of, for example, a few kilograms can usually achieve a high degree of uniformity by hand mixing in a bowl or bucket. Larger seed lots can be mixed by pouring them onto the floor and mixing them manually by shovelling or raking from side to side a few times. When the lot has been manually mixed, it can be divided into two or four equal parts. Two parts are put into a container and the lots are then simultaneously poured into a larger container (Fig. 7.2). The procedure may be repeated once or twice (Willan 1985).

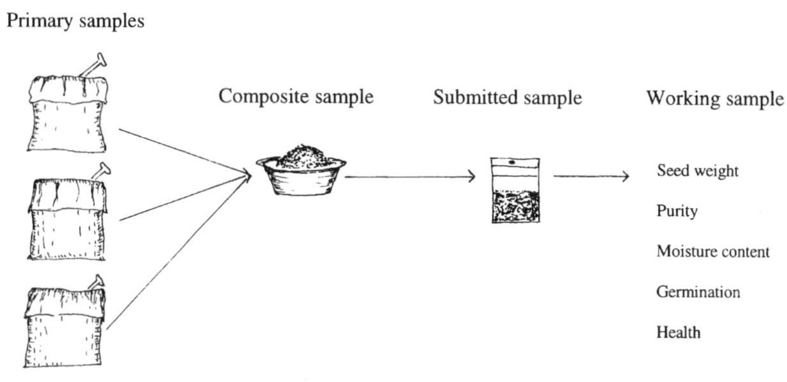

Fig. 7.3. a Procedure of sampling. Primary samples are drawn from the seed lot and mixed into a composite sample. The composite sample is reduced to a submitted sample to be forwarded to the seed laboratory for testing. In the seed laboratory, working samples are drawn for the individual test. The same working sample may be used for more than one test if the test is not destructive, e.g. purity test, followed by tests for germination or moisture content.

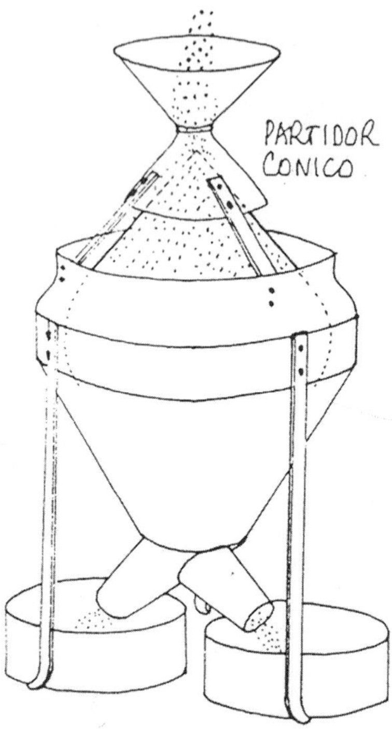

Fig. 7.3. (Continued) **b** Conical divider used for dividing composite samples into submitted samples. (M. Robbins)

When the sample size has been reduced by subsequent dividing, the now-small quantity is mixed by hand and further divided, e.g. using a mechanical seed divider. Seed dividers are mostly used for subdividing submitted samples during seed testing (see later), but can also be used in this connection for drawing samples from smaller seed lots of small and smooth seed.

Seeds stored in several containers and bags, yet belonging to the same seed lot, are mixed separately, samples are drawn from each container and then the samples are mixed before testing. Each part of the seed lot should contribute with a proportionally equal amount of seeds to the sample, i.e. a container with 20 kg of seeds should contribute about twice as much to the sample as one containing 10 kg.

7.4.1.2
Drawing Subsamples

Another way of compensating for stratification or unequal distribution of seed in a seed lot is to draw subsamples from different positions in the seed lot. This

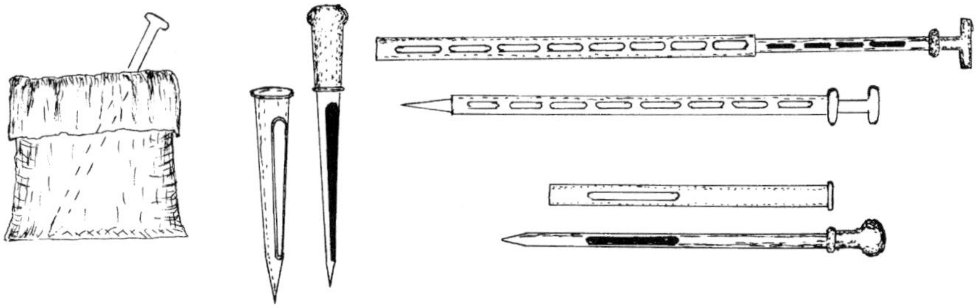

Fig. 7.4. Various types of triers used for sampling in large seed lots

is much used for agricultural seed, where there is a large quantity (often many bags), and mixing in practice is not possible. Sampling is here drawn by probes or triers, which is a device consisting of two tubes, one fitting outside the other as a sleeve (Fig. 7.4). (ISTA 1986; Edwards and Wang 1995). In practice, triers have limited applicability in forest seed testing because they are designed for large quantities of small and smooth seed. They are good for pines, most of the smaller-seed legumes (*Acacia, Cassia, Albizia*) and similar seed but cannot be used for large, winged and otherwise irregular seed. Most seed handlers find it easier to draw subsamples ('primary samples') manually by taking a handful, a cupful or another reasonable uniform quantity from different positions of the seed lot.

7.4.2 Reduction of Sample Size for Testing

Each small quantity of seed taken out from a single position in the seed lot makes up a *primary sample.* All the primary samples taken from different parts of the lot are then bulked or mixed into what in seed testing terminology is called a *composite sample.* Usually this sample is several times larger than the sample actually needed for testing. The quantity of seed necessary for seed testing depends on species and seed size (some tests are done on weight, some on number). For official seed testing, the ISTA (ISTA 1986; Poulsen et al. 1998) has issued prescriptions for the quantity of seeds needed by the seed laboratory to carry out standard tests. This makes up the *submitted sample.* The submitted sample is further reduced in the laboratory to a *working sample* according to the quantity required for the individual test (ISTA 1986; Poulsen et al. 1998).

The composite sample is reduced to the submitted sample during several divisions, each under observation of the same strict rules of maintaining it as representative for the whole lot. Because the quantity of the composite sample is relatively small, it requires little effort to maintain the samples in a homogeneous state. In its simplest form, seeds are spread on a plane table in an even layer. The sample can then be divided in halves once or several times by a ruler or other straight tool (Peterson 1987). Mechanical dividers (Fig. 7.3) are convenient accessories for unbiased reduction of sampling size. They can be used for most seed types other than large and winged types.

Tests are carried out on working samples from the submitted sample. Each test uses two or more subtests, which allows the mean and variance to be calculated. Errors in sampling methods prior to working samples will not be revealed in seed testing unless two different samples are submitted for testing.

To carry out basic tests of purity, seed weight, moisture content and viability/germination analysis roughly 2,500–5,000 seeds are needed, depending on seed size (ISTA 1996; Poulsen et al. 1998). However, for very small seeded species, a sample size of less than 1–5 g is impractical, although it may contain many more seeds than actually required. For large-seeded species, reduction of the sample size to a minimum of 500 seeds is acceptable (Box 7.2). The ISTA (ISTA 1986; Poulsen et al. 1998) proposes that samples submitted to the laboratory be twice the size of the total required working samples. Examples of the weight of some submitted samples are listed in Table 7.1.

Where figures are not available, the quantity can be calculated on the basis of the number needed for the two types of 'destructive tests' (tests in which the seed cannot be reused for another test), viz. germination test and moisture content test times two (in case a test has to be redone). The number used for each germination test is usually four replication times 50–100 seeds, and the moisture content test is normally carried out on 5 or 10 g of seed except for seed from large-seeded species, where a bigger quantity is needed.

7.5 Purity

Purity is, in common terms, an expression of how 'clean' the seed lot is, i.e. how much is seed and how much is something else. The purity changes during processing as inert matter and debris are removed from the seed lot. For large seed collected manually, the purity is often close to 100% and a test has little meaning, so purity tests for these seeds are often omitted (Hor 1993).

The purity of a seed lot indicates the percentage of pure seeds of the target species, and the percentage of inert matter and other seeds. Impurities consist of any non-seed material (leaf, flower, fruit fractions, soil, etc.), small fractions of seeds of the actual species as well as seeds of other species. The ISTA (1996, 1999, 2006) specifies the pure seed fraction may contain:

Box 7.2

Testing large seeds

Large seeds impose two genuine problems, their size and their usually recalcitrant physiology. 'Large' is a relative measure – in a flexible context it means something larger than a pigeon egg or an acorn.

Exact sampling devices such as probes are not useful in testing large seeds. Sampling is thus done manually. Large seeds with large appendices have a higher tendency to stratify.

Most large seeds are indehiscent fruits like nuts and samaras, or stones of drupes. Dispersal structures of nuts and samaras often remain firmly attached to the seed. This implies some difficulties when defining pure seed because wings, cupula and other appendices often make up a significant part of the seed. Very bulky appendices are often removed manually, e.g. breaking off large wings. Seeds with or without wings differ significantly in terms of seed weight and parameters where seed weight is used for calculations, e.g. moisture content.

Purity usually has little meaning for large seeds, where seeds are handled individually rather than as a bulk. Impurities are usually break-off parts of the fruit appendices, and their weight is very small compared with that of the seed.

The moisture content in large seeds is usually high because most large seeds are recalcitrant. Where dry appendices, e.g. dipterocarp wings, make up a significant part of the seed, the total moisture content appears to be relatively low. Moisture content analyses which contain a sufficient number of seeds to account for individual seed variation require a large volume of seed. Variation between individual seeds in a seed lot often varies significantly, e.g. owing to differences in maturity. A seed lot of *Dipterocarpus tonkinensis* in North Vietnam showed variation from some seed with 23% moisture content and others 42%. At least 25–50 seeds are usually necessary to get a representative sample, which falls within the permitted statistical error. An average seed of, for example, *Dipterocarpus tonkinensis* weighs about 100 g – a 50-seed sample thus weighs about 5 kg, which is a very large sample for standard seed testing. Two such samples would take up more than the normal drying capacity of an oven. In practice, smaller amounts of seeds are tested. Seed moisture meters are not applicable to entire large seeds because of their limited volume capacity, but the moisture meters can still be used for ground fractions, provided the devices are appropriately calibrated.

Germination of large seeds requires large germination capacity. The germination method and medium are in practice restricted to sand trays, and standard trays contain, for a 'pigeon-egg'-sized seed, about 10–15 seeds only. Height of the tray must be at least 25 cm, another enlarged standard.

Table 7.1. Examples of the weight of samples submitted for seed testing, given that each submitted sample contains twice the minimum weight of the working sample (approximately 2,500 seeds). (From Poulsen et al. 1998)

Species	Submitted sample (kg)	Species	Submitted sample (kg)
Acacia nilotica	1.100	*Dryobalanops oblongifolia*	24
Acacia Senegal	0.550	*Gliricidia sepium*	0.835
Acacia tortilis	0.420	*Khaya nyasica*	2.500
Afzelia quarzensis	25	*Khaya senegalensis*	1.600
Cedrela odorata	0.165	*Swietenia macrophylla*	2.400
Ceiba pentandra	0.500	*Tamarindus indica*	3.600
Dalbergia melanoxylon	0.840	*Ziziphus mauritiana*	3.500

1. Intact seeds of the actual species as well as dead, shrivelled, diseased, immature and pregerminated seeds.
2. Achenes[3] and similar fruits (e.g. samaras), with or without perianth and regardless of whether they contain a true seed, unless it is apparent that no true seed is contained.
3. Fractions of broken seeds, achenes, etc. which are more than half of the original size. However, seeds of, for example, legumes and pines which have the entire seed coat removed are regarded as inert matter.

For practical purposes, pure seed may be redefined as any seed which is likely to germinate plus entire seeds even if they are most probably dead (Fig. 7.5). The most important thing is consistency in the definition of the classification. Any seed regarded as 'pure seed' should be included in the moisture content and germination test. If an apparently dead seed is included in the pure seed fraction, it must also be included in the germination test as it will otherwise be an error in the calculation of number of viable seeds per kilogram, which is the ultimate figure of interest for the seed user.

During purity analysis, each 'pure seed' fraction (items 1–3 above) is separated from the working sample. Purity is expressed as the weight percentage of the pure seed fraction over the total weight of the working sample:

$$\text{Purity} = \frac{\text{weight of pure seed (g)} \times 100}{\text{total weight of working sample (g)}}\ \%.$$

'Other seed' in terms of weeds or different cultivars often occurs in crop seed (Table 7.2). The standard of separating impurities into inert matter and other

[3] Achene is the fruit of grasses.

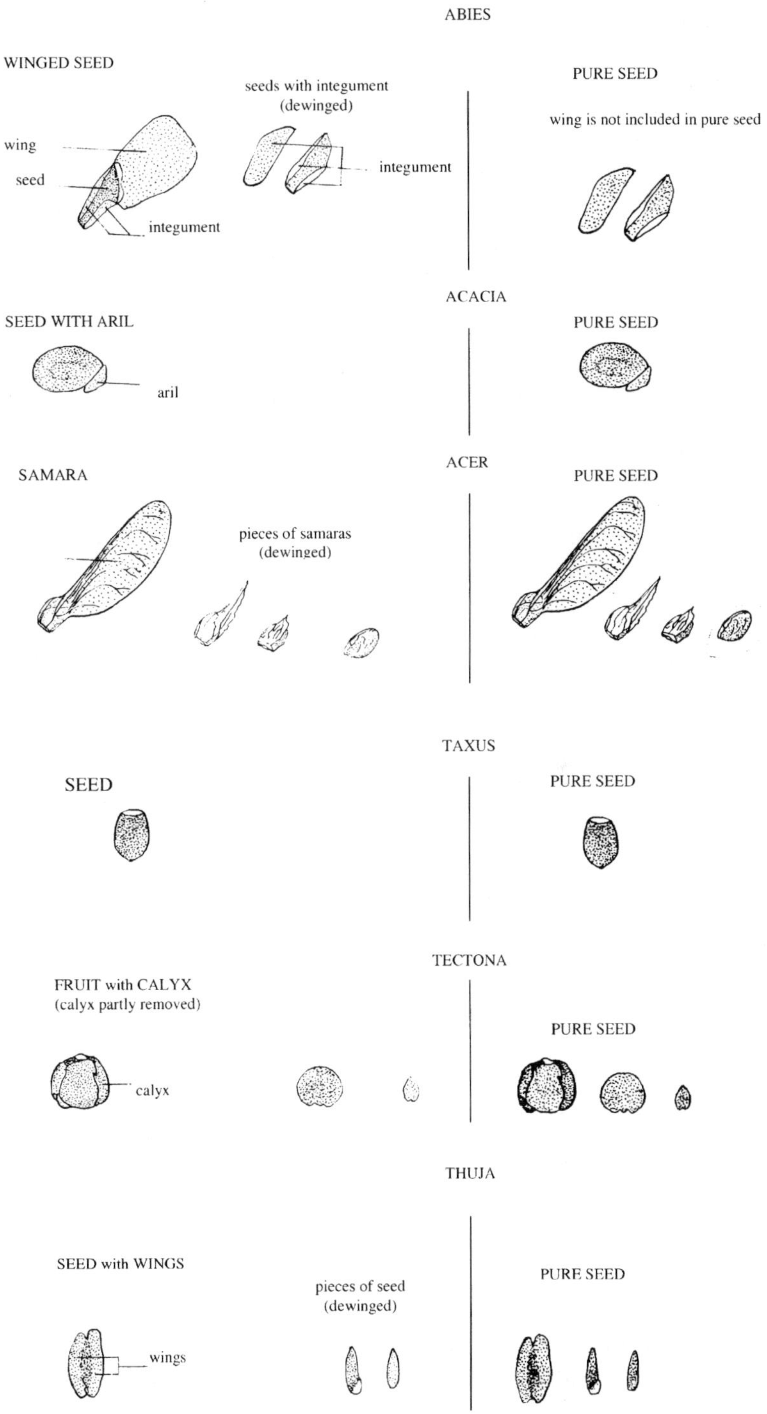

Fig. 7.5. Examples of 'pure seed' definitions. Pure seed is a seed that contain the morphological structures necessary for germination and protection of essential structures. Dispersal appendices like wings, hairs and arils are not physiologically necessary but are often closely attached to the seed. (From ISTA 1991)

Table 7.2. Example of fractions indicated in a purity test for *Pinus merkusii.* 'Other seed' occurred in this seed lot because some comes of another pine were accidentally mixed with the seed lot. In most cases the other seed fraction is not relevant. (Figures from Vietnam seed test)

	Weight (g)	Percentage
Working sample	60	100
Pure seed	54	90
Other seed	1	1.7
Inert matter	5	8.3

seed is thus relevant for agricultural crop seed (Poulsen et al. 1998). In forest trees, contamination with other seeds may occur during collection if crowns are entangled with those of other species or if there are seed-bearing epiphytes in the trees. Some bulk collection methods such as vacuum collection often result in contamination by several other species. Contamination can also occur during processing, e.g. if machines are not properly cleaned between processing of different species. However, as such contamination is relatively rare and often does not have greater implications, separation into two impurity classes is generally not included in a 'routine test'. Seeds with arils, wings or other attachments can change the purity status, if the seed appendices fall off during storage: as long as the appendices are attached to the seed, they are part of the seed; when they fall off they become inert matter.

7.6 Seed Weight

Seed quantity is typically indicated in weight, while seedlings are planted by numbers. Seed weight is thus primarily a conversion number. It is a necessary figure when calculating the number of expected plants from a certain quantity of seed and hence the seed demand for a given planting programme. Further, seed size may be correlated with vigour and hence may be an indirect measure of potential performance. It has been shown that high seed weight is often correlated with rapid germination and good seedling establishment (Griffin 1972; Sorensen and Campbell 1993). Seed weight can be indicated in two ways, viz. the number of seeds per kilogram or (for small seeds occasionally per 100 g) or the weight in grams of 1,000 seeds (Poulsen et al. 1998). The two figures are sometimes directly convertible from one to another. However, it should be noted that the weight in grams of 1,000 seeds as used in seed testing always refers to pure seed, sometimes for precision indicated as 1,000 *pure* seed weight (tpsw). When the number is converted to the number of seeds per kilogram, it also refers to *pure* seeds. Examples:

1. The 1,000-seed weight of *Eucalyptus camaldulensis* is 1.5 g. The number of seeds per kilogram is 1,000 seeds/1.5 g×1,000 g, i.e. 666,000 seeds. If the seed lot contains impurities (purity less than 100%), the figure should be multiplied by the purity percentage to give number of seeds per kilogram.
2. *Pinus caribaea* contains 3,500 *pure* seeds per kilogram. The 1,000-seed weight is 1,000 g/3.5, i.e. 285 g.

In standard seed testing, seed weight is usually calculated for eight replications of samples of 100 seeds (Fig. A.1 in Appendix 2). For very large seeds, the calculation is conveniently based on a smaller number in each sample, yet with the same number of samples. The high number of replicates in seed weight testing is necessary to account for the often high variation in this analysis (see variance calculation). The figure expresses the variation in seed weight within the sample.

Seed weight is subject to large variation within and between samples and seed lots. Variation between seed lots can be caused by genetic, developmental and environmental factors. Variation can also reflect differences in the measuring basis like pure seed and moisture content.

Seed weight analysis uses the same criteria as the purity test for what may be included as 'seed' in the calculation. Seed weight can thus vary a lot according to what is the pure seed basis, e.g. the seed with or without wings or other appendices. For species where there is a difference in processing, extraction can thus not always be compared directly. For example, the 1,000-seed weight of *Pterocarpus indicus* could be 300 g for entire fruits, 200 g if fruits are dewinged and 10 g if seeds are extracted. In *Sindora cochinchinensis*, seeds without arils weigh about half of those with arils.

The moisture content can influence seed weight since moist tissue has a higher density than dry tissue. For example, the seed weight of dewinged *Swietenia macrophylla* seeds roughly increases from above 2,300 seeds per kilogram at 5% moisture content to 2,400 seeds per kilogram at 9% moisture content. The difference is considerable in recalcitrant seed where the initial moisture content at harvest is often well over 50% and the moisture content after drying may be less than half, which also corresponds to a halving of seed weight.

7.7 Moisture Content

Moisture content is an indirect quality parameter since it is known that it has a crucial influence on storage and longevity. Analysis with a high or a low moisture figure can thus suggest a different storage fate. High demand for exactness is relevant for agricultural crops used for consumption, because it influences nutrient quality. Less exactness can be accepted in connection with reproductive material of forest seed. Seeds may absorb or desorb moisture according to the balance with atmospheric humidity (relative humidity), so to eliminate error caused by varying humidity, seeds should be packed in waterproof material as quickly as possible after sampling and should maintained within this packaging until the working sample for moisture content determination has been taken out. Moisture analysis should be done as quickly as possible to prevent errors caused by absorption from the air (Karrfalt 2001).

The conventional method of moisture content testing is the oven-drying method as described in, for example, ISTA (1999, 2006). This *direct* method can also be used for calibrating moisture meters for *indirect* measurement of moisture content. The indirect methods provide very quick results, which can be used as a guide during seed handling, e.g. to determine the necessity for further drying (Chap. 3).

The moisture content of a sample is the loss of weight when it is dried in accordance with the prescribed rules. It is expressed as a percentage of the weight of the original sample (ISTA 1996, 1999, 2006). This is the fresh-weight basis. Moisture content measurement contains the following components (Fig. 7.6):

1. The container (heat resistant) with or without the cover is weighed (M1). It is important to be consistent in the weighing with or without the cover before and after drying. If the cover is included, it is important to use the same cover in both weighings, as covers differ in weight. In practice the cover and container may be identified by a number or a letter.
2. Seeds are ground or cut into smaller fractions before drying to ensure that moisture can escape from the interior. Cutting may be omitted in small, thin-coated seeds. When making routines for new species, it is advisable to test individual samples with/without cutting or grinding in order to establish a practice for future testing of the species.

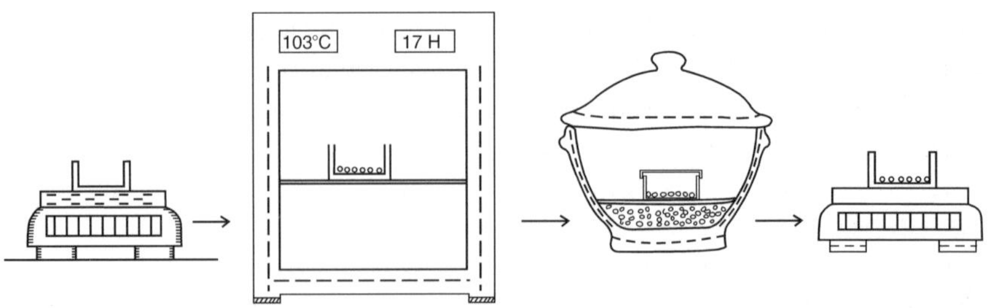

Fig. 7.6. Moisture content test. From *left to right*: weighing, oven-drying, cooling in a desiccator, weighing

3. Seeds are placed in the container and weighed together with the container (with or without the cover) (M2). The weight of the sample should be roughly 5 g. The sample should contain roughly at least ten to15 seeds, so the weight may be reduced for small-seeded species and increased for large-seed species samples.
4. Seeds are placed in an oven at 103±3°C for 17±1 h. This will remove all water from the seeds. For practical laboratory routines, moisture content analysis would be initiated in the afternoon, allowing the samples to be taken out next morning after approximately 17 h. Samples should be removed from the oven as soon as possible after the oven has been switched off in order to avoid moisture absorption during cooling.
5. If samples can be weighed immediately, they are taken directly from the oven to the scale. If there is a significant delay until weighing can take place, the containers with seeds are covered and placed in a desiccation chamber with (dry) silica gel to avoid reabsorption of moisture from the atmosphere.
6. After cooling, the seeds plus container (with or without the cover) are weighed again (M3).

The moisture content (fresh-weight basis) is calculated as follows:

$$\text{Moisture content} = \frac{(\text{M2-M3}) \times 100}{(\text{M2-M1})}\ \%.$$

Different parts of the seed can have different moisture contents (Sacande et al. 2004). Some seeds contain less moisture in the seed coat and pericarp than in the embryo and endosperm. Hence, processing may influence the moisture content both directly in terms of drying rate and indirectly in connection with extraction and possible dewinging. For example, the moisture content of a seed sample of *Khaya senegalensis* may be lower if the seeds are not dewinged before the moisture content test, since in that case the entire dry wing contributes to the seed weight, yet little to the total moisture content.

The method anticipates that the total loss of weight is caused by evaporation of water. In practice other volatile compounds such as oil and resin are also lost during drying, which, in seeds rich in these compounds, contributes to an overestimation of the moisture content. Despite this potential source of error, the oven-drying method is still used as a standard for these seeds, but the seed handler should be aware of the likely overestimation of the real moisture content when testing oil- or resin-rich seeds (Poulsen 1994).

The above methods all refer to calculation of moisture content on fresh-weight basis. Moisture content expressed as loss of moisture in percentage of dry weight (dry-weight basis) is sometimes used, especially by some

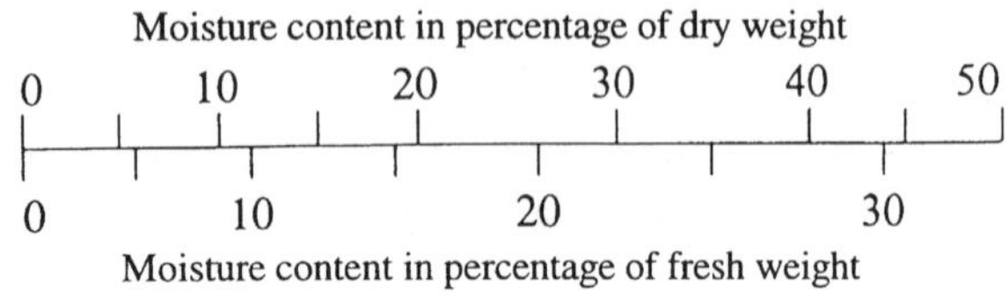

Fig. 7.7. Approximate conversion scale of moisture content calculated on a dry-weight basis converted to moisture content on a fresh-weight basis. (From Willan 1985)

researchers. A conversion scale is shown in Fig. 7.7. It should be noted that percentages above 100 may occur when calculations are performed on a dry-weight basis, while such figures are obviously impossible when calculations are performed on a fresh-weight basis.

7.8 Viability and Germination

The two words 'viability' and 'germination' are sometimes used synonymously – in seed testing there is, however, an important distinction. Viable means alive and a viability test indicates the percentage of alive or potentially germinable seed in a seed lot. A germination test indicates how many seeds germinate. Since germination is the ultimate target for the seeds, a germination test gives the best direct indication of the physiological quality (Box 7.3). A viability test is thus an indirect test and is only valid if there is a close correlation between viability and germination. Yet, viability tests are in some situations even better indicators of potential nursery germination than germination tests themselves. Viability tests can be better than germination tests in the following situations:

- Where seeds have a very short viability. The duration of a germination test is typically 2–6 weeks. For short-lived recalcitrant seed, significant loss of viability may take place during the test period. For such seed, the germination percentage obtained from the test is not valid for the seed lot from which it was taken because the viability of the seed lot has declined during the test period (Poulsen 1996).
- Where germination is delayed or suppressed by deep dormancy. If pretreatment has been insufficient to overcome dormancy, germination may be low even if seeds are viable. This indicates an insufficient pretreatment in the germination test.

Box 7.3

Dead or alive?

Seed ageing or deterioration is often a long progressive series of cytological and biochemical events that ultimately lead to the death of the seed (Roberts 1972, 1973b). Death is, by definition, irreversible, but many events leading up to the ultimate death are reversible. Repair of, for example, cell membrane damage is, for instance, a natural event in germination of seeds. Many seeds contain necrotic tissue but it does not necessarily influence viability. Dead storage tissue does not influence embryo structure, and damage to, for example, cotyledons, either by necrosis or by insects, can often be overcome or the tissue can be regenerated.

Whether a seed is able to recover from ageing also depends on the environmental conditions during germination. Optimal conditions enhance the turnover and repair mechanisms of aged seed, while the same recovery may fail under poor conditions.

The progressive nature of ageing and the ability to repair and regenerate, a feature very unique to plants, make it complicated to 'declare a seed dead'.

Since the ultimate objective of a seed is to germinate, this ability should set the criteria as to what may be considered a live seed. A seed that will not germinate under optimal conditions should be considered dead. The line is, however, not very sharp, as it is not always clear what is 'optimal'. Deteriorated seed may germinate slowly and produce poor and abnormal plants under normal conditions and not germinate at all under field conditions. Such seed should, from a seed quality perspective, be considered dead. Germination and vigour (stress) tests may reveal such seed.

Seeds that do not germinate in a normal germination test are not necessarily dead. Dormant seeds are fully viable but germination is constrained by some impeding factors, which must be overcome for germination to proceed. Viability tests such as TTZ stain are viable for dormant seed – the cells are alive and the reduction process to red formazan takes place. Such staining will also occur in immature seeds because they are alive but they cannot germinate.

- Where fast test results are required or germination is slow (some species take several months to germinate) the duration of a germination test may be inconvenient. Where a seed lot is to be disposed of soon after collection, there is often not enough time for a germination test.

A viability test may also be carried out for more practical reasons, e.g. if proper standard germination facilities are not available, e.g. temperature regulation. Because they are quicker, viability tests are usually significantly cheaper.

Although prescribed test procedures aim at creating concurrence between the two types of tests, some divergence may occur. For example, seeds deemed viable may not be germinable because of an advanced stage of deterioration (reduced vigour) or dead tissue in vital parts of the embryo. Another example is tetrazolium staining, which indicates live tissue (see later). Young (immature) seeds may stain normally by this procedure although they have not achieved germinability. On the other hand, viability tests are not always inferior to germination tests. In the two situations listed above, a viability test is preferred. Viability tests are also used as a supplement to germination tests in order to examine the character or quality of seeds that have not germinated during the standard test.

7.8.1 Viability Tests

Viability tests include methods in which seeds are visually assessed (do they look alive?) and methods in which at least some life processes are measured. The visual methods use the rationale that if seeds do not look obviously dead, they must be alive. None of the viability tests actually prove that seeds are germinable, only that they are (most likely) alive.

Viability tests include tests with cutting, tetrazolium, X-ray, excised embryos and hydrogen peroxide, which are described in the following subsections. Of these methods, only tetrazolium, hydrogen peroxide and excised embryo tests actually prove a life manifestation, in the first case as the activity of a metabolic enzyme complex, in the last as directly observable embryo development. It should be emphasised that all types of viability test are subject to some subjectivity in the interpretation of results. Viability tests are generally less applicable to very small seeds such as those of eucalypts, and in case of excised embryos, the method is practically impossible (Boland et al. 1980, 1990). Examination under a stereomicroscope greatly improves the possibilities of studying cross-sections of small seeds.

7.8.1.1 Cutting Test

Exposing and examining the interior of ungerminated seeds gives a clue to their condition (Fig. 7.8). This is widely used during seed handling to examine, e.g. maturity, insect damage and health. A cross-cut will show if seeds are empty, if they have an embryo, if the interior is underdeveloped or showing other distinct signs of damage of if they are damaged by insects. A cutting test will usually not show if seeds are aged. Desiccation-sensitive seeds are quite

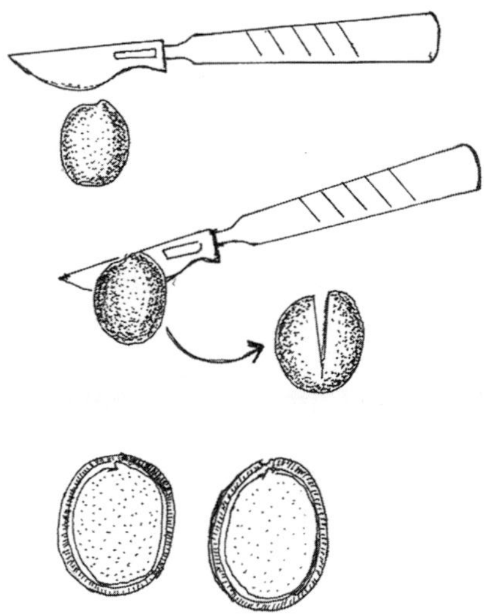

Fig. 7.8. A cutting test reveals the conditions in the interior of the seed

easy to cut; legume seeds, pyrenes or other types with a hard seed coat or pericarp can easily be damaged by cutting. It is advisable to soak seeds before cutting. Cutting tests are used to examine the conditions of non-germinated seeds after completion of a germination test. Aged seed will often appear rotten after a germination test, because they have lost their inert protection. Intact seeds with no apparent damage after a germination test are dormant.

7.8.1.2 X-radiography

X-radiography is a quick test to differentiate empty, underdeveloped, insect-damaged or physically damaged seeds from morphologically intact and healthy seeds by the aid of X-rays (ISTA 1996, 1999, 2006). A thorough description of the principle and practice in X-radiography in tropical tree seeds is found in Simak (1991) and Saelim et al. (1996). X-rays are electromagnetic waves with wavelengths of 0.05–100 Å[4] (visible light approximately 4,000–8,000 Å). The seeds are placed between the X-ray source and a photosensitive film or paper. When seeds are exposed to X-rays of low energy (longer wavelength, approximately 1 nm), an image (radiograph) is created on the film/paper. Photographic processing converts the radiograph into a visible picture. Since X-rays are non-destructive, seeds examined by the X-radiography method may also be used in direct germination tests.

[4] Wavelengths of light are usually indicated in angstroms or nanometres: 1 Å=0.1 nm, or 1 nm=10 Å.

X-radiography basically reveals the same type of damage as cutting tests, just without cutting. It is thus particularly relevant where cutting tests are difficult to carry out or interpret. The method does not require presoaking and does not harm the seed. Some situations where X-radiography is applicable are:

1. Small seeds, which are difficult to cut. Magnification of the photographic picture makes interpretation easier.
2. Empty seeds in pines, eucalypts and others, where the seed develops to full size even if it contains no embryo.
3. Insect-infested seeds where no entry hole is visible, e.g. legume seeds infested by bruchids or conifers or eucalypts infested with chalcids, e.g. *Megastigmus* spp. (Fig. 7.9b).
4. Hard fruit structure, e.g. drupes or samaras, where the pericarp or endocarp bears no sign of the presence or condition of the enclosed seed(s). X-radiography may reveal both the number of seeds in such fruits and their condition (Chayiyasit et al. 1990).
5. Hard-coated seed, e.g. legumes. X-radiography may reveal both the condition of the seed and possible multiple embryos.
6. Seeds where internal mechanical damage to the embryo may have occurred, e.g. during processing. Embryo damage can, for instance, be damage to the radicle, the plumule or attachment of cotyledons.
7. Seeds with shrunken or underdeveloped embryos, e.g. immature seeds (Fig. 7.9b).

Comparison between X-radiography and germination tests has mostly shown good correlation (Chaichanasuwat et al. 1990); however, the X-radiography test misses out some types of deterioration and therefore tends to overestimate viability compared with germination tests (Bhodthipuks et al. 1996a). Laedem et al. (1995) found good correlation between X-radiography and germination tests in *Dalbergia cochinchinensis* and *Pinus kesiya*, both of which had a high seed quality, while there was poor correlation for *Pinus merkusii* in which the germination percentage was low.

Application of specific contrast chemicals, e.g. $BaCl_2$, $AgNO_3$, NaI or KBr, to the seed before X-rays enhances the possibility of evaluating the viability of the tissue. Because these chemicals stain differently in live and dead tissue, the X-ray contrast method gives a different image of live and dead seed or seed tissue, similar to the tetrazolium test (Saelim et al. 1996; Simak 1991). Some experience is required for the interpretation of results. X-radiography is especially used in medicine, where advanced equipment has been developed during the last decade. The new technology has made X-radiography faster, better and

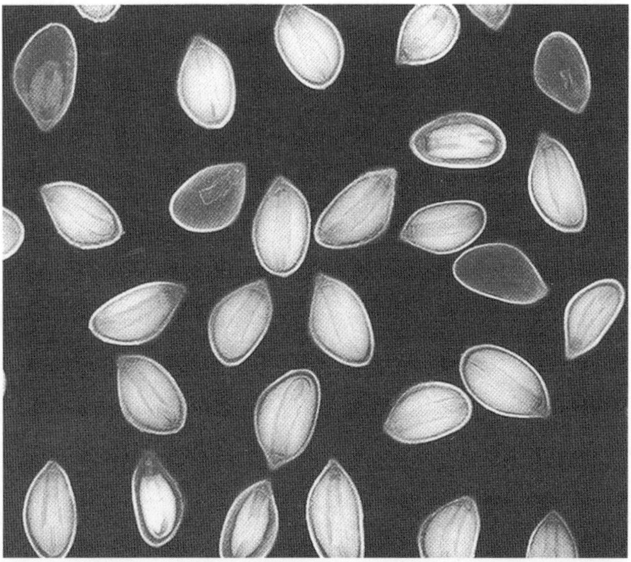
a

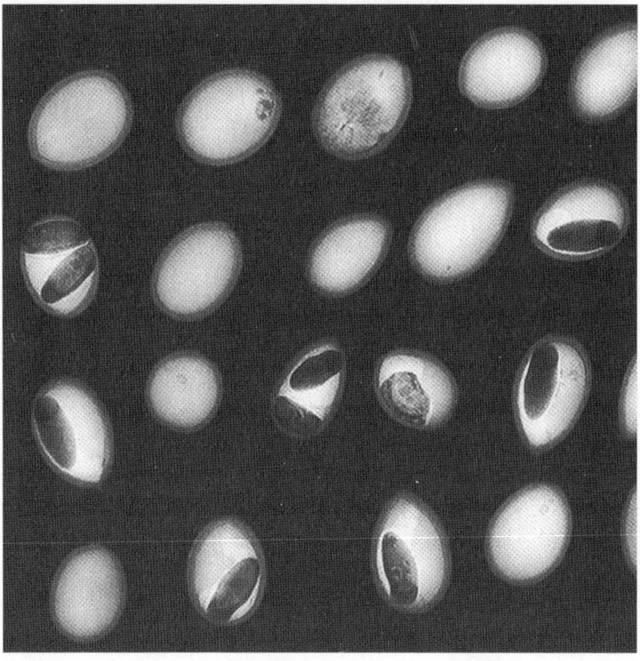
b

Fig. 7.9. X-ray radiographs used for seed quality analysis. **a** *Pinus kesiya*; some seeds with rudimentary embryos and some empty seeds. **b** *Albizia procera*; seeds infected with bruchid beetles. (From Saelim et al. 1996)

cheaper, but the equipment is quite expensive for seed testing (Karrfalt 2001; Craviotto et al. 2004).

7.8.1.3
Topographical Tetrazolium Test

This test is a further development of the cutting test, but the test uses a biochemical reaction to prove active life processes are occurring. Topographical refers to the differentiated examination of different areas of the embryos (Moore 1985; Yu and Wang 1996; Enescu 1991).

The principle of the tetrazolium test is as follows. Dehydrogenases are a group of metabolic enzymes in living cells which release hydrogen during the reduction processes in metabolically active cells. The hydrogen is able to reduce an applied pale-yellow solution of 2,3,5-triphenyltetrazolium chloride or 2,3, 5-triphenyltetrazolium bromide (TTZ) to a stable, bright-red triphenylformazan; hence, the formation of red formazan is an indication of dehydrogenase activity, which in turn is an indication of metabolism and hence viability. Because staining of tissue is local, it is possible to distinguish living (coloured red) and dead (colourless) parts of the seed (Fig. 7.10). Where dead (necrotic) tissue occurs only superficially in cotyledons, whereas the radicle stains normally, the seeds may still be viable. On the other hand, even small patches of necrotic tissue in the vital part of the embryo normally means that the seed would not be able to germinate. Interpretation of the staining pattern has been described for a number of species (Moore 1985; Yu and Wang 1996; Enescu 1991).

The tetrazolium test is especially useful as an alternative to the germination test for species that require long periods of pretreatment to overcome dormancy (e.g. several temperate species), but the test is also widely used as a quick

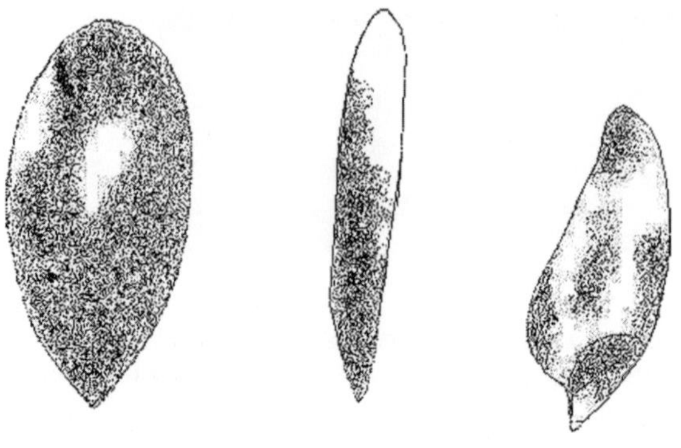

Fig. 7.10. Seed stained by tetrazolium. Bright-red areas (*dark areas* in this figure) indicate live tissue, pale areas are dead or necrotic tissue. Necrotic tissue in the radicle or embryo axis normally means that the seed is not viable, while some necrotic tissue in the cotyledons may not affect germination, and the seed is thus still viable

test for species with no or less complex dormancy. The method is in principle applicable to all seed types but does have some limitations:

1. As for all viability tests, interpretation of the results becomes extremely difficult for very small seeds.
2. The amount of chemical needed for very large seeds makes the test uneconomical for some species.
3. Immature but non-germinable seeds will stain red because of the metabolic activity.
4. Species with natural red cotyledons and embryo axis at maturity, e.g. Vietnamese *Madhuca pasquieri*, make it difficult to distinguish tetrazolium-stained areas from the natural colour of the seed.
5. Physically damaged but not necrotic tissue may stain normally.
6. Seeds with no embryo and inbred seed may stain normally although they show no or poor germination in germination tests. Inbred seeds may form abnormal seedlings, which would be disqualified in a germination test.
7. Seeds infected by fungi or bacteria may stain because of the metabolic activity of the microorganisms and not the plant cells. However, such fungi-infected cells generally stain dark brownish-red, not bright red as live sound plant cells do.
8. Slightly stained seeds or parts of seeds are difficult to interpret. Viable tissue should be bright red and pink ones are usually disqualified. However, there are transition areas where interpretation becomes quite subjective.

Seed embryos are likely to stain whether they are dormant or not; therefore, the result of the tetrazolium test is likely to include the three classes in the germination test: normal seedlings, abnormal seedlings and live but not germinated seeds (including hard seeds) (Poulsen et al.1998). Yu and Wang (1996) found good concordance between the tetrazolium test and germination in comparative studies of different viability tests on several tropical tree species.

7.8.1.4
Excised Embryo Test

In some seeds with deep, complicated or unknown dormancy, seeds that require long-term pretreatment and seeds with very slow germination, germination is restricted, delayed or impeded because of inhibitors in the non-embryonic structures of the seed (seed coat, endosperm, albumen). Such seeds can be made to germinate faster by removing the surrounding structures by excising the embryo.

The embryo is manually excised from the seed under aseptic conditions, placed on filter or blotting paper and incubated in a germination cabinet at the physiological optimum, e.g. 20–30°C, depending on the species. The result of the excised embryo test is the germination percentage under incubation (Johnson and Chirco 2005; ISTA 1999).

The excised embryo test is a transition form to a true germination test, since the embryos are evaluated on radicle development, which is essentially an early germination event. However, the germination process is concluded before the seeds develop into seedlings that could be evaluated for normal growth, as is done during normal germination tests.

7.8.1.5 Hydrogen Peroxide Test

Hydrogen peroxide (H_2O_2) has several effects on germination, both in breaking physiological dormancy and in speeding up the germination rate (Puntarulo et al. 1988). In seed testing it is used as a transition to a normal germination test, but where germination of the seeds is evaluated only after the first stage of radicle protrusion (Fig. 7.11).

The H_2O_2 test method is illustrated in Fig. 7.11. Seeds to be tested are initially soaked in a 1% solution of the chemical for 8–12 h. A small piece of the seed coat at the radicle end is then removed and the seeds are incubated in a H_2O_2 solution for a period of about 7 days. The solution is changed after about 3 days. Incubation is carried out under dark conditions as the chemical is sensitive to light. Seeds are considered viable when the radicles emerge from the cut end (Laedem 1984; Bhodthipuks et al. 1996b).

7.8.2 Germination Test

Germination potential is most directly determined in a germination test. Germination tests are carried out under optimal germination conditions of temperature, moisture and light, and with appropriate pretreatment to overcome possible dormancy for the species in question. These conditions are for most species listed in testing handbooks (ISTA 1999; AOSA 1997). Under such conditions everything that can germinate should germinate. The results do not necessarily reflect the germination in the nursery, where stress factors typically cause a lower germination percentage.

Standard tests are subject to strict prescriptions to pretreatment methods and germination conditions (ISTA 1999; ASEAN 1991). There are three main types of pretreatment:

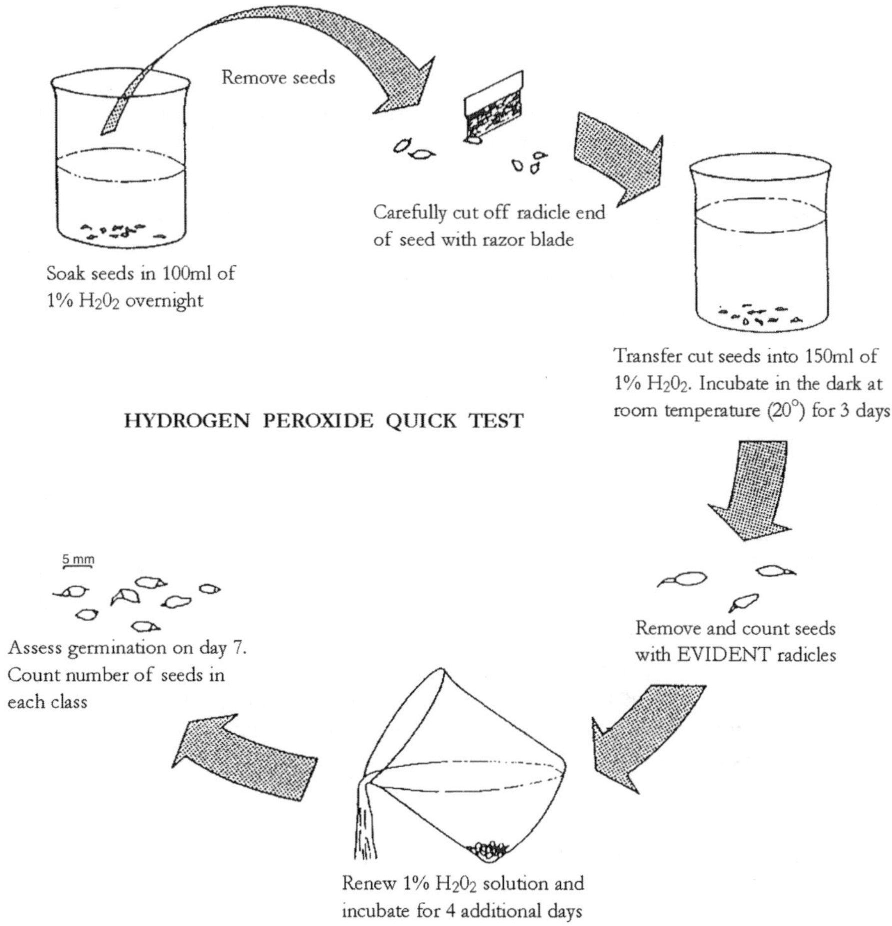

Fig. 7.11. Working procedure for hydrogen peroxide viability test. (From Laedem 1984)

1. Hard seed coat (physical dormancy), e.g. in Leguminosae. Pretreatment by scarification of the seed coat (or pericarp), for example by a wire burner or a file.
2. Thermodormancy, e.g. in highland northern species. Pretreatment by cold exposure of imbibed seed for a variable length of time, (depending on the species) in a refrigerator.
3. Chemical inhibitors, e.g. in many fleshy fruits. Pretreatment by thorough washing in running water prior to germination.

The germination conditions as prescribed by the ISTA include the following variables:

- Temperature (level and regime, e.g. constant day and night or fluctuating)
- Light (with or without light or period of day/night cycles, e.g. 12-h day + 12-h night)
- Substrate [sand (S), top of sand (TS), top of paper (TP), between paper (BP) and pleated paper (PP)] (the abbreviations in parentheses correspond to those used in ISTA prescriptions; see Fig. 7.12).

It is important that above variables fit the optimum for the species. The germination test should reflect the quality of the seed lot, not the germination conditions. Technically, temperature regulation is performed via thermostats connected to different cooling or heating systems. Germination cabinets have inbuilt temperature regulations but limited space capacities for operational seed testing of many seed lots and species. Germination rooms can be provided with the same temperature regulation as cabinets, e.g. via air conditioning or heating systems. Germination rooms without temperature regulations will inevitably imply some fluctuations. The problem is especially critical in seasonal climates with large annual fluctuations. Germination of seeds tested during winter months with low local temperatures will be very different from germination during hot summer months. Temperature regulation can also be necessary when testing germination of seeds from different climates. Highland species will often not perform well under testing in a lowland laboratory under ambient conditions.

Temperature regimes differing significantly from the ideal may cause germination to fail altogether, germination may be slow or seedlings can be abnormal. Cold temperature, e.g. cool winter periods, tends to slow down

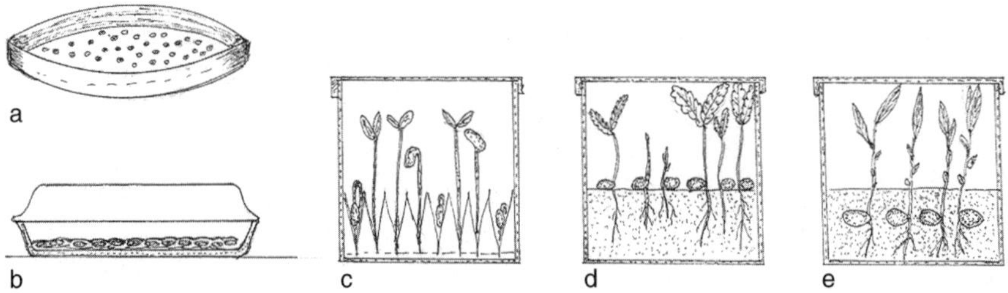

Fig. 7.12. Germination under different test conditions. **a** top of paper (TP), **b** between paper (BP), **c** pleated paper (PP), **d** top of sand (TS), **e** in sand (S). The abbreviations in *parentheses* refer to ISTA abbreviations

germination and frequently causes infections by fungi which tend to have a relative advantage over slowly germinating seed (Fig. 7.18).

Tropical countries have approximately 12-h light–12-h dark cycles with some annual variation depending on latitude. Daylight through windows follows the seasonal cycle. Additional light sources are regulated by timers. Relatively few species have a critical requirement for light regimes. Even photodormant seeds will germinate as long as they get some light. Dark-demanding species are best germinated under a dark cover.

Sowing substrates primarily have the purpose of providing sufficient water for the germinating seeds. When seeds have germinated, the different substrate and substrate arrangement provide support for the seedlings. Most small-seeded species are conveniently germinated on moist blotting or filter paper, or even tissue paper in transparent germination boxes or petri dishes (top of paper, TP). Germination development can easily be observed and counted. The method does not, however, allow development into the seedling stage, especially if germination is in flat petri dishes. In germination boxes, there is more space for vertical development, and germination in folded paper (BT) gives some support to the seedlings during development. Larger seeds are germinated in a sand medium, either on top of sand (TS) or covered with sand (S). Sand may here be river sand, vermicolite or any other material with good water-holding capacity, yet able to drain off excess water. Soil is not suitable in a laboratory because it always carries pathogens. It is important that substrates are absolutely free of pathogens, i.e. sand must be washed and heated in an oven to kill any possible organisms.

During germination there must be sufficient moisture, yet not so much that the seeds will suffocate. Where seeds are germinated on filter paper, connection of the filter paper to a water source will allow a continuous water supply during germination without the risk of overwatering. It is advisable to keep control the moisture content in the sand by applying a definite amount of water to a definite amount of sand. In this way, the moisture content is controlled and the same conditions can be created for all samples in a test. Germination boxes with perforated bottoms placed in water allow both excess water drainage and water absorption by capillary traction. Germination boxes or trays are always covered with a lid in order to reduce loss of moisture. If there is no continuous water supply, and if the germination period is long, occasional watering may be needed. Water should be clean, ordinary tap water. During germination the seedlings rely entirely on their own nutrient resources and no nutrient should be applied.

Fungi are occasionally a problem in germination tests and may interfere with the result. Seeds may be disinfected by a short (less than 10 min) dip in a 1% solution of sodium hypochlorite ($NaClO$). Fungicides should be avoided in germination tests since some types interfere with germination.

Germination is carried out on the 'pure seed' fraction, which on one hand excludes all 'other seeds' and inert matter, and on the other hand includes large, damaged seeds.

Seeds are sown on the appropriate substrate after possible pretreatment. The separation must be sufficient to ensure that the seeds and early seedlings do not touch each other; possible infection easily spreads from one seed to another in a moist warm substrate. Sowing in rows allows easy counting, another practicality to consider.

The germination/test time varies from about 1 week in some small-seeded and fast-germinating species to several months in some extremely slowly germinating species. There is a certain flexibility in time requirement. The ISTA rules also indicate the days of the first and the last count in order to standardise the duration of the test period for selected species. Germination is defined as 'the emergence and development of the seedling to a stage where the aspects of its essential structures indicate whether or not it is able to develop further into a plant under favourable conditions in the soil' (ISTA 1996). That means for seeds of trees a root system, shoot axis, cotyledons and terminal bud. The exact criteria of evaluation vary slightly between species, e.g. in eucalypts a seed is considered to have germinated when the radicle has developed normally and the cotyledons have emerged from the seed coat and have unfolded (Boland et al. 1980). In species with dehiscent cotyledons, there should preferably be one or two permanent leaves before evaluation.

In practice, the germination test is often terminated once the radicle has emerged and has a definite length, sometimes defined as the length equal to the length of the seed, or an exact measurement, e.g. 1 cm for most species. It is important to define the exact germination criteria, especially when they are used for calculation of germination speed.

Germinated seeds are counted regularly during the prescribed germination period from the indicated first count to the final count. Counting may be carried out once a week, for species with rapid germination every 2 or 3 days. Germinated seeds are removed from the germination tray once they have been counted (Fig. 7.13). This has three rationales:

1. To facilitate subsequent counting
2. To avoid germinants being counted more than once
3. To minimise the risk of possible fungal spread

Both 'normal' and 'abnormal' germinants are counted, recorded and removed during the period. At the end of the period, all non-germinated seeds are examined. Notice that sometimes more than one seeding may appear from a seed (Box 7.4). The final test result is grouped into the following classes:

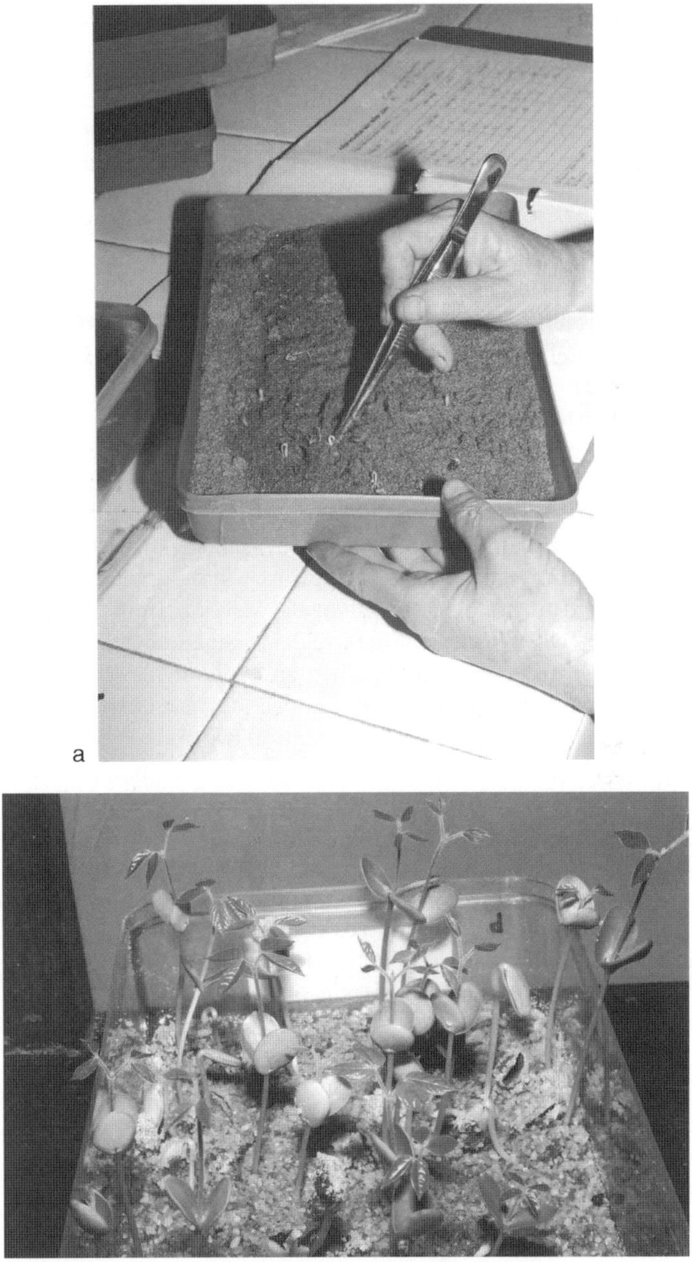

Fig. 7.13. **a** Counting of germinated seeds. **b** Germination of *Inga* species in sand. Seedlings are counted and evaluated after unfolding of the first pair of persistent leaves

Box 7.4

One seed – more plants: polyembryony and multiseeded fruits

Sometimes several plants appear from one sown unit. This can principally be caused by one of three reasons:

1. If the sown unit is an indehiscent fruit and there is no mechanical extraction of the seed, the fruit can produce several plants, viz. one for each morphological seed. This is a common phenomenon in pyrenes (stones of drupes), nuts and samaras. The plants are siblings and genetically not necessarily more related than any other seeds from the mother tree. Usually a fruit is formed if one ovule is fertilised, the other locules may remain empty (Fig. 7.14).
2. Morphological seeds can sometimes contain two or more sexually produced embryos. The genetic relationship depends on whether the embryos are produced before or after fertilisation
   (a) Simple polyembryony (in gymnosperms called archegonial polyembryony) occurs when two or more ovules are fertilised by different pollen (Dogra 1967). This can occur in several ways. If more than one ovule (up to four) develops from the meiotic division, the ovules are genetically different from each other, i.e. the plants will be genetically as different as independent seed. If several ovules develop from the one surviving mother cell after meiosis, the ovules are genetically identical.
   (b) Cleavage embryony occurs when an ovule divides/cleaves immediately after fertilisation and gives rise to two identical offspring. The offspring are genetically similar (clones/'identical twins').
3. Non-sexually produced embryos can be derived from apomixes or somatic embryogenesis. The former is seed development without fertilisation; the latter is embryo formation without previous sex cell formation. Somatic embryos are natural clones, i.e. genetically identical offspring.

Different types of polyembryony may occur in the same seed. This may include both sexual and somatic/adventitious embryos (Costa et al. 2004; Martinez-Gomez and Gradziel 2003). Polyembryony and multiseeded nuts or stones can give several plants from the same sown unit and thus a germination percentage over 100 (Fig. 7.14).

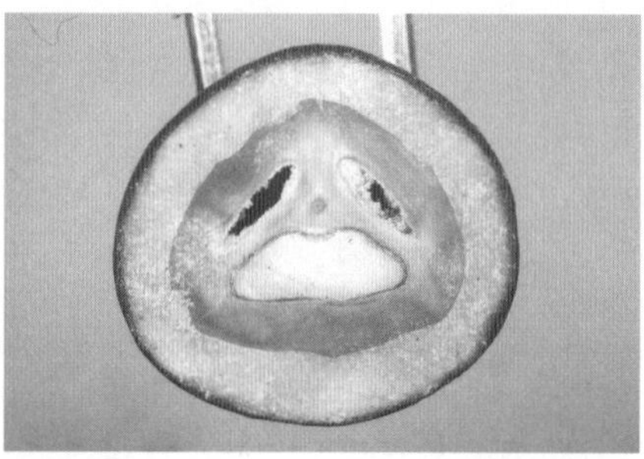

Fig. 7.14. Transverse section of *Canarium album* stone. The pyrene can contain up to three seeds but in most fruits one or two cavities are empty and the fruits thus only produces one seedling

1. *Normal germinants.* The cumulative number of seeds which have developed into seedlings of normal and healthy appearance with all essential structures of a seedling. This also includes seedlings where possible damage is caused by secondary infection.
2. *Abnormal germinants.* The cumulative number of seeds which have germinated during the test period but in which the seedlings show abnormal or unhealthy appearance, e.g. lacking essential structures such as cotyledons, or being discoloured or infected by seed-borne pathogens (primary infection).
3. *Ungerminated seeds.* Seeds which have not germinated by the end of the test period. These are grouped into the following the subclasses:
   (a) Hard seeds, which are seeds that remain hard because they have not imbibed (normally because of insufficient pretreatment)
   (b) Fresh seeds, which are seeds that have not germinated although they appear firm and healthy
   (c) Dead seeds, which are seeds that are soft, or showing other signs of decomposition
   (d) Other seeds, e.g. empty seeds (seeds without embryo)

Seeds in categories a and b may be germinable but dormant. Their correct status may be further determined by a viability test (e.g. cutting or TTZ). If the number of viable but not germinated seeds is high, a new germination test following a new pretreatment may be appropriate. However, germination failure can also be due to, for example, inbreeding, and in such case the problem is inert and cannot be overcome by pretreatment.

The final evaluation of the germination test is reported as the germination percentage or germination capacity, which counts 'normal germinants' (class 1).

7.9 Other Seed Testing

Several types of tests are available to elaborate or document seed physiological quality where standard tests are considered inadequate. None of these methods are carried out as routine tests by seed laboratories, and unlike for the tests described in the previous section, there are no strictly adopted standard procedures for conducting the tests and evaluating the results. However, in some cases results from germination or viability tests can be used for evaluating both seed vigour and health.

7.9.1 Vigour Test

Standard germination tests indicate germination under a set of optimal germination conditions. These tests are unable to detect quality differences between seed lots which show a large discrepancy between germinability in the laboratory and germination under field conditions. The vigour test is a more sensitive test which aims at detecting such differences. The term 'vigour' refers to *relative* strength or power, where a germination test indicates a plus/minus germination. Vigour is defined by the AOSA (1983) as 'those seed properties which determine the potential for rapid, uniform emergence, and development of normal seedlings under a wide range of field conditions'. The ISTA uses the definition of Perry (1981) as 'the sum of the properties which determine the potential level of activity and performance of the seed or seed lot during germination and seedling emergence'. Seeds which perform well are termed 'high-vigour seeds' (Perry 1981).

The rationale of vigour testing is that seeds undergoing natural ageing lose vigour at a faster rate than they lose viability (Fig. 7.15). Or even seeds that germinate under optimal germination conditions may have undergone some degree of ageing or deterioration, which affects their total physiological quality (Fig. 7.15). It is known that progressive ageing encompasses damage to cell membranes, chromosomes and other cell constituents. Minor damage is repaired during the initial phases of germination, but the greater the damage, the more complicated and the longer it takes to repair. Biochemical methods can detect some types of deterioration. In connection with germination, deterioration or decline of vigour may be manifested in reduced germination speed and reduced capacity to germinate under suboptimal conditions. Suboptimal conditions can logically not be standardised and thus only have meaning as relative values.

7.9.1.1 Germination Speed

The result of a germination test is the germination percentage (germination capacity), which states the percentage of a sample that germinated during the test period, but not whether germination occurred during the first or the last part of the test period. A seed lot where seeds germinate fast is considered a better quality seed lot than a seed lot with delayed germination. The speed of germination (or the velocity of germination or germination energy) can be calculated from the current germination record and is thus not a separate test. However, if the speed of germination is to be calculated, more frequent records are necessary than if only the germination percentage is recorded. In fast-germinating species, germination should be recorded daily or every

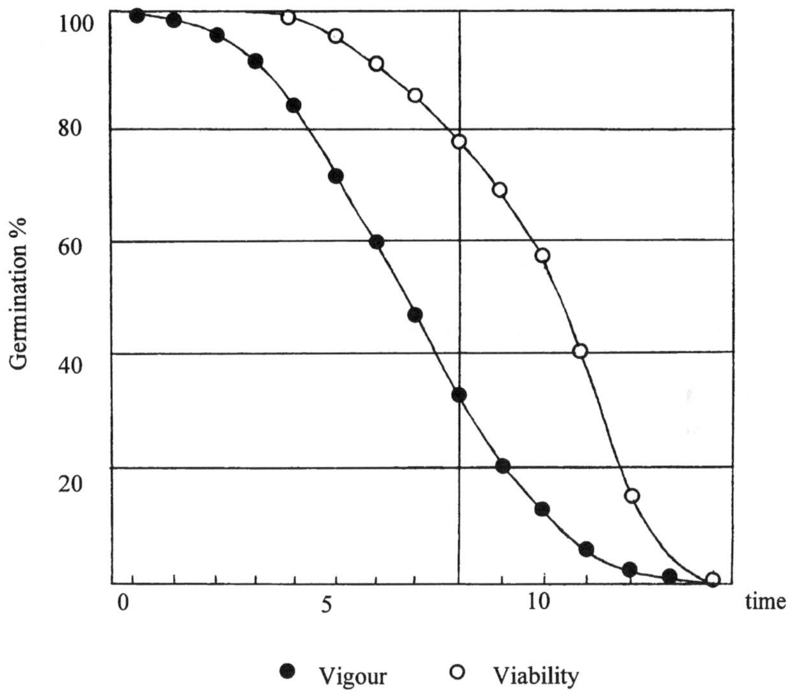

Fig. 7.15. Relation between viability and vigour over a period of time. The relation may be expressed as germination under a certain set of conditions after a certain ageing period. The viability curve represents germination under optimal conditions, while the vigour curve expresses germination under stressed conditions. For example, after 8 months' storage, germination under optimal conditions is about 80%, while under stressed conditions it is only about 30%. (Redrawn from Duangpatra 1991)

2 days; in slow-germinating species, weekly recording is sufficient. For calculation of the germination speed, it is important to establish very clear criteria for germination recording, i.e. when germination is considered completed. An example of records of two seed lots with different germination speeds is presented in Fig. 7.16. There are various alternative ways to calculate the germination percentage:

1. As the number of days to reach a certain definite germination percentage, e.g. 25%. This is 8 days for seed lot 1 and 11 days for seed lot 2 in the example.
2. As the germination percentage after a certain definite germination period. In the example, the germination after 14 days is 63% for seed lot 1 and 50% for seed lot 2.

3. As the percentage of tested seeds that germinate within a given period, and shorter than the total test period, e.g. 7, 14 or 21 days, depending on species.
4. As the percentage of tested seeds that germinate up to the time of peak germination, which is the highest number of germinants appearing in a given 24-h period. The peak of the records in Fig. 7.16 occurs at day 9 for seed lot 1 and at day 11 for seed lot 2.
5. As the number of days required to reach 50% of the final germination percentage. It is 7 days for seed lot 1 and 11 days for seed lot 2.
6. As the average germination speed over the full test period, based on daily counts.

7.9.1.2 Conductivity Test

Immature seeds have incomplete cell membranes and when immature seeds are placed in water various cell constituents will leak into the water. Leachate conductivity, measured with a conductivity meter, can be used as a measurement of the maturity stage (Sandeep-Sharma et al. 2003). As seeds age, cell membranes and other constituents disintegrate. During the early stages of imbibition, the cell membranes reorganise and damage is repaired. Until repair is complete, various leakages will take place. Delayed repair or failure to overcome such membrane

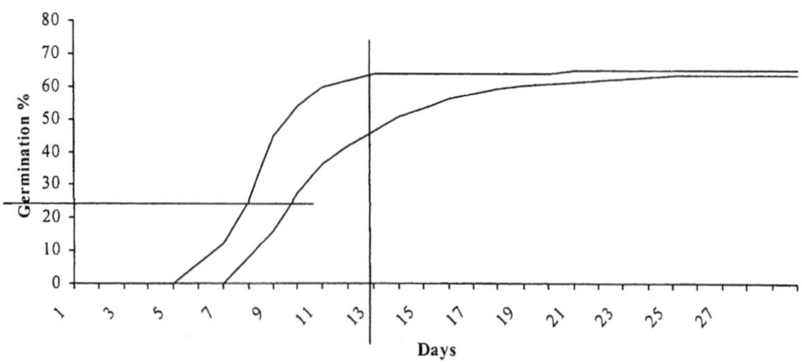

Fig. 7.16. Cumulative germination of two seed lots of *Pinus kesiya* during a 4-week-germination test period. The two seed lots have the same ultimate germination percentage but different germination speed (vigour). The highest germination per day (peak germination) in the cumulative curve is the time where the germination curve is steepest. As appears from the figure, the difference in germination speed may be expressed as different germination after a definite period of time, or the time to reach a definite germination percentage. (original data)

damage causes increased leakage of electrolytes from the imbibing seeds. Leachate conductivity of the water in which the seeds are imbibing is thus a measurement of ageing. Since high-vigour seeds are able to reorganise their membranes more rapidly and repair possible damage to a greater extent than low-vigour seeds; the relative magnitude of electrolyte leakage is an indication of vigour. Low conductivity indicates low electrolyte leakage and thus high vigour; high conductivity accordingly indicates low vigour (ISTA 1995).

7.9.1.3 Accelerated Ageing

Accelerated ageing is a stress test with two main applications in practical seed handling: (1) to predict the potential storage life of seeds and (2) to assess the vigour of a seed lot. Accelerated ageing is based on the assumption that if seeds deteriorate at a certain predictable rate under a given set of storage conditions (mainly as a function of temperature and humidity), then deterioration will occur much faster under suboptimal conditions of increased temperature and/or humidity. The basic assumption is that the same process of deterioration which takes place during a natural (slow) ageing period will occur during a short period when seeds are exposed to unfavourable conditions (Delouche and Baskin 1973; Elam and Blanche 1990; TeKrony 2005). Natural deterioration is thus simulated and 'compressed' into a short convenient test period. Under such conditions, high-vigour seed lots will show only a slight decline in germination, while low-vigour seed lots will decline markedly after exposure to accelerated ageing.

Accelerated ageing has proven to be a useful method to compare parameters related to seed deterioration; however, there is some divergence between natural ageing and accelerated ageing. For example, microflora (fungi) and the repair mechanism of cell organelles are two factors that apparently prevail more under accelerated ageing conditions than under natural ageing (Priestley 1986).

7.9.1.4 Stress Test

A number of methods have been used to evaluate seed and seedling performance under stressed conditions, and are all attempt to simulate single stress factors occurring under field conditions. These tests are germination tests carried out under suboptimal conditions and hence differ from normal germination tests. The type of suitable stress factor depends on the species and, except for the exhaustion test, the factor most likely to be encountered in the field. The methods of conducting stress tests are described thoroughly in ISTA (1995) and AOSA (1983). The methods are mentioned only briefly here.

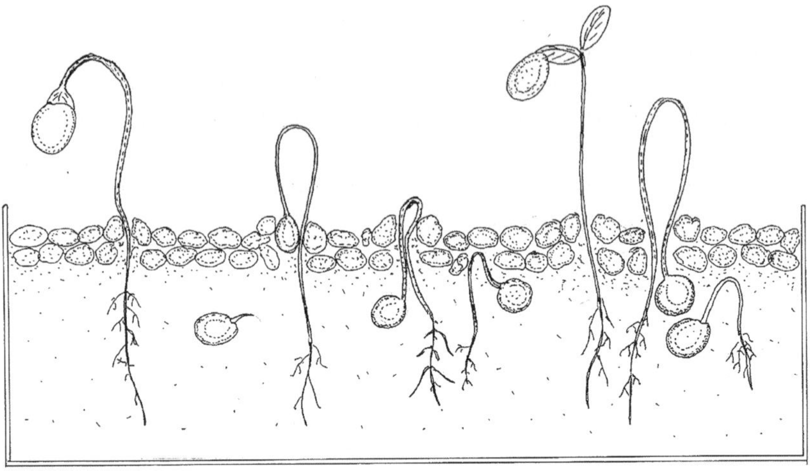

Fig. 7.17. Vigour stress test during seed germination. Seeds are germinated under a layer of gravel, which restricts expansion. Vigorous seedlings will grow through the gravel layer, while less vigorous seedlings are not able to overcome the impediment

During the *Hiltner* test, the ability to overcome physical stress is evaluated by germinating the seeds under a 3–4-cm-thick layer of crushed brick stone or gravel (Fig. 7.17). A cold test evaluates the ability of seeds to germinate and grow under low temperatures. This test is frequently used for temperate species but is also suitable for tropical and subtropical highland species. High temperature and water stress are other variable factors likely to reflect a difference in vigour.

The *exhaustion* test is based on the principle that seeds germinated in darkness do not carry out photosynthesis but rely entirely on nutrients derived from the seed. The germinants become etiolated, and after a specified test period the dry weight of the seedlings is measured. Seedlings derived from high-vigour seeds have the highest dry weight (Poulsen 1994). Obviously this method is not applicable to seeds that require light for germination.

7.9.1.5 Seedling Evaluation

The standard germination test only distinguishes between normal and abnormal germinants. Variations in seedling size and vigour are likely to occur within the category 'normal germinants'. Since initial growth is highly influenced by the seed, evaluation of seedling vigour, expressed, for example, as dry weight or evaluated in size classes, is in turn an expression of seed vigour. Comparison of different seed lots must obviously be carried out under strict observation of standard germination conditions and the duration of the test period. The latter implies that seedlings must not be removed during the test as is customary during normal germination evaluation.

Ageing is progressive and progressed ageing leads to the death of the seed. Minor repairs occur as a natural event during imbibition. Progressive ageing which affects the multiplication mechanism may be irreversible, i.e. it will affect the plants with some abnormal growth during the rest of their lives and will also affect next-generation seed. However, ageing affecting seed germination is not necessarily permanent. Slow starters may fully recover once new cells have formed.

7.9.2 Seed Health Testing

Seed health is indirectly revealed during viability, germination or vigour tests since infected seeds are often unable to germinate or appear non-viable when examined by X-rays, the TTZ test or other viability test methods, or they germinate slowly and produce poor seedlings. However, in some instances a more thorough examination of the presence and type of seed-borne pests and pathogens is relevant. Especially in international transfer of seed, where there is a risk of introducing seed-borne organisms together with seeds, a special health test may be required. Methods of seed health testing are described in Richardson (1990), and the ISTA rules (ISTA 1996, 1999, 2006) provide general guidelines on health testing.

The level of seed health testing varies from simple assessment of the infection rate by visual examination of the seed sample under a stereomicroscope, to thorough examination and species identification after incubation.

Assessment of the insect infestation rate may be carried out as part of X-radiography or a cutting test as described earlier. Where identification of insects is required, it is often necessary to acquire adult specimens. Since the insects present inside the seeds are often in the larval or pupal stage, incubation under conditions that promote their development may be necessary.

Fungal spores present on the surface of the seed may be detected by microscope examination of an aqueous suspension after washing the seeds in a small quantity of water (Desai et al. 1997); however, most fungal examinations requires preincubation under warm moist conditions. Incubation is normally conducted on blotting paper, sand or agar plates. Sand and blotting paper are used where germination is desired. After a few days' incubation, fungal growth may be visible on seed coats or as symptoms appearing on the seedlings (Fig. 7.18). It should be stressed that certain pretreatment methods, e.g. sulphuric acid or hot water, used for breaking physical dormancy should not be used in seed health testing because possible fungi will be killed by such treatment and hence this will interfere with the result. In the agar method, seeds are placed on the surface of a sterile nutrient agar gel during incubation. The fungi will grow and form a colony on the agar plate, and the fungi may be identified by colour and type of growth (Desai et al. 1997).

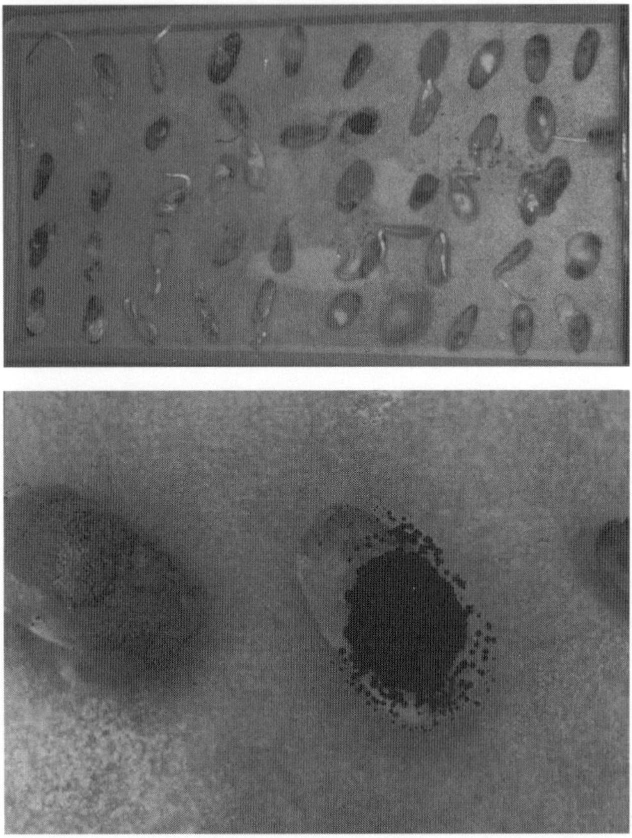

Fig. 7.18. Fungal attack of germinating seed. The fungi (*Penicillum* spp.) here attacked necrotic tissue after burning treatment. Seeds that germinated fast at a higher germination temperature were free from fungi. The better germination conditions helped seeds overcome attack by the fungi

Seed Supply and Distribution 8

8.1 Introduction

Market mechanisms have proven a quite efficient tool for development of, for example, agricultural production in most countries. Market mechanisms work when there is a good balance between consumers, producers and developers. For example, customers who can and will pay for improved products provide an economic incentive to producers (here farmers) to produce this product, which in turn provides the basic conditions for researchers to develop even better products (Martinussen 1999). Most production can be put into this simplified triangle, although the network is usually more complicated. Real systems contain many constraints and impediments, e.g. delay in improving products and limited resources. It has also been observed that development is often dependent on an economic 'kick-start', i.e. input of external resources and incentives that can initiate the cycle.

Supply of high-quality forest seed has proven reasonably effective, on a large scale, for industrial species and for species with short rotation. These systems are characterised by fulfilling the requirement for basic market mechanisms already mentioned. They are relatively simple networks, where the distance from the seed producer to the consumer is short (sometimes the same company). National forest seed supply suffers from several constraints, because it often has to work under conditions where common driving market mechanisms are not good or sufficient regulators. Some impediments are:

1. Skewed balance between production and use. Quality seed production of forest trees requires large production units, because trees require space. Fulfilling the requirements of the number of mother trees and the distance between them sets some high limits on the area of seed sources. For lesser-used species, production is far higher than demand. Field testing is also space-demanding.

2. High diversity in species. Tropical countries are generally species-rich, and if a reasonable diversity is to be maintained, many species must be included. With the aforementioned large production units, overproduction and thus wasted resources are inevitable.
3. Many seed users are resource-poor. Small-farmers are often unable to pay the production price, which includes high production, development and distribution costs.
4. Seed demand for many users is very low and rare. Long-rotation species need replanting at long intervals.
5. Distribution channels for forest seed are underdeveloped. Tree planting on farms is generally a relatively recent activity, and with the constraints of small quantities needed, good distribution systems for small end users/farmers have not been developed.
6. The time frame for production, improvement and testing is very long. It takes a tree generation of sometimes 20 years or more to develop a new cross of a long-rotation species. It usually takes several years to check whether the tree really produces better in the field. Under these conditions it is difficult to implement legal complaint processes, which in turn discourage quality production.
7. Tree planting/afforestation often contains several elements which are not directly measurable or economic, e.g. environmental aspects. It is sometimes difficult to get environmental funds for good-quality seed.
8. The definition of 'good-quality seed' is often very blurred, speculative and poorly documented. This pertains especially for lesser-used species, where there are no established seed sources, no trials and no other documented proof of quality.
9. Regulations and legislation to promote use of particular types of 'quality seed' are often poorly implemented and subject to corruption. Regulations are also often not in context with reality. It is often seen that the regulations' definitions of good-quality seed are simply not available for a number of species.

Seed supply contains three elements, viz. production, procurement and distribution. Compared with the many potential species growing and grown in the tropics, very few are currently distributed by seed suppliers (Kindt et al. 1997). This implies firstly that the number of available species is limited for seed users. Secondly, it implies that if species are not available from seed suppliers or grown, they often come from random sources and are often of poor quality. Seed supply systems should aim at overcoming diversity problems, i.e. extend the number of species and extend the distribution distance. Far more species can usually be supplied than are actually asked for, and the key issue pertaining to seed supply systems becomes 'how to get good seed to where it is needed'. Good seed here means quality seed in the broad sense: the best available which is suitable for the site.

Except for a handful of commercial/industrial species it is unlikely that a seed supply system can work efficiently without government interference. Efficient here means fulfilling diversity, quality and ubiquitous availability. Public interference should aim at concentrating on issues where market mechanisms are insufficient, e.g.:

1. *Research and development.* There must be a close link between the results of public seed research and the use of the results, e.g. by private seed suppliers.
2. *Decrees and regulations.* Regulations may include minimum standards of documentation, seed source use and other quality parameters.
3. *Organisational framework.* National systems usually consist of a mosaic of centralised and decentralised systems. A formal organisation may include representatives for both seed producers and users.
4. *Quality control.* This includes expertise in particular fields, e.g. certifying board for seed sources.
5. *Price regulation.* Pure market mechanisms tend to favour easy species at the expense of more difficult ones, e.g. recalcitrant species, and the result may tend to narrow diversity and favour species which are cheapest during establishment but not necessarily the preferred species in the field. Price regulation may include subsidies to less profitable species, to improve quality or to distribution chains.

Development constraints, which are part of the conditions in many tropical countries, become particularly evident in this part of the seed chain. Insufficient infrastructure and communication hamper efficient distribution. Demand and supply are often un-coordinated, with the result of many wasted resources at both ends. That is one reason why suppliers tend to concentrate on a few reliable species, where the demand is stable and the risk small. Forest seed supply is to a large degree built on confidence. The time frame of forestry makes effective control in particular of genetic quality very difficult. At the same time genetic quality is usually neither well defined nor well documented. Often reforestation and afforestation types which are not profitable in the short term are left with the public sector. That does not necessarily guarantee the use of good-quality materials. Confidence in the public sector is often lacking, as political systems with too much room for personal benefit in the public sector are unfortunately also prevalent in several tropical countries. Rules, regulations and control measures can, with sufficient public commitment, be made efficient. However, where corruption is a problem it seems that the forest seed sector is easily affected. One reason is the long time frame; another is the skill needed to implement regulations. It often hampers implementation of otherwise wise rules and regulations.

8.2 Distribution Patterns for Forest Seed

Supply models consist of three components: producer, distributor and user (Fig. 8.1). The number and geographical coverage, and their interaction make up the national seed supply systems. The key of the models is seed flow. Seed is here implicit seed of high physiologic and genetic quality and representing a broad range of species. Seed supply systems usually consist of different flow systems.

In large private and public forestry companies, the whole seed supply system, including product development and research, is included in one company. Companies with their own internal seed production and tree improvement are typically paper pulp factories and other forest industries with rapid turnover. Some private forest seed sectors have little involvement with the national tree seed supply system – they do not supply seed outside their company; they do not buy seed from outside. This closed type of seed supply inevitably implies some duplication and maybe waste of resources, yet the efficiency and site matching appear to compensate for these possible drawbacks.

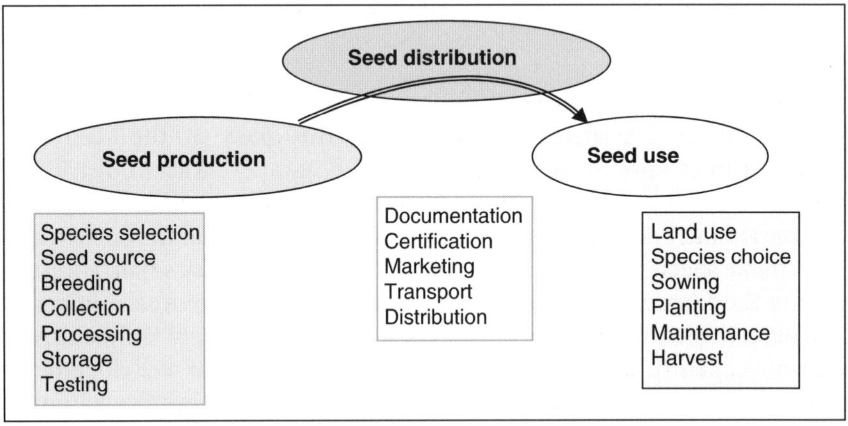

Fig. 8.1. Model of a seed supply system. The system consists of producers, distributors and users, each with various support functions. In, for example, companies with one or few species and their own test backup, the whole seed supply system may be a closed unit. In national systems, the seed supply system is a network of many stakeholders with different roles and functions

Historically, seed supply systems have been targeted large plantation programmes and have thus concentrated on few species, few stakeholders and large quantities (Graudal and Lillesoe 2006). Seed supply is here to a large extent commercial, with seed prices covering both procurement cost and research and development backup. Commercial seed suppliers may also service smaller customers with the main species. Typically, however, they concentrate on relatively large customers and do not have a direct link to farmers. Quality aspects are upheld as long as there are economic incentives linked to the output, e.g. the future harvest. However, many afforestation activities are carried out by private stakeholders on a contractual basis and without quality and diversity obligations. If such institutions do not have any long-term interest in the area, normal budgeting would favour the cheapest possible planting material.

National seed supply systems typically consist of a large network of larger and smaller stakeholders and often with a strong public/government share. The strong government interference is justified because the forest seed sector contains a number of necessary, yet unprofitable aspects, e.g. watershed management, rural development, resource poor farmers, forest rehabilitation and conservation. The time frame of forestry makes, e.g. improvement and field research of slow-growing species unprofitable. Donor input is in many cases an important stakeholder in the system, because donors tend to work completely independently of market conditions and with niche approaches such as diversity and gene conservation.

Today much tree planting in the tropics takes place on farms, and in some countries, the farmers constitute the most important group of tree planters (Simons 1997). Natural forests tend to have two fates: either they are exhausted by overexploitation or they become protected forests under some category (national park, protected forest, reserve, etc.). In any case natural forests as a tree and tree product resources and reserves tend to get out of reach of farmers. Wood and non-timber forest products consequently tend to be produced more on-farm as a part of the farming system, i.e. various types of agroforestry. Agroforestry is an old method of mixing tree and crop production (Nair 1993). Agroforestry includes traditionally a number of species with different roles on farms. It is thus significantly more species-diverse than other types of plantations. As the farming area in most tropical countries is generally increasing, e.g. owing to extension into what used to be marginal land, the area for on-farm planting in contrast to the area for natural forest resources is increasing. Finally, there is a tendency for the political development being towards providing better land tenure security for farmers so that long-rotation crops become feasible. All these aspects tend to promote on-farm tree planting.

Because of the historical background of seed supply linked to large plantation programmes and their tree improvement efforts, there has been a historical reason and tradition to concentrate seed supply in central units, e.g. national tree seed centres (Graudal and Lillesoe 2006). A number of technical factors support at least some volume in forest seed production. Good-quality seed is always produced in a relatively large quantity since any seed source must have a certain minimum size (to ensure outbreeding and a reasonable genetic base). Processing, documentation and other basic setups require a relatively large flow, and legal aspects are easier to control in larger units. These facts tend to favour large centres.

In contrast to this is the requirement of using seed sources close to the planting site and close to seed users. Appropriate source–site matching inevitably implies that seeds should be collected from a site that is reasonably similar to the user site. Transport to and from central large units is often a wasted resource. It has also proven difficult for large central centres to provide small quantity seeds directly to small end users such as farmers (Nathan 2001). This aspect tends to favour a more decentralised approach with many small seed units dispersed over a region. Although regional 'subunits' have been a prevailing model for national seed supply in large countries with regional site differences, the 'farmers approach' tends to emphasise decentralisation to a higher degree.

8.3 Commercial Distribution

Commercial distribution of tree seed is dependent on customers who need the product and are willing to pay the required price for it. Since large-scale afforestation is often subject to strong political priorities, these priorities strongly influence seed demand. Seed suppliers must thus be aware of the trends and priorities of the national forestry policy. 'Emotional' political trends typically shift between environmental approaches of favouring indigenous species and a production approach favouring fast-growing exotics. Seed suppliers range from small and medium-sized suppliers relying on domestic and often local customers in the vicinity of their headquarters/centres, to large suppliers often specialised in a few species such as teak, pines, eucalypts or mahoganies. International customers tend to be erratic/occasional. Commercial seed supply is a business with the same base factors as determine trade of other products. Some of the specific elements are listed in the following subsections.

8.3.1 Market Analysis

Market/customer analyses are much used by new as well as established seed suppliers. Such analyses should in particular address the following issues:

1. *Potential customers, their species preferences, niche species, amount of seed required.* End users have different priorities and demands, and it is always good to be able to supply the product a customer needs. In addition, you may influence his/her choice. Many seed users tend to ask for what they think is available. Preferences should thus not necessarily limit development of something new.
2. *Other seed suppliers, their profile and specialisation.* Other seed suppliers can be competitors, colleagues or partners. Competition may be an incentive for improvement, but it is also a loss for the one who loses – sharing can be a gain for both parties. Being aware that seed demand does not increase drastically in the short term, other suppliers' geographical coverage and species selection give some clue of the chances of establishing or expanding a niche.
3. *Potential market development/environment, e.g. new upcoming projects and political preferences.* Forest seed demand tends to develop over long periods of time. Afforestation alone tends to target uncritically any tree species, while political indications of environmental planting, rehabilitation or the like suggest some emphasis on higher diversity.
4. *What customers are willing to pay for.* Genetic quality is especially difficult unless there is already an established practice, a high degree of confidence and customers have a commercial interest in such quality. Many customers may be unwilling to pay for the particular measures taken to secure high genetic quality, e.g. seed source establishment and management and individual tree selection, as the result may only be seen in the distant future. The confidence of customers has to be extremely well established since the (genetic) quality can rarely be proven. Genetic improvement and good quality are always good trade words. However, advertisements strongly arguing for superior genetic quality may easily create expectations beyond what is reasonably realistic.

The result of a general survey or market analysis of customers and present seed supply should end up with an estimate of market share, i.e. how much of the present market that is likely to shift to a new seed supplier. In this should also be included the likelihood that tree-planting activities may increase or change character in case a new seed supply becomes available (Raae and Christensen 1997).

8.3.2 Product Development, Diversity and Species

Analysis of present tree-planting activities with a breakdown to species will show how much seed is actually needed from various species. A simple adjustment would be procurement of the most frequently planted species. However, in order for a new supplier to be successful, he/she must usually present something new and better than what is already available. The seed supplier may influence tree planting by making seed of new species and provenances available plus by ensuring a generally high seed quality. Suppliers must be aware that the cost of seed does influence some customers strongly. However, genetically improved seeds harvested in seed orchards is not necessarily too expensive even for farmers, provided they can buy them in small quantities. If seed orchards are available, it rarely makes sense to collect seeds from lower-grade sources unless these sources represent another type of genetic material, e.g. better matching to the potential planting site.

8.3.3 Seed Pricing

The price of seed is set by two key considerations, viz. what is the actual procurement cost plus necessary/reasonable profit, and what are customers willing to pay. On average the latter must at least be as high as the former – otherwise the trade will soon run into trouble. The best profit is where the difference between the ability or willingness to pay and the procurement cost is highest.

Procurement costs inevitably contain a high variation even within species because of differences in seed collection, processing, treatment and storage costs. In particular, climbing is very labour intensive and can thus influence procurement cost significantly. Seeds from large collected lots generally have a lower relative cost than those from small lots, as, for example, transport cost to and from seed sources is independent of quantity, and processing of larger quantities is usually more efficient. Excess collection will, however, add to storage cost and, if excess seeds are not sold, will be wasted effort.

The procurement cost of seed collected in natural forest is often high because of distance and difficult collection procedures. The collection cost in seed orchards is for the same reason often low. On the other hand, seed orchard seed may need to account for high improvement costs, whereas collection in natural forests or plantations is free or with some low fee to the seed source owner. In most low-grade seed sources such as plantations and natural stands, seed is a by-product. Calculation of the time (man-days/man-hours) during handling and the duration of seed storage helps in calculating the approximate procurement cost. An example of specieswise procurement costs from a number of Central American species is given in Table 8.1.

Seed procurement costs calculated per seed lot or species form a useful base guideline for price setting. Yet, a number of expenses cannot be allocated to particular seed lots but are part of the overall procurement expenses and must as such be added to the seed price. These include, for instance, equipment,

Table 8.1. Breakdown of seed procurement costs for some Latin American species. (From CATIE 1998)

Código	Especie	Procedencia	No. Personas	Horas/ día	Período	Distancia km	Costo Transp. $	No. Sacos	Peso frutos	Peso semillas	Costo rec. $	Costo proc. $	Costo equipo $	Costo adm. $	Costo total $	Costo kg $
068/97A	*Swietenia macrophylla*	Pocosol Guanacaste	2	8	7-10/01	845	228.15	14	379.0	25.7	172.00	67.65	210.74	128.5	806.64	31.38
070/97A	*Cedrela odorata*	Turrialba	2	6	6,7/02 24/3	307	82.89	2.5	57.4	3.3	129.00	10.70	27.06	16.5	266.15	80.65
012/97C	*Gliricidia sepium*	Naranjo	2	12	8-10 y 23-25/4	712	192.24	11.2	234.3	22.0	258.00	33.88	180.4	110.0	774.52	35.20
009/97D	*Erythrina berteroana*	Dota	2	8	29,30/04 5-7/05	477	250.83	8	37.9	6.1	172.00	35.99	50.02	30.5	539.34	88.41
072/97A	*Cedrela odorata*	Abangares	3	9	31/03-4/4	670	180.9	7	170.0	8.0	169.59	19.43	65.6	40.0	475.52	59.44
006/97D	*Pithecelobium saman*	Abangares	3	6	31/03-4/4	149	40.23	7	126.8	8.2	113.06	143.07	67.24	41.0	404.60	49.34
073/97A	*Tabebuia rosea*	Abangares	2	4	8-12/04	119	32.13	11	158.6	17.9	86.00	56.00	146.8	89.5	410.43	22.93
074/97A	*Tabebuia crysanta*	Abangares	2	2	8-12/04	50	13.50	1	15.3	0.88	43.00	23.47	7.21	4.4	91.58	104.07
013/97B	*Enterolobium cyclocarpum*	Abangares	2	4	8-12/04	788	212.76	11	190.0	47.5	86.00	21.97	389.5	237.5	947.73	19.95
071/97A	*Cedrela odorata*	Matambú	2	4	18-21/03	807	217.89	4	97.1	5.2	86.00	18.30	42.64	26.0	390.83	75.16
008/97D	*Erythrina poeppigiana*	San José	3	6	15,16/04	312	84.24	2	34.0	14.6	113.06	7.18	119.72	73.0	397.20	27.20

Codico: seed lot code
Especie: species
Procedencia: provenance
No. personas: number of persons
Horas/dia: hours / days
Periodo: period
Distancia km: distance, km
Costo Transp.: transport cost
No. sacos: number of sacs
Peso frutos: weight of fruits
Peso semillas: weight of seed
Costo rec.: cost of collection
Costo proc.: processing costs
Costo equipo: cost of equipment
Costo adm.: cost of administration
Costo total: total [procurement] costs
Costo kg.: Cost per kg

housing premises, vehicles and basic salary for staff. Seeds that deteriorate or for other reasons are not sold involve procurement costs, which must be covered by the seed that is sold. New investments and profit are other items that must be considered in the total price setting.

The procurement cost varies for each seed lot and from one year to another – sometimes the variation is considerable. Large price fluctuations are usually not appreciated by customers, and for administrative reasons it may also be inconvenient to handle too many price levels, e.g. with a breakdown for each species. Some seed suppliers prefer to compile species into price groups. The Australian Tree Seed Centre's pricing system operates with four main price classes, based on rarity of the species, ease of collection, relative abundance of the seed (these three relate to procurement costs) and the demand for particular species or provenances. The relative cost is reduced according to the seed quantity ordered, with larger quantities being relatively cheaper per unit (ATSC 1995). The costs of administration and processing of documents in connection with shipping are usually the same no matter whether the seed order is large or small; therefore, seed suppliers usually add a fixed handling fee to each invoice together with the individual freight cost (ATSC 1995; Gunn 2001).

The pure marketing pricing works in a free, open and commercial market. Many forest seed markets are far from free and are market-regulated, and many systems contain whole or semisubsidised elements. Central command economies like that of Vietnam used to have fixed prices for seed of each species. Public seed suppliers may work semicommercially but with government support for research, base facilities or salaries. Such systems will obviously influence the price level adopted by possible commercial, private suppliers.

Seed prices based on customers' willingness to pay are more 'supply and demand' driven. This implies that rare and highly demanded species and provenances may be priced relatively highly (and with high profit), while other species may be sold at prices just covering the procurement expenses. Political or strategic considerations may include low (and subsidised) prices for species that are to be promoted. These could be indigenous species, high-yielding varieties or new introduced species. The philosophy of subsidies is to overcome price bottlenecks if long-term benefits (including environmental) are anticipated.

8.3.4 Marketing

Species choice and the quantity of seed needed differ from one customer to another and the market strategy must be adjusted accordingly. The aim of the marketing strategy is to make the whole range of potential customers aware that seed is available for sale. The information is disseminated via advertising that may take different forms. Different customers are reached by different types of information and approach. Some examples are:

1. *Seed catalogues.* These provide the essential information about each species, normally listed in columns informing on provenance, geographical coordinates of the seed source, altitude, purity, viability, quantity in stock and seed price or price group (Fig. 8.2). This information suffices for estimation of the right quantity of seed and determination of the best seed source (provenance). The catalogues or seed lists are usually only distributed and updated annually and the figure of quantity of seed in stock is thus subject to change. Recalcitrant seeds are usually not kept in stock for long periods and seed catalogues should rather state that seeds of the particular species are available on request during a specific period. Seed catalogues may be distributed to all major customers, such as large nurseries, large private tree planters, e.g. wood industries, donor-funded projects plus former customers. Potential overseas customers should also receive catalogues. A disadvantage of catalogues is that they are relatively expensive both to print and to distribute. Some catalogue information like stock and viability may also be out of date quickly after distribution.
2. *Direct communication.* For main customers it may be advisable to keep in regular contact in order to ensure planting targets are met, the species required are supplied, the required time of supply is kept, etc. Two-way communication allows mutual adjustment of plans, schedules, etc. by both the customer and the supplier.
3. *Advertisements.* Advertisements in newspapers, in publications, on radio, on TV, etc. are designed to draw attention to the topic rather than to provide specific information. They must be followed up by requests from those listening to or seeing the advertisement. The messages in advertisements are normally short and rather inexact. Forestry magazines and network newsletters will often reach most target groups.
4. *Internet.* The development of the Internet has opened up a whole new area of marketing possibilities. Homepages contain almost endless options for information dissemination and bring together the strengths of many other types of media, viz. an inviting appeal, cheap distribution to anywhere and easy updating of information. Web-based seed catalogues have thus replaced many printed and manually distributed seed catalogues from larger seed suppliers. The main limitation and the reason why advertising cannot be limited to this type is customers access and active searching. Despite very rapid development of networks and use of computers in most parts of the world, many seed users, in particular small end users, are cut off. In addition, information is only available if users search and know how to search. A homepage is thus not a replacement for an advertisement.

Nombre científico Nombre común	CODIGO FUENTE	RN	TIPO FUENTE	PROCEDENCIA	PAIS	ALT. (m)	LAT. LONG.	TEMP. (°C)	PRE C. (mm)	GERM. (%)	TPO (**)	SEM/ VB (kg)	US $ (kg)
Jacaranda mimosifolia Jacaranda	BL075		FI–P	San José	CR	1040	09-57 84-07	20	1883	86		24571	60
Leucaena leucocephala Leucaena	1805			Chiquimula	GT	380	14-41 89-36			77	*	17101	30
Leucaena leucocephala	4104		FI–P	Liberia, Guanacaste	CR	150	10-37 85-27	26	1653	95	*	12780	30
Leucaena leucocephala	4262		FI–P	Nicoya, Guanacaste	CR	160	10-08 85-20	26	2232	86	*	14583	30
Leucaena leucocephala	4474		FI–P	Nicoya, Guanacaste	CR	130	10-08 85-27	26	2232	99	*	19583	30
Leucaena leucocephala	4507		FI–P	Nicoya, Guanacaste	CR	130	10-08 85-27	26	2232	100	*	18346	30
Leucaena leucocephala	4549		FI–P	Nicoya, Guanacaste	CR	130	10-08 85-27	26	2232	85	*	16285	30
Leucaena leucocephala	4592		FI–P	Nicoya, Guanacaste	CR	130	10-08 85-27	26	2232	70	*	17315	30
Leucaena leucocephala	BL049		FI–P	Nicoya, Guanacaste	CR	160	10-08 85-20	26	2232	98	*	17980	30
Pinus tecun-numanii Tecunumani	BL036			San Rafael del Norte	NI	1100	12-54 85-47		1800	45		57443	250
Pinus tecun-numanii	BL037			Las Camelias	NI	1000	13-46 86-18		1600	80		46411	250
Pinus tecun-numanii	BL038			Yucúl, Matagalpa	NI	1100	12-54 85-47		1600	69		57808	300

Fig. 8.2. Example of a page from a seed catalogue from Costa Rica. (From CATIE 1997–1998)

The design of advertisements and other public relations material to specific target groups implies both a consideration of details and the language used. For example, catalogues distributed to overseas customers must be in English, French, Spanish or other widely spoken language, and botanical (Latin) species names should be used rather than local names. Further, in international catalogues prices should be indicated in convertible currencies, like US dollars. These basics are obviously the same for Internet distribution. Information addressed to local communities, NGOs, etc. should use local species names, language and currency.

Small end users such as farmers are best addressed by existing distribution channels. Agricultural departments usually have a fairly well developed distribution system for crop seed, fertilisers, pesticides, animal fodder, etc. Vegetable seed is usually used in smaller quantities but is mostly distributed through agricultural dealers. Traditionally, agricultural distributors have not been much concerned with tree seed distributions:

1. Agricultural seeds are sown – trees are (usually) planted. It is more common that farmers buy (or acquire) plants rather than seeds.
2. Farmers are often unaware of genetic seed quality aspects of trees. They would thus rather collect seeds from a single tree in a neighbour's field or a roadside than buy from a shop.
3. The quantity of seeds used by each farmer is very small. A farmer with a small land holding may typically want to plant one or a few trees on his land. If there is good germination, he will get the plants he needs from a few seeds. If the plants grow well, they will take up their space for years ahead and need no replacement. An individual farmer may thus need a couple of seeds every 10 years! Such demand is just too small to maintain a normal market.

Seed distribution in small quantities to farmers imposes a real problem. On one hand, there is a demand which must necessarily be fulfilled. On the other hand, the market is small and insecure. Small-quantity distribution of tree seed via agriculture distributors was tried on a pilot basis in Nepal in 2002–2003 (Nathan et al. 2005). Seed of five species widely used for cattle fodder were distributed in small bags containing 50 and 500 seeds, respectively (Figs. 8.3, 8.4). The bags were supplied with propagation instructions and basic information on the origin/seed source. Brochures with more detailed information were distributed together with the seed bags, and distributors (Agrovet dealers) had some more detailed information about how to grow the trees. For the five species in the pilot project the outcome was quite positive – all seeds were sold and most of them sown. This was in an area which had not had access to seed

Fig. 8.3. Commercial seed distribution in small bags in Agrovet in Nepal. Small agriculture suppliers are distributed throughout the country and are regularly visited by farmers. This type of distribution chain has been shown to be appropriate for small orthodox seeds. (From Nathan et al. 2005)

before. Although seeds necessarily become relatively expensive because of packing material, including a printed attractive display, it appeared that price was no constraint in Nepal – the seeds became affordable to a wide range of farmers.

8.3.5 Managing Seed Stock and Sale

The seed stock serves as a buffer from which seeds are removed when demand is high (around sowing time), and where seeds are stored when supply is high (harvest) and demand is low. Seed lots should generally be dispatched in the same order as they enter (first in, first out), but obviously dispatch of seed should primarily attempt to meet customers' preferences for particular provenances. Further, for long-lived orthodox seeds, for which there is no significant difference in viability of fresh and stored seed, it will often be more appropriate to supply freshly harvested seed and hence avoid storage altogether, rather than to prepare all seed for storage by reducing the moisture content drastically.

Fig. 8.4. Packing for shipment of seed. Individual seed samples are packed in sales packages of moistureproof material. The total consignment is packed together in a large plastic bag and shipped in a cardboard box (transport package). (P.Andersen)

Several computer programs for handling stock and sale are available. Computer registration becomes a great help when many species, seed lots and orders are handled. The Microsoft Access database program is user friendly and easy to set up and manage.

8.3.5.1 Seed Orders

Orders should be as precise as possible to ensure the right seed is delivered. Orders may refer directly to the seed lots listed in the seed catalogues, or they may indicate the location of the planting site and conditions. In the latter case, it is left to the seed supplier to find the most appropriate seed lot suited for the planting site. Customers may indicate the exact quantity of seed in their order, or they may indicate the proposed planting area or the number of seedlings required, from which the seed supplier should calculate the quantity of seed required on the basis of purity, viability, etc. A seed order may also state the time the seed is wanted. The latter is important where orders are placed a long time in advance and where collections are made according to orders. In some cases a customer may prefer to receive the seed consignment as near as possible to the sowing time, e.g. where storage facilities are not available, or where fresh seed is preferred. In such cases where orders are not dispatched immediately, it is important that confirmations of orders are sent to the customers and reservations made for the quantity ordered. Orders should be listed according to the dates of their receipt. If the supply is short, e.g. owing to poor seed setting, customers who have placed their orders first must be given priority.

8.3.5.2 Labelling and Shipment Documents

Proper labelling is part of the basic seed documentation system. It is a good routine to use double labels for any seed lots. One label is fixed outside the bag, the other is put inside. This also holds for consignments of several species: a copy of the invoice may be fixed to the packet and another copy put inside.

Most seed suppliers use carbon copies or copy blocks of three to four different colours for invoices, e.g. a white copy is kept by the accountant, a blue copy is mailed to customers in advance of the consignment and a yellow (plus red) copy is sent with the seed. Labelling of seed lots and information to the customer includes basic information such as species, provenance, country, date of collection and seed testing results (date indicated) (see later).

8.4 Dispatch and Shipment of Seed

Transport of seed from supplier to user is typically undertaken by postal companies, shipping companies or airlines, which are paid and are responsible for safe delivery but not for the maintenance of the content. Modern transport time is relatively fast – yet conditions during transport can be fatal for sensitive material. Prolonged storage in airports often happens during international transfer. Deterioration in transit is often experienced during unaccompanied road transport where seed may lie in border transit stores for weeks or months, subject to both adverse climatic conditions and rodents (Campbell 1983). Requirements for legal documents can seriously delay shipment to the final destination. This is particularly encountered in international trade where slow and bureaucratic procedures in airports can delay release of cargo in the importing country. Special attention should be paid to any accompanying inoculants of microsymbionts, which are often both sensitive to storage environment and possible compulsive phytosanitary treatment.

Seed suppliers and customers mostly have no control of such procedures; however, being aware of possible delays, the seed supplier should prepare the seed consignment so that it will keep as long as possible, and complete possible documents exactly to speed up the delivery process.

8.4.1
Packing Material

Packing should protect the seed from both mechanical and environmental damage (Lauridsen et al. 1992). Mechanical damage to the seed itself is unlikely as seed coats usually provide sufficient protection against, for example, pressure damage. A higher risk is that frequent handling will cause tearing and other damage to the package, with the result that the seeds fall out. Packing material must be strong enough to resist damage during ordinary handling. Both moisture and temperature can be detrimental. Moisture damage is prevented by packing seeds in moistureproof material such as sealed polyethylene bags. This is, however, only suitable for completely dry seeds. Desiccation-sensitive seeds and seeds which are not dried below the respiration point should not be packed in completely moistureproof material. The temperature inside transparent plastic bags can rise dramatically if the bags are exposed to direct sunlight because of the so-called greenhouse effect: shortwave solar rays pass through the material easily, whereas longwave heat waves are retained. A combination of respiring seed and high temperature inside bags can initiate an accelerating deterioration process. Heating is prevented by storing transparent bags in some lightproof material, e.g. paper, and using some insulation of the seed consignments such as double-lined envelopes and corrugated cardboard (Fig. 8.5).

Seeds stored in small portions in laminated plastic bags with CO_2 (Chap. 4) may be shipped without repacking. This type of packing may also be used after weighing out desired quantities of seed according to seed orders. The CO_2 in laminated sealed plastic is absorbed by the seeds and hence functions as a vacuum packing. It is very convenient to handle and resistant to damage, but is only practical for relatively small seeds and quantities. Large seeds and large quantities of seed may be packed in gunny bags, wooden boxes, metal tins, drums or the like. The volume and the weight of packing material may be worth considering, especially for air shipment.

Small seed bags with few seeds have several other purposes except from being transport packages. The bags are used as a display and they are addressed to farmers with limited knowledge of tree seed handling. The bags thus also contain information about the tree and how to handle and germinate the seeds. For larger 'traditional' consignments, the latter is provided by separate seed lot documents. Some properties of various packing material for small bags are listed in Table 8.2.

Fig. 8.5. Tree seeds in small bags used in Nepal. The bags contain a picture of the tree and information about germination and tender on the back. For vegetable seeds, some companies insert a small 'window' so the seeds can be inspected. (From Nathan et al. 2005)

8.4.2 Seed Treatment

Use of pesticides should be minimised as their use could cause health risks for the customer, and legislation on the use of pesticides differs between countries. Treatment of seeds with a compound banned in the receiving country may cause importing problems. Seed treated with special dangerous and environment-damaging chemicals such as DDT and other chlorinated hydrocarbons are likely to be rejected in many countries with strict environmental legislation. If pesticides are used, the customer should be informed of the particular remedy (Willan and Barner 1993). Toxic pesticides should never be used for consignments for non-skilled customers, e.g. farmers.

8.5 Seed Documentation

Computerised systems have made the former tedious work of seed registration and distribution much easier. The revolution of computer software has provided user-friendly systems allowing users to create and modify systems according to their own liking and requirement. Geographical information systems (GIS) are especially suited to site–source matching aspects, e.g. location of seed sources and planting sites, delineation of planting zones/seed zones, and identification of possible 'holes' in geographical coverage of seed supply systems.

Table 8.2. Material used for small sales bags. The material must be durable and protect seeds against any environmental damage, e.g. moisture (desiccation – wetting), temperature (overheating), light, insects and fungi. The material must be strong enough not to break or tear and it must be easy to work with, e.g. it must be possible to close the bags and write on the outside. Notice that small tears in bags treated with insecticide or fungicide powder can be annoying and potentially dangerous, e.g. if the bags are handled in the same place as food items. Transparent material makes it easy to see and check seed conditions (e.g. if there is mould), but since also seed insects use their eyes to locate seed, transparency does a this drawback. Finally, the price of the material and possible necessary operation equipment is fairly important. All materials are available in different qualities

	Paper	Plastic	Aluminium foil
General handling properties	Easy to handle. Does not require special equipment. Is easily made into any size. Closed with glue. Easy and cheap to print and write on	Available in many small sizes from factories. Specially designed bags with zipper closing. Airtight closing requires relatively thick material and a sealing machine. Direct writing with special pens or print. Information can be put on a paper sticker. Transparency makes it easy and convenient to check the content	Usually used with a plastic lamination. Special design not readily available. Closing with special sealing equipment. Direct writing with special pens or special print. Information can be put on a paper sticker
Resistance to mechanical damage	Many qualities available. Good-quality paper is quite strong at normal handling	Many qualities with different thickness available. Good-quality material is very resistant to damage	Tears easily; therefore, often combined with plastic lamination
Permeability to air and moisture	Permeable to moisture. Water-resistant cover will reduce permeability significantly, but not enough for, e.g. CO_2 treatment	Airproof and moistureproof. Strong material impermeable to gas and applicable to CO_2 treatment	Impermeable to both. Can be used inside plastic or paper wrapping
Light and heat	Practically lightproof – slight permeability will prevent heating	Transparent. Direct sunlight will cause the content to heat up via a greenhouse effect	Lightproof
Insect and fungi	Most insects are able to tear through paper material but seeds will escape visually oriented insects	Insects can see seeds and penetrate bags made of thin plastic	In combination with plastic lamination, virtually insectproof

Databases are the core of seed documentation where virtually any relevant seed source and seed lot information can be stored.

Storage of electronic data has made tremendous progress as computer hard disks and CD-ROMs have increased data storage capacity manyfold in just a few years. Technical progress in the information technology has thus made technical management of documentation much easier. Unfortunately this implies a few pitfalls: outdated and no longer relevant seed documentation, e.g. old seed tests, tends to be kept, and data entry tends to be more comprehensive than reading capacity. In other words, seed records and registers are getting filled up with redundant and immense quantities of data, which is almost as bad as no record at all, because the records are not used for what they should be used for: to increase the use of good-quality seed.

Seed documentation is principally and in the short term information provided by the seed supplier to the customer encompassing all relevant product information, viz. genetic information (seed source records, origin and collection records) and physiological information (seed testing records). The records also have a longer and more scientific purpose, namely to help develop the national seed supply, tree improvement and conservation system. In addition, seed suppliers keep internal management records such as handling records, seed storage and dispatch records and customer registers both for trade purposes and as a part of their own reference system to improve seed technology.

Documentation to customers must be reliable, exact and short. Behind it is a comprehensive evaluation, assessment and monitoring system, which must be systematic and transparent. Filed detailed internal records serve as documentation of summary information provided to seed users. Widely accepted standards on nomenclature and procedures should be adopted. Yet, seed documentation also contains a large amount of confidence. A claim that collection has been done from 25 widely spaced individuals can rarely be refuted. However, rumours that collection is not done according to the documentation can have a serious backslash because it is also difficult to prove an improved standard.

As a link between supplier and end user, seed information becomes particularly important when seeds are dispatched to end users who are not in direct contact with those who collected and handled the seed. This is particularly so in international transfer of seed.

Seed documentation is often a requirement from official authorities and part of the whole seed legislation system. The objectives are twofold:

1. To protect customers. A 'minimum standard' may be set for any invisible documentation (including also, for example, food quality).
2. To regulate transfer over borders with possible spread of pest and diseases (importing country) and national genetic resources (exporting countries).

8.5.1 Documentation and Certification

Seed records contain notes of facts of different aspects of seed handling. Seed documentation is a compilation of the records into some standard formats such as seed source, seed testing or seed collection documents. Seed documentation is carried out by the seed supplier. A certificate is a piece of information accepted and approved by an official authority. It is thus also a legal document, a guarantee, which can be used in case of a juridical dispute. Certification pertains both to the truth of the content and to a certain standard of the method. For example, a certified seed test certifies that the test procedures live up to the high (usually international ISTA or AOSA) standard and the results are thus reliable (though not necessarily good). Certification of a seed source contains an additional aspect, viz. approval. A certified seed source has been assessed according to a set of standard criteria, the certifying body guarantees that the records are true and valid (could, for example be based on progeny trials) and on the basis of these facts the authority approves the seed source for seed supply. If not, a certificate will not be issued.

Certifying authorities can be permanent national seed bodies such as government institutions or institutions authorised and accredited by the government. This is typically the case for laboratory certification such as seed quality tests. Seed source approval is more often carried out by a board/committee consisting of experts from different institutions. The latter can be reasonable in order to avoid conflicting interests since seed sources often have different functions.

Certification schemes encompass different quality aspects:

1. *Genetic quality.* The lowest-grade quality certificate is 'certified origin' (equivalent to source-identified material); the highest is tested material. The latter is seed collected from trees of proven genetic superiority from seed orchards with controlled pollination and a subsequent progeny test.
2. *Physiological quality.* Where seed testing is carried out according to the ISTA (or AOSA) rules (Chap. 7) by special accredited seed testing laboratories, an ISTA certificate of seed quality may be issued. An ISTA seed quality certificate is hence an assurance that seeds have been tested according to the rules.
3. *Health and diseases.* Governments often require imported seeds to have been certified by an official authority, stating that the seeds are not infected or carrying diseases. Such a 'phytosanitary certificate' states that seeds have been examined and 'to the best of our knowledge' have been found free of special pests and fungi (Fig. 8.11).

Many countries have adopted their own certification schemes. Larger uniform standard modes of certification are adopted by, for example, the USA and OECD countries (Mangold and Bonner 2001).

8.5.2
Accession Numbers

A seed coding and numbering system creates a quick reference system across all handling procedures. Accession or reference numbers are identity numbers assigned to seed sources and seed lots for the particular purpose of referring to seed quality. Seed sources have different geographical names; their number and code as seed sources pertain only to that function. For seed lots, the number is the only reference identity. Although new software can deal with full names, codes or abbreviations are often convenient in databases – they save space in writing and make printout of summary tables with several columns on standard paper easier.

In seed trade, supplier and customer codes are applicable (Lauridsen 1994). Codes and reference numbers must be unique in the sense that two objects cannot have the same identity number. Although this is a common-sense observation, experience shows that frequent errors occur, in particular when different people or authorities assign identities. Once a system has been implemented, it can be quite complicated to change it. It is therefore quite important to consider carefully any potentially upcoming confusion. For example, a seed source may change status from a selected stand to a seed production area, after a culling. If the seed source reference name contains a code for the type of seed source, the identity number needs to be changed after upgrading. This requires in turn that the change is smoothly communicated to seed source users. Reference codes may contain letters and numbers, sometimes both.

Letters are often used as abbreviations or acronyms, for example, for species, seed source types or geographical names. It is important to be clear on what it refers to. SW could, for example, refer to southwest, south Wales, Swietenia or Schima wallichii, depending on the context!

Species codes are often convenient in connection with seed sources or seed lots. Seed lots always consist of one species only, while natural forest can be a seed source for several species. A species code can, in the latter example, be confusing. Most species can be identified by the first three letters of the genus and species epithets, respectively. SWIMAC can thus easily be recognised as *Swietenia macrophylla*. In a few species this system can result in confusion. *Eucalyptus microtheca* and *Eucalyptus microcorys* have the same initials and could rather be coded as EUCTHE and EUCCOR, respectively.

Regional codes are convenient in large geographical areas. Regional codes are used both technically as a quick identity reference and administratively in order to manage different seed source certifying authorities. Where large administrative regions exist, they are conveniently used as identity numbers – it is important that boundaries are exact. For example, Vietnam has been divided into nine ecological regions, viz. northwest (NW), northeast (NE), central-north (CN), Red River (RR), north-central (NC), south-central (SC), central highland (CH), southeast (SE) and southwest (SW). The regions have been adapted to follow provincial boundaries. Subdivision into regions has the advantage of avoiding number confusion; all seed sources in the central highland in Vietnam thus start with CH.

Seed source type can be referred to by an acronym, e.g. SSO for seedling seed orchard and SPA for seed production area. However, because of the aforementioned occasional change of status/upgrading, this author prefers to avoid linking seed source type to the seed source identity number.

Numbers should always be two to three digits starting with 1 (01 or 001) for each continuous series with a unique starting point (for seed sources typically 001 onwards, for seed lots years, e.g. 06-001 for the first seed lot of species XX in 2006). In many cases it is necessary to handle units before they are assigned a permanent number. This can be preliminary registration of a seed source or handling of a seed lot in the field right after collection. Preliminary numbers are here convenient, but it is advisable to use a different system from the permanent codes and numbers. A preliminary number could, for example, consist of the initials of the member of staff in charge.

Coding or numbering of seed sources is subject to much confusion because seed source systems often follow a national system, and since two sources must necessarily not get the same number, the process is frequently delayed by bureaucratic procedures. The official code must then be communicated back to the seed suppliers, who use them in their seed documentation system. In most countries, seed source registration and numbering are regionalised, i.e. provinces or regions ascribe numbers for seed sources within their region.

Many different national systems exist. The National Tree Seed Programme (NTSP) in Tanzania used, for example, a system consisting of three components for seed sources (Rasmussen 1992):

1. Two capital letters, referring to one of the three regional seed centres responsible for the seed source.
2. A three-digit serial number referring to seed source identity within the particular regional seed centre.
3. One capital letter referring to the type of seed source, e.g. Z (seed collection zone), I (identified stand), S (selected stand), A (seed production area), P (provenance seed stand) and O (seed orchard).

The seed source number MO149S thus refers to a seed source located in the area of the Morogoro regional seed centre, it is number 149 of their seed sources, and is a selected stand.

Since seed sources may be used by different suppliers, the registration systems must be adopted nationally. Seed lot identification and registration systems can be adapted to the preferences and need of individual seed suppliers.

The NTSP in Tanzania uses a three-component seed lot accession number system: (1) the seed source number (according to the system described already), (2) a two-digit number indicating the year of collection and (3) a capital letter indicating the number of the collection in the particular seed source. The seed lot accession number MO149S/91B thus refers to the second seed collection (indicated by 'B') in 1991 (indicated by '/91') from the aforementioned Morogoro seed source number 149, which is a selected stand (indicated by 'S') (Rasmussen 1992). The Danida Forest Seed Centre used a system where seed lots were numbered in sequence as they entered the seed bank, the year of accession being indicated. A seed lot with identity number 5320/92 means a bulked seed lot number 5320 received in 1992 (Lauridsen 1994). The Australian Tree Seed Centre uses a continuous five-digit number for each species (ATSC 1995).

Codes and reference numbers for seed suppliers and customers may include letters for the country code, region, project, etc. Accession numbers are particularly useful when linking different databases or registers, e.g. a list of all seed lots of a species, a list of all seed supplied to a certain customer or a list of all seed sources in a region. Once a reference code or accession number has been allocated, it should be entered on all seed forms, labels and databases pertaining to the particular object.

8.5.3 Documentation Systems

Most modern seed documentation systems are based on computer databases, and the possibilities and limitations of software influence the setup. A seed documentation system consists of a series of forms which are usually filled in the field or laboratory and later entered into the computer database. The purpose of the computer is primarily to ease data management, such as making cross-references, and writing out summary sheets. Both aspects must be taken into consideration when designing the documentation forms. The relevant information is the same as for the primarily manual systems depicted by Bowen (1980), Rasmussen (1992), Lauridsen (1994), ATSC (1995) and Willan (1985).

In databases each field represents a specific type of information, and sorting of data according to fields is a powerful tool in databases. When designing the fields it is thus relevant to consider which type of data management could be relevant later and from which criteria data should be managed.

An example of data management for seed sources could be a list of a given species, prioritised by seed source type (e.g. starting with seed orchards and

Box 8.1

Site–source matching – seed zones and geographical information systems

Ecological adaptation within species suggests that species growing under a certain set of ecological conditions will produce offspring that are also most adapted to the same type of environment. Provenance tests have confirmed this theory. However, a shortage of trials of most species has made it necessary to adopt some more generalised concepts of site–source matching. Seed zones are the largest units in seed collection. The concept was developed to serve as a broad guideline for transferring seeds for national plantation programmes (Barner and Willan 1985). Seed zones are based on climatic factors (temperature, precipitation), physiographic structure (topography, geology, soils) and geographical elements (vegetation). It is envisaged that the genetic variation of a species within a seed zone is less than that between two different zones (Albrecht 1993). However, as species occupy different niches, and have patchy distribution, the generalised seed zones make it difficult to predict species growth from these zones alone. Advances in geographical information systems have provided much more flexible tools for site–source matching. Geographical information systems are digitalised mapping programs with wide application in seed source management. Used in site–source matching, potential planting sites can be identified for a given species, provenance or seed source depending on the actual trial information available (Booth 1996; Javanovic and Booth 2002).

ending with plantations), and listed for locations. An example of seed lot data management could be a list of seed lots collected at a particular seed source, and listed according to collection time. A link between the two databases could be matching a given planting site with the best matching seed source with second priority to seed source type (genetic quality) (Box 8.1). Since details on seed source ecological factors do not appear in the seed lot database, the link to seed sources is required.

The order of fields in databases is not important, but it can be practical with a certain system. The layouts of some forms, which are easily set up in a Microsoft Access database, are described in the following pages. The examples are limited to a few relevant forms.

8.5.4 Seed Source Records

The objective of seed source information is to provide information on accessibility to seed of high genetic quality. Seed source information should thus include the following information:

1. *Growth site information, primarily climate and soil.* Information on climate would typically be obtained from the nearest weather recording station with similar climate (beware that large differences can occur in areas with strong gradients or patchiness, e.g. in mountains) The Global Positioning System (GPS) has made it possible to locate seed collection sites with great precision (Box 8.2). Terrain and soil information may be taken from observations/measurements on the site or from secondary information from, for example, official topographical and soil maps. In forestry the name 'provenance' refers to a relatively well identifiable place on a map, e.g. a lake, a hill or the nearest large town, where the stand occurs (Box 8.3).
2. *Genetic quality information.* Seed sources are classified according to the national system (which should preferably be harmonised with the international system). The type of stand gives information on the genetic history and sometimes quality. Neither natural stands nor most plantations have been subject to selection for quality. A natural stand often has a wider genetic base than a plantation unless the natural stand is a small isolated group of trees, e.g. a small relict of a previous large stand. Both natural stands and plantations can be upgraded to selected or certified seed sources or seed stands. The latter involves some selection since inferior individuals must be eliminated to upgrade the average genetic quality. Where a seed source has been established on the basis of genetic results from an improvement programme, the seed source is referred to as a seed orchard. Both the generation (first, second, etc.) and the mode of establishment (clonal or seedling) are indicated in the seed source record (see also the discussion on seed sources in Chap. 2). The fill-in form version contains a list where the selected type/s is/are 'ticked off'/marked. The database is a multiple-choice list, where only the selected type will appear (note that it is not possible to make more than one choice in the database system). Information on possible progeny/provenance tests is referred to (it is possible to link this system to a trial database).
3. *Stand description.* This information is a supplement to the genetic information, but information about, for example, seed productivity can be derived. Some countries and organisations prefer photographic documentation on this point. Pictures visualise, for example, stand conditions and can in some cases be a good supplement; however, they has a few drawbacks and limitations:
   (a) Pictures of forests are always quite difficult to take because of light and angle conditions. Photographers thus tend to take their pictures where they can get the best shot, which is typically the open glades or hillside positions. This may give a wrong picture of the average of the stand.

Box 8.2

Finding a position anywhere – the GPS

The Global Positioning System (GPS) is a modern satellite-based navigation system, originally developed and intended for military applications. Since it was released for civil use in the 1980s, it has become the universal system for positioning and navigation.

GPS is a satellite system, based on a network of 24 GPS satellites placed into orbit by the US Department of Defence (Fig. 8.6). The satellites circle the earth in 12 h in a very precise orbit and transmit signal information to earth, where it can be received by GPS receivers (Fig. 8.7). The receiver uses triangulation of the signals to calculate an exact position. Receiving signals from at least three satellites, the receiver can determine the user's position to latitude and longitude. Altitude determination (three-dimensional position – latitude, longitude and altitude) requires reception from at least four satellites.

GPS receivers have numerous applications, which are all based on calculations from the determination of a position. Modern GPS receivers display the position on the unit's electronic map. When the user moves, the GPS receiver can calculate track and speed. Many other functions, like sunrise, sunset, distance to a certain destination and trip distance, are inbuilt functions in many GPS units. Particular interest in forestry is a planimeter function, a function which can calculate the area of, for example a seed source after walking around it.

Fig. 8.6. The GPS satellite system consists of 24 satellites that are orbiting the earth at about an altitude of about 10,000 miles. The satellites travel with a speed of about 11,000 km/h and make two circuits in less than 24 h. The GPS satellites are powered by solar energy and small rocket boosters on each satellite keep them flying in the correct path

(Continued)

Box 8.2

Finding a position anywhere – the GPS—Cont'd.

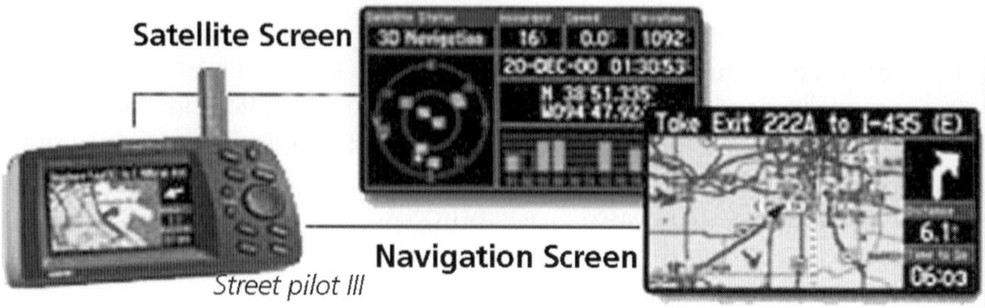

Fig. 8.7. The GPS receiver is a small, portable or handheld unit with an antenna, a screen and various function bottoms. Modern GPS receivers have USB ports that allow connection to computers and thus data transfer to further management in the computer. GPS receivers are battery powered. (Text and picture source: Garmin)

GPS receivers work under any weather conditions, though certain atmospheric factors can affect their accuracy. Speed of operation and accuracy vary with brands but are continuously improving. Modern GPS receivers will display positions after 10–15 s. They use up to 12 parallel channel receivers, which make them accurate to within 15 m on average. With use of additional ground-based signal systems, which are operating in some areas, GPS signals can be corrected to an average of 3–5 m. Such accuracy is sufficient to locate single 'mother trees' in a seed source. Both speed and accuracy may be influenced by 'shading' objects like buildings or canopy trees. This problem has, however, been reduced significantly as GPS receivers have become increasingly sensitive to signals

Box 8.3

Provenance or origin

The provenance name is, by definition, the place the seed is collected no matter whether it is from a natural stand or a planted source. Since trees carry their genes with them, exotic trees may rather reflect the growth conditions of the original growth site than that of their new site, where they are actually growing. Adaptation takes place over several generations, and introduced populations develop land races which can possess special and distinct characters. Both the original place where trees came from and the actual growth site thus have an impact on seed quality.

For example, seed of *Eucalyptus camaldulensis* in Salima, Malawi, is necessarily a Salima provenance. Eucalypts are not native to Malawi and must thus have been introduced. The origin of eucalypts in Malawi could be Petford in Queensland, Australia. The origin is thus Petford.

Sometimes there have been several links from origin to the provenance, i.e. the Salima seed mentioned above could have been collected in Dedza, Malawi, where there might have been a few generations that have developed a special land race for the site.

   (b) Pictures are static and conservative. When seed sources change – positively or negatively – pictures should be changed.
   (c) Pictures cannot be handled in databases in the sense of retrieving specific data from them.
4. *Seed production and harvest.* By appointing a stand as a seed source, it implies that the biological preconditions for productions are there, i.e. the stand is of mature age and produces a regular seed crop. Key phenological data as well as the production potential (estimate) will give potential seed collectors some clue of when to harvest and how much can be expected to be harvested from a given source. Notes on the applicable harvest methods will assist the collector during a possible seed collecting expedition.
5. *Accessibility and collection permits.* Rules and regulations for collection depend on ownership or formal administration right. Advanced seed orchards are usually the property of a specific supplier who may have more or less the monopoly of using the particular source. Seed is typically collected and supplied by the one authority only. For lesser-used species, protected forests make up the largest portions of seed sources. Most protected forests are subject to strong protective measures which can interfere with seed collection, e.g. some types of collection techniques (spurs, cutting of branches, etc.).

Seed source information is recorded on forms like the example in Fig. 8.8.

Much information on seed source forms has only relevance to the seed collector or supplier. That pertains to, for example, information on ownership, accessibility, size, age and productivity of the stand. Also details on flowering and fruiting periods, potential local labour availability and other pure collection details are of little interest for the seed customer, while they are highly relevant for the seed supplier. Only information that refers directly or indirectly to seed quality is transferred to the form sent to the customer. The Access database links specific selected fields to another database sheet with information to be provided to the seed user.

Summary seed source lists with selected relevant information can be created directly from the databases. Listing may be done according to species or seed source type. The latter is sometimes more appropriate where seed sources contain and serve as the source for more species, e.g. many mixed natural forests. Access databases can easily shift between seed source listing and species listing. Eventually, seed sources are conveniently plotted on a national or regional map indicating seed source reference number. Using a GIS, seed sources can be overlaid on the maps (Fig. 8.9).

SEED SOURCE INFORMATION FORM

Species information

Seed source reference No. ________________ Seed zone
Provenance name
Species (botanical): ______________________________________ Species code:
Common name:

Seed source classification:
☐ Unclassified, ☐ Seed zone, ☐ Identified stand, ☐ Selected stand, ☐ Seed production area,
☐ Provenance seed stand, ☐ Seed orchard

Location description

Seed source location: _________________________ District/county
Region/state ____________________ Country
Geographical coordinates: Latitude: ______ ° ______ 'N/S, longitude: ______ ° ______ 'E/W
Altitude: _________ m.a.s.l
Koeppen climatic code*: ☐ Af, ☐ Am, ☐ Aw, ☐ Aw1, ☐ Bsh, ☐ Bsk, ☐ Bwh, ☐ Bwk,
☐ Cfa, ☐ Cfb, ☐ Cw, ☐ Cw1, ☐ Cs, ☐ D

Rainfall regime: ☐ Summer, ☐ Uniform, ☐ Winter, ☐ Bimodal
Mean annual rainfall (mm):
Length of dry season (<60mm) (indicate months):
Mean annual temperature (°C):
Mean daily min. temp coldest month (°C): _______ , Mean daily max. temp hottest month (°C):
Absolute min. temperature (°C):
Other information:

Site description

Terrain: ☐ Flat, ☐ Hilly, ☐ Mountainous, ☐ Ridge top
Slope: ☐ Flat or gentle (<5%), ☐ Intermediate (5-10%), ☐ Steep(11-45%), ☐ Sheer
Aspect: ☐ North, ☐ East, ☐ South, ☐ West, ☐ Level
Soil type: ____ pH:

Stand description

Total area: ______ hectares
Type of stand: ☐ Unknown ☐ Natural stand ☐ Plantation, Planted year
☐ One species ☐ Mixed species, Associated species:

Accessibility

Distance to nearest forest station:
Accessibility road, ☐ 2WD, ☐ 4WD,
Walking distance from nearest road accessible by 4WD
Collection permit: ☐ Required ☐ Not required

Seed production

Flowering period _______________ Fruiting period
Harvestable fruit production (estimated): _____ Kg, or seed production (estimated): ________ Kg

Labour availability

Name(s) of nearest village:

Distance from seed source to nearest village:
Available labourers:

Other Information:

*** Koeppen climatic codes**

Af:	Permanent humid	Cfa:	Humid subtropics, east side of continents, incl. montane.
Am:	Monsoonal, short dry season	Cfb:	Temperate, maritime
Aw:	Subhumid, drier than Am	Cw:	Highland, subhumid
Aw1:	As Aw but bimodal rainfall	Cw1:	As Cw but bimodal rainfall
Bsh:	Semi arid, hot "steppe", "Sahel"	Cs:	Mediterranean.
Bsk:	Semi arid, warm to cold	D:	Temperate continental, also tropical & subtropical montane
Bwh:	Arid, hot desert		
Bwk:	Arid, warm to cold		

Fig. 8.8. Example of seed source registration form. Key information should preferably be entered into a database system where database functions such as different type of listing, updating and linkage to geographical information systems can be used

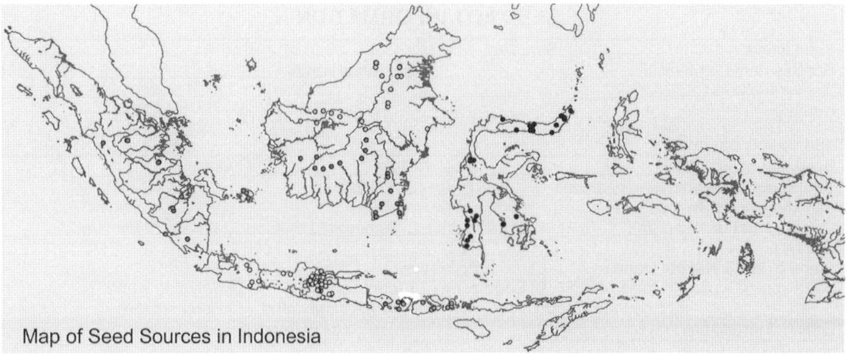

Fig. 8.9. Example of a country map of seed sources, here Indonesia. Digitalised maps are linked to databases of seed sources, and a 'click' on the seed source dot will display information about the particular seed source. Many seed sources tend to be located near tree breeding stations with good access, as in this example in central Java

8.5.5 Seed Lot Information

Labels attached to the seed lot indicate species, provenance, quantity and seed lot number. Additional information may be provided on seed lot forms containing information considered relevant for the seed user, e.g. genetic information and seed testing records (Fig. 8.10). In addition, the form may contain brief recommendations on handling from receipt to sowing.

8.6 Rules and Regulations

Rules and regulations should assist seed distribution in the sense of maximising the use of good-quality seed and minimising the use of random undocumented seed. Regulations should help keep track of seed distribution, set standard requirements and protect seed users from being cheated. Legislation may use incentives or prohibitions ('the carrot or the whip'). Often both types are used. Compulsion and prohibition are used to hinder the absolutely unacceptable (cheating, setting minimum standards). The area between denotes the spectrum of what is allowed; within this area incentives can be used to direct development towards the wanted.

All rules and regulations must aim at promoting seed quality. Yet, as we have seen, seed quality is often the 'best available', which is a blurred, yet realistic term, but rather hard to transfer into regulations. Therefore, and because regulations are difficult to control in practice, prohibition and compulsion are poor tools in seed regulations. Incentives, in addition to being more positive

SEED LOT INFORMATION

Seed lot no. ______________ Supplier
Species name (botanical): ________________ Provenance name: __________________
Common name: ___________________________
Country : ________________________________

Seed source information

Seed source location: ________________ Region/state __________ Country
Geographical coordinates: Latitude: ______° ____'N/S, longitude: ______° ____'E/W
Altitude: ______ m.a.s.l
Mean annual rainfall (mm): ______ Rainfall regime: ☐ Summer ☐ Uniform ☐ Winter ☐ Bimodal
Soil type: ____ pH
Stand type: ☐ Natural stands ☐ Plantation
Seed source type: ☐ Unclassified, ☐ Seed zone, ☐ Identified stand, ☐ Selected stand,
☐ Seed production area, ☐ Provenance seed stand, ☐ Seed orchard
Other information:

Collection data

Collection date:
Genetic representation: Number of parent trees collected from: _______
Average spacing between parent trees: ________________
Phenotypic selection of seed trees: ☐ Yes ☐ No
Selection criteria: ☐ Height, ☐ Straightness, ☐ Branching habit, ☐ Health, ☐ Others

Test results

Date of (latest) test	Germination percentage
	Viability:
Purity: ____________%	Measured by: ☐ TTZ
	☐ Cutting
Moisture content: ________________%	☐ X-ray
1000 grain seed weight: _________	☐ Other:
	No. of viable seeds per gram:

Seed treatment	Pretreatment:
Seeds treated with: ___________	☐ Scarification, method and duration:
Date of treatment ___________	☐ Stratification, method and duration

Recommended seed handling before sowing

☐ Soaking in water, duration:	Date: ___________
☐ Leaching, duration	Signature
☐ Manual extraction, method:	
☐ Other	
Inoculation: ☐ Mycorrhiza, species/ type:	
☐ Rhizobium, species/ type:	
☐ Frankia, species/ type:	

Fig. 8.10. Seed lot information form. The form contains extracts from various records such as seed sources, collection and seed testing. Via the seed lot number, additional information can be retrieved from the original data if needed

than prohibition, have also the practical effects on control mechanism in that they change where the burden of proof lies. Breaking of a rule must be proven by the authority, deserving an incentive must be proven by the applicant!

Unintentional side effects refer to effects on areas not intended to be regulated or where regulation may have negative effects. As an example, seed testing is usually considered to contribute to an improvement of seed quality (provided it is followed up by the action of removing seed lots with poor test results). A rule compelling seed testing is beneficial in an environment and for species where it is feasible and practical to test all seed lots. That is, it requires a relatively well functioning network or infrastructure, so that seed can be tested quickly; and it does not encompass recalcitrant seed, which will deteriorate parallel with the test. If a rule compels seed testing and it is not possible to implement it, it tends to incriminate/illegalise all seed supply, and this would have a very negative effect on tree planting. Another example is strict rules on seed transfer, e.g. between seed zones, may exclude harmless use of seed at a planting site, which happens to be on the other side of a generalised (and sometimes rather randomly created) zone border, but that is otherwise rather similar to the seed source. Further, it may restrict necessary experiments of using different seed sources at different sites. Rules and regulations can be used to promote and support an ongoing process, but they can also lock and set back a process. In the first example, where tests cannot be implemented, it forces seed producers to either break the rule (and in widest consequence be ignorant of any such rules) or to stop seed producing altogether, none of which was intended. In the second example, it places restrictions on where restrictions should not be used.

A seed source certification scheme is good provided a real difference can be documented between different candidate sources, that evaluation is based on sound scientific criteria, and that uncertified sources can still be used locally or where they are the best considering site–source matching.

In strongly regulated areas like most of Europe, Australia and North America basically any movement of seed is registered, both as a quality measure and, as far as trade is concerned, for use by tax departments. In such systems, where communication is fast and there is an open reflection from controlling civil servants to higher authorities, 'overlegislation', i.e. with undesired side effects, can easily be adjusted, based on feedback. In less controlled systems, and where seed control may not be the highest priority, regulations may tend to be empty and uncommitting.

Regulations may be used in a 'softer' way for standardising procedures, e.g. seed source evaluation, seed testing and seed lot numbering.

8.6.1 Target Group

It is a general consensus that any rule and regulation should apply to everybody unless they are specifically exempted. In forest seed regulations, exemptions are often necessary in order not to incriminate all small-scale, local, unauthorised but important seed supply. Specification of target groups and exceptions may be included either directly in the text or in notes to the decree. For example, 'seed suppliers with an annual turnover of xxx (currency) must assign seed lot number and documentation to all seed lots'. Or, 'seed suppliers with an annual turnover of less than xxx (currency) are exempted from the rules of seed lot number and documentation'. In some countries the main suppliers are authorised. Authorisation may be compulsory for all seed suppliers with a certain turnover, or it may be granted upon application. Authorisation would usually be encompassed by a special set of rules including privileges and duties, e.g. authorised seed suppliers are allowed to (have permission to)/must (are compelled to) use approved seed sources. Rules for authorised suppliers may include seed lot number assignment, use of seed sources, seed testing and pricing. If authorisation includes special privileges, it may accordingly be taken away from a supplier who does not fulfil standards or breaks the rules. Regulating mechanisms may be annual, biannual or rarer assessments and renewal of authorisations.

Local seed distribution occurs within or between neighbouring farms (one farmer sowing his own seed or giving it to his neighbour). It may qualitywise be poor, yet such small transactions should not be encompassed by seed decrees as it would hinder an important local exchange. With regards to quality, most local seed supply is probably poor. There could accordingly be good reasons to reduce or eliminate small dealers selling randomly collected seed on markets or stalls via a prohibiting decree. However, it should be carefully observed if this seed flow can be fulfilled by another and better source – random seed may be better than no seed at all – or random quality trees better than no tree planting at all!

Formulation of rules and regulations depends very much on the political system and traditions in the country, and whether seed supply is mostly a governmental or a private business. Experiences have shown that it is difficult to apply the seed-quality aspect and ensure sufficient diversity in both purely public and purely private businesses. Government systems may be less dependent on income generation (provided they have sufficient core funding) and can afford long-term objectives without short-term profit. However, control is often weak when one government institution controls another, and there is no real consequence of breaking rules. Firms in the private sector depend on short-term profit and will give in on quality if they are not profitable and not subject to rules and efficient control. A strongly regulated system with use of incentives and subsidies appears to be the most efficient for quality aspects.

8.6.2 Legislation on Seed Quality

Rules and regulations contain the same set of standard quality parameters as contained in seed documentation. Quality aspects include the following key areas:

1. Physiological quality, i.e. how seeds germinate and possible pests and diseases
2. Genetic quality, i.e. seed source and improvement aspects
3. Site–source matching, i.e. seed distribution aspects

Regarding key area 1, legislation should set the standard on measurement of quality parameters such as seed weight, seed moisture content, purity, germination and quantity and species of pests and diseases. Decrees may also contain a certain minimum standard, generally with a breakdown for species and species groups, below which seed should be disposed of. International rules and standards of seed testing are issued by the ISTA and in the USA by the AOSA. Legislation may adjust the standards to local conditions, cf. the above remarks on what is possible, e.g. what type of laboratory facilities are available. Legislation should also contain a statement on which laboratories are authorised to conduct the test. Finally, legislation may indicate which type of seed and which type of transfer need a physiological and phytosanitary test.

Test of physiological quality use the same methods as for agricultural seed and are sometimes encompassed by the same framework as these. This is not wise as agricultural seeds are quite different and have different specific quality aspects, i.e. because they are often improved varieties, are used for consumption and large samples are available for testing.

Regarding key area 2, assurance of genetic quality is probably the most problematic aspect in tree improvement since the concept is relative and the proof is difficult. The rules should include a definition of seed source categories, the seed source numbering system and seed lot assignment. International standards on seed source categories are practical both for international trade and for coordination. However, it is important that the categories indicate a difference, and that the difference is defined. Most seed sources in developing countries may from an international standard fall under the category identified stand or extensive seedling seed orchard (ESSO) because they do not live up to the standard required for, for example a selected stand, a SPA or management and test for a SSO. Yet it may thus be practical for domestic purposes to redefine concepts according to what is prevalent.

Once categories have been defined, seed sources can be classified within the system. A decree should give the authority to a board of experts to classify and approve seed sources. This system should be flexible enough to allow frequent adjustment or updating of the national list of seed sources. Seed sources disappear, for example, owing to logging or are degraded by encroachment or other environmental damage. New sources appear, either as new selected or established sources. Some sources change category, e.g. a selected stand is upgraded to a SPA by phenotypic thinning, or an untested seed orchard is upgraded to a tested orchard after a genetic test and rouging. A decree should be issued on how often updating of the seed source list takes place. Again it is important to be realistic on how such updating will work in practice.

The board of experts should have well-established terms of reference including the number of members, the appointment/selection procedure and authority.

Regarding key area 3, the theoretical aspect for site–source matching regulations is to avoid transfer of material from a seed source to a site where it is not adapted and, in some cases could become weedy. Special considerations apply to national transfer, and are discussed later. Whether there should be regulations on seed transfer within a country is subject to debate and obviously the first consideration should be is there a problem, how big is it and how is it easiest solved without interfering and unnecessarily incriminating most of the country's seed trade. There are many both biological, technical and legal problems in making regulations on national tree seed transfer. Biologically, the problem is that some species are quite adaptive and tolerate planting in very different conditions, while others have quite a narrow margin. Some species have a screwed lopsided tolerance in the sense that they perform well under 'better' (e.g. warmer) conditions and poorer under worse (e.g. colder) conditions. In some cases, transfer of planting material will have an immediate benefit, e.g. because the material is removed from local controlling factors (a factor well known from cultivation of exotics) but may be sensitive to rarely occurring climatic extremes such as a short cold spell. Technically, there is the problem that sites and sources are always ecologically different. Climatic parameters of both site and source have usually been extrapolated from the nearest climate station, which may actually be quite far, and the documentation of which climatic parameter (rainfall, temperature in coldest month, etc.) could be crucial is usually not available. Seed zones are sometimes used as a practical tool to compare ecological conditions over large areas. They are normally based on a number of parameters such as climate, soil and vegetation and as such they give a good picture of regional differences. They have, however, three key problems:

1. Information often comes from very few and scattered stations and is then extrapolated and interpolated to represent a very large area.
2. Significant within-zone variation gets blurred because the zones must be practical to handle. Species distribution is often according to the microclimate, e.g. occurring as small 'islands' or 'niches' over a larger area which dilute at the boundary of their ecological and geographical distribution.
3. Boundaries between zones are arbitrary and do not necessarily reflect species parameters. In areas with geographically long gradients and transition zones such as a flat area from the coast inland, an arbitrary change in, for example, rainfall parameter of say 100 mm may shift the border 100 km. Classification which includes many species becomes very insecure and interpretation inconclusive; hence, seed zone maps should ideally be drawn for each individual species.

There are legal impracticalities as well. A breeding programme typically concentrates on one or two promising provenances. Following a site–source matching concept, an improved seed source may only be used in a limited area, i.e. a large part of the species' growth area is kept out of reach of improved material.

In conclusion, decrees should only deal with cases and issue restrictions on transfer for species where problems have occurred and are well documented. It is rare, however, that failure or non-adaptability is very well documented and conclusive, and in most cases there should simply be technical recommendations, not legal restrictions.

8.6.3 Legal Authorities and Implementation

All countries have a system of different levels of decrees and ordinances, how they are passed and who can sign them. An analysis of the legal system is beyond the scope of this book. However, legislation on forest seed is often vague and poorly implemented. A sound legal framework is considered of major importance in order to implement good-quality seed supply. Seed people are the technical experts, who will often propose or advise on legislation and ordinances on forest seed.

1. Forest seed and agricultural seed should be treated as two groups with regards to testing and documentation. This inevitably implies a lot of difficult aspects for transition/agroforestry species; however, it is never more complicated than a list of species can be made.
2. The higher up in the system, the more difficult passing of legislation and a later change, but probably the stronger the implementation. Ministerial decrees should give a framework only. Ordinances or guidelines should be issued by lower authorities. The hierarchy of laws is that no rule or ordinance from a lower authority must contradict that of a higher authority, i.e. provincial rules must not contradict national rules.
3. The law of hierarchical authorities sometimes creates areas of potential conflict, e.g. between higher provincial authorities and lower national authorities. On higher decrees, a national decree cannot be overruled by a provincial decree. Boundary areas are often interpretation and implementation.
4. Laws frequently have boundary areas. Economic legislation, for instance, affects all areas of economic transactions and trade. Specific legislation on seed quality aspects must not conflict with such overriding laws.

5. Rigidity and flexibility are two common conflict areas in legislation. Countries with a poor control system and risk that anything unspecified will be interpreted to the benefit of individuals tend to make legislation very detailed. However, it should be remembered that forest seed quality is almost always debatable and needs to contain room for interpretation and adaptation to species and field conditions.
6. Formulation and feedback mechanisms. These are to ensure that all legislation has a local anchorage and is in line with what is possible to implement.
7. Laws and ordinances must have local anchorage and be implementable. Any comprehensive new regulation should be consulted by involved authorities and stakeholders during formulation. Guidelines to and possible training for implementing authorities should follow official regulations.
8. Rules and regulations which cannot be implemented, or where breaking or not following the rules and regulations has no consequences, can result in the respect of the entire seed sector being lost. Prohibition and incentives must be followed by action: if a rule is broken or an ordinance ignored, there must be a concrete system of action, e.g. fine, withdrawal of licence or the like.

8.6.4 Export and Import Regulations

Seed export is subject to the same general requirements of clearance as holds for any goods. However, special legislative restrictions on seed export exist in some countries. In India, any seed leaving a state has to be officially cleared, stating that it is seed surplus to the needs of the state in question (Campbell 1983). Increasing concern regarding the national right to genetic resources occasionally restricts unofficial trade over national boundaries. Seed exporters should be aware of any national restriction and legislation. Normally any consignment for export has to be inspected and cleared by official authorities before being exported. This is to ensure that the consignment does not contain illegal items like drugs or protected goods. Export agencies or authorities (postal service, airlines, freight companies, etc.) will in addition require appropriate documentation as demanded by the importing country, most commonly a phytosanitary certificate (Fig. 8.11), in order to avoid problems at the point of delivery.

Most international transfer of seed is subject to restriction and legislation in the importing country. The most common type is phytosanitary legislation,

Name and address of exporter (1) EXPORTER NAME EXPORTER ADDRESS EXPORTER ADDRESS SUBURB STATE POSTCODE AUSTRALIA	No. 0D6990001 Australian Government Department of Agriculture, Fisheries and Forestry **PHYTOSANITARY CERTIFICATE** Plant Protection Organisation of the Australian Government (5)
Declared name and address of consignee (2) CONSIGNEE NAME CONSIGNEE ADDRESS CONSIGNEE STATE CONSIGNEE COUNTRY	Import Permit No. (6) CN1234 1 MARCH 2006 Place of origin (7) QUEENSLAND — Code 4750 Country of final destination (8) CHINA — Code CN To: The Plant Protection Organisation of (9) CHINA
Declared means of conveyance (10) AIR / CA175 Declared point of entry (11) BEIJING	

Distinguishing marks and container nos (12)	No. and description of packages (13)	Name of produce/ quantity declared (14)	Botanical name of plants (15)	Commodity code (16)
LOT NO: 552X	10 BAGS	SUNFLOWER SEED	Helianthus annuus	12060000

Total no. of packages (17)	Total mass (18)
10	100 KGM

This is to certify that the plants, plant products or other regulated articles described herein have been inspected and or tested according to appropriate procedures and are considered to be free from the quarantine pests specified by the importing contracting party and to conform with the current phytosanitary requirements of the importing contracting party, including those for regulated non-quarantine pests. (19)

DISINFESTATION AND/OR DISINFECTION TREATMENT (20)

Date (21)	Treatment (22)	Chemical (active ingredient) (23)	Concentration (24)

Duration and temperature (25)	Additional information (26)

Additional Declaration (27)

1) Fiveangled dodder (Cuscuta pentagona) is not known to occur in Australia,
2) Crown gall of grapevine (Agrobacterium vitis), Chrysanthemum foliar nematode (Aphelenchoides rizemabosi), Stem nematode (Ditylenchus dipsaci) and Tobacco ringspot virus are not known to occur on sunflower seed in Australia,
3) A representative sample has been inspected and found free from Dodder (Cuscuta chinensis and Cuscuta europaea).

Name of Inspector (28) A INSPECTOR	Inspection Date (29) 27/6/06	Code 4750
Name of Authorised Officer (30) B BEACHAM	Inspection Date (31)	Code 4000
Place of issue (32) BRISBANE		Code 4000
Signature (34) B Beacham		

AUSTRALIA

Stamp (33)

An officer of the Department of Agriculture, Fisheries and Forestry (35)

SAMPLE

Fig. 8.11. Example of a phytosanitary certificate used in the USA. A phytosanitary certificate is a clearance document stating that a consignment is free of pests and diseases. The certificate is often compulsory in connection with import

which is valid in most countries. The purpose of phytosanitary rules is to avoid the risk of introducing dangerous seed-borne pests and pathogens into a country where they are not already found. Such pests may easily find their way to, for example, related native species, which may not have any resistance against the disease. In other cases the purpose is to keep exotic species free from pests found in the native country, but not in the country in which the trees are to be grown. In most countries a phytosanitary certificate will be required for import of forest seed. The certificate is issued by an accredited authority in the exporting country and states that the seeds have been examined and found free from pests and diseases. It is thus an official guarantee from the exporting to the importing country. The certificate will also state whether the seeds have been subject to fumigation or chemicals, and which type.

Customs authorities in the importing country may or may not accept phytosanitary certificates as a guarantee of freedom from pests and pathogens. If not, the seed consignment will be required to go through the quarantine regulation, which implies that the seeds are reexamined and/or retreated with a pesticide. Import treatment is unfortunately routine in many countries, and serious delays may occur on that account. In addition, application of seed pesticides in both exporting and importing countries may be highly damaging for the seeds as most seed dressings are phytotoxic in large doses and the effect is cumulative (Willan 1993).

General import restrictions imposed on any commercial product may also affect forest seed. Some countries require particular import permits. In addition, the importing party may need to pay duty on the consignment upon arrival. The duty is normally calculated as a certain percentage of the invoice amount (where freight and handling fees are normally excluded). However, small non-commercial seed lots, e.g. for research or trials which are sent free of charge, are normally exempt from customs charges, which usually makes clearance and release from ports much quicker (Willan 1995).

Import regulations can cause a serious delay to delivery, which may ultimately lead to reduced seed quality because of deterioration in transit. Bottlenecks vary from one country to another. Frequent trade and transfer with the same customer, through the same ports and using the same shipping agent will often help speed up the procedures of clearance. It is advisable that seed suppliers keep a file of country regulations of import and necessary arrangements for shipment. For neighbouring countries, various alternative transport modes and routes may be considered, e.g. ship, air, road or rail. Freight companies specialised in international transfer of goods often make efficient arrangements. Both exporting and importing parties can take measures to facilitate clearance and shorten the transit time:

1. Several administrative issues may be cleared without (i.e. before) the physical presence of the seed. For example, the importing party should make necessary prearrangements for settling particular import restrictions before the seed is submitted. A letter of confirmed order, including price (and type of currency) may be submitted by the supplier before the seed is shipped (at this point some suppliers claim part of the payment). The confirmation letter may help the importing party to obtain the necessary clearance.
2. The seed supplier should provide the necessary documents required by the importing country. Copies of, for example, the phytosanitary certificate and the freight document may be faxed or mailed to the customer prior to shipment. The freight document should specify thetransport (e.g. airline and flight number) and the expected arrival date.
3. The consignment should be properly labelled, indicating the addressee, contents, quantity, date, possible treatment etc. A copy of the phytosanitary certificate should accompany the seed where required.
4. A short message addressed to the shipment and custom authorities may state 'sensitive to high temperature', 'urgent expedition' or the like to help to draw the attention to the sensitive nature of the content and hence avoid unnecessary delay or transit damage.

Appendix 1: Seed Processing Table – Species List

Table A.1. Seed processing table – species list

Family/genus / species group	Prevailing fruit type – description	Extraction procedure
Apocynaceae • *Alstonia* • *Wrightia* • *Dyera*	Dehiscent, dry, often long and slender double follicles with many seeds	Drying will cause fruits to split open. Seeds fall out by themselves or with minimal mechanical impact
Anacardiaceae • *Spondias* • *Dracontomelum* • *Swintonia* • *Gluta* • *Mangifera*	Drupe with fleshy, often edible mesocarp. Mesocarp fibrous in, e.g., *Mangifera*. In *Swintonia* and *Gluta* drupes remain attached to a 5-winged placenta formed from persistent petals	Depulping by ingestion or soaking followed by stirring or high-water pressure, or mechanical depulping. Seeds are not extracted from the pyrene. Removal of wings not necessary as they will fall off during wet processing
Araucariaceae • *Araucaria* • *Agathis*	Dehiscent cones, often large. Disintegrate at maturity	Drying causes cone scale and seeds to separate from the central cone axis. Cone scale removed by sifting and/or winnowing; fine cleaning by flotation
Bignoniaceae • *Marchamia* • *Fernandoa* • *Stereospermum* • *Millingtonia* • *Spathodea*	Long slender dehiscent follicles/pods – in some species up to 80 cm. Winged seeds attached to central columella	Sun-drying causes dehiscence. Seeds usually fall off or out readily or with little mechanical impact. If extracted manually, fruits are discharged by the same procedure
Bombacaceae • *Bombax* • *Ceiba* • *Coelostegia* • *Durio* • *Adansonia*	Large woody capsules. Dehiscent with woolly seeds in *Bombax* and *Ceiba*; dry edible pulp in *Adansonia*. Indehiscent with arillate seed in *Durio*	Dry extraction from fruit followed by removal of testa appendices. In *Bombax* and *Ceiba* mechanical deflossing or burning of seed hair. In *Durio* removal of aril by depulping procedures, e.g. high-water pressure or, in edible species, by soaking off the edible pulp. Hard pulp in *Adansonia* removed after soaking

(*Continued*)

Table A.1. Seed processing table – species list—Cont'd.

Family/genus / species group	Prevailing fruit type – description	Extraction procedure
Boraginaceae • *Cordia*	1 seeded drupe	Pyrene extracted by wet extraction, e.g. high water pressure after softening
Burseraceae • *Canarium* • *Commiphora* • *Boswellia*	Drupe with fleshy pulp and hard endocarp containing up to 3 seeds	Wet extraction for removal of pulp, e.g. high-pressure water after softening. Seeds are not extracted from the pyrene
Casuarinaceae • *Casuarina* • *Allocasuarina* • *Gymnostoma*	Dry dehiscent multiple fruits, 'conelike', spherical to oblong, opening by slots	Drying makes fruits open. Tumbling usually suffices to make seeds fall out. If trapped, seeds can be extracted after disintegration of the whole fruit, e.g. threshing
Celestraceae • *Kokoona*	3-valved woody capsule	Drying until dehiscence, then mechanical raking, shaking or tumbling to remove seeds
Combretaceae • *Combretum* • *Terminatia* • *Anogeissus*	Mostly dry winged fruits, in *Combretum* with 4 angular wings, in *Terminalia* with 1 wing surrounding the seed (wing much reduced in, e.g., *T. catappa*, making fruit drupelike)	Extraction reduces storability and is generally avoided. To reduce bulk, fruits can be dewinged by rolling between wire-mesh screens. In *Combretum* seeds may be extracted by manually splitting open the wings before sowing
Cupressaceae • *Cupressus* • *Fokienia* • *Libocedrus*	Dehiscent cones with central cone scales that open upon drying	Cone scales open upon drying and seeds are released by gentle tumbling. Usually no dewinging
Datiscaceae • *Octomeles* • *Tetrameles*	Dehiscent capsules with many seed	Extraction by drying and shaking. The volume of fruits and that of seeds are always small and the tiny seeds easily spill out. To avoid loss, opened fruits can be shaken thoroughly manually in a pail with a closed lid and extracted through a fine masked sieve
Dilleneaceae • *Dillenea*	Dehiscent follicles making a star-formed compound fruit surrounded by enlarged fleshy sepals that split open at maturity	Fruits split open upon drying; seeds extracted manually. Fleshy sarcotesta removed by wet extraction, e.g. high water pressure or wet tumbling

Table A.1. Seed processing table – species list—Cont'd.

Family/genus / species group	Prevailing fruit type – description	Extraction procedure
Dipterocarpaceae • *Anisoptera* • *Dipterocarpus* • *Dryobalanops* • *Hopea* • *Parashorea* • *Shorea* • *Vatica*	Nuts with 2 or 4 (occassionally5) large wings originating from persistent sepals. Fruits contain usually only 1 embryo. Fruits usually large, including wings from 3–20 cm. Most species have desiccation-sensitive seed	Manual removal of wings sometimes done to reduce bulk and ease sowing. Sensitivity to desiccation and their short storability makes fast sowing mandatory
Ebenaceae • *Diospyros* • *Euclea*	Berry with persistent sepals and from 1 to a few seeds. Most species with fleshy pulp, but species with dry pulp occur in dry areas	Fleshy pulp removed by normal wet extraction, e.g. water pressure or wet tumbling. Dry fruits keep well when sun-dried. Pulp must usually be removed before sowing to remove germination inhibitors
Euphorbiaceae • *Aleurites* • *Bridelia* • *Croton* • *Hevea* • *Macaranga* • *Trewia* • *Clutia*	Dehiscent capsules. Seeds usually small	In this group of euphorbia, seeds can be extracted by any dry extraction procedure, i.e. drying until dehiscence and tumbling or other mechanical impact to separate fruits from seeds
Euphorbiaceae • *Aleurites* • *Bischofia* • *Drypetes* • *Endospermum* • *Trewia*	Drupes or berries, usually small, often with sticky, milky pulp	Stones or seeds extracted wet after softening by soaking or initiated decomposition. Bleach or some mild liquid soap help remove sticky residual pulp
Fagaceae • *Castanopsis* • *Quercus* • *Fagus* • *Lithocarpus* • *Castanea* • *Nothofagus*	Nut with enclosing, dehiscent or open cupula. Usually large	The dehiscent cupula in *Fagus* and *Castanea* open by slight drying. The cupula remain firmly attached to the fruit in some *Lithocarpus* and *Quercus* species. Wetting and slight drying help soften the attachment, but many Fagaceae are desiccation-sensitive. Cupula must often be removed manually
Guttiferae • *Calophyllum* • *Mesua* • *Cratoxylum*	*Callophyllum* has a drupe fruit with fleshy/fibrous mesocarp. The fruit in *Cratoxylum* is a woody capsule	Wet or dry extraction for fleshy and dry fruits, respectively. Residual pulp of fleshy mesocarp removed by tumbling in sand or by brushing

(*Continued*)

Table A.1. Seed processing table – species list—Cont'd.

Family/genus / species group	Prevailing fruit type – description	Extraction procedure
Hamamelidaceae • *Altingia* • *Liquidambar*	Semidehiscent, casuarina-like, compound, dry fruits, which open by apertures	Apertures open upon drying and seeds may be extracted by tumbling. If seeds are stuck inside the fruit, it is necessary to disintegrate the fruits, e.g. by threshing or in a hammer mill
Juglandaceae • *Carya* • *Engelhardtia* • *Juglans*	Dry drupes or nuts, in *Engelhardtia* with wings	Dry exocarp/mesocarp removed manually or, for some species, by tumbling in a cement mixer with abrasive material or in brushing machines with hard brushes
Lauraceae • *Cinnamomum* • *Cryptocarya* • *Eusideroxilon* • *Litsea* • *Machilus*	Most genera with 1 to a few seeded berries. In *Eusideroxylon* fruits are large drupes (up to 15-cm long). In *Cryptocarya* fruits are surrounded by a persistent flower tube. *Cinnamomum* often have a persistent placenta	Fleshy pulp removed by wet extraction, e.g. high-pressure water after soaking. Some species have fragile seed coats, which are easily damaged by mechanical handling
Leguminosae – Caesalpinaceae • *Erythrophloeum* • *Intsia* • *Pelthophorum* • *Senna* • *Brachystegia* • *Delonix* • *Bauhinia* • *Baikaea* • *Sindora* • *Afzelia*	Dehiscent/semidehiscent pods. Often large, woody and thick. Seeds often remain enclosed in the fruit until after dispersal. In *Sindora* and *Afzelia* seeds have large and thick arils	Mature fruits will split up upon drying. However, owing to the thickness of the pod and the woody character, drying for a long time, occasionally using an artificial heat source, is necessary. Pods that remain closed can be split open manually by a few blows with a club. Arils are easiest to remove immediately after extraction when they are still soft. Strong drying for dehiscence has the drawback of hardening the aril. A few hours' soaking immediately after extraction facilitates manual removal of the aril
Leguminosae – Caesalpinaceae – • *Cassia* • *Tamarindus* • *Dialium* • *Koompassia*	Indehiscent often round pods, 30–70-cm long. Seeds surrounded by a sticky substance	Drying and then thrashing or pounding to crush the fruits. Seeds usually separate readily from the fruits. Residual pulp removed by washing with addition of sodium hypochlorite. Seeds cleaned by sifting following winnowing

Table A.1. Seed processing table – species list—Cont'd.

Family/genus / species group	Prevailing fruit type – description	Extraction procedure
Leguminosae – Mimosaceae • *Acacia* (some) • *Albizia* • *Paraserianthes* • *Xylia* • *Leucaena* • *Calliandra* • *Gllericidia*	Dehiscent thin pods, usually with many (4–16) seeds. Seeds usually remain attached to half of the pods during dispersal. Australian acacias frequently with funicle developed into an aril	Sun-drying until dehiscence. Shaking or thrashing used to extract seeds – the strength and method depend on the strength of funicle attachment. Arils detached by threshing, in brushing machines or by biological means (e.g. ants)
Leguminosae – Mimosaceae • *Prosopis* • *Inga* • *Pithecellobium* • *Acacia* (e.g. *A. nilotica*)	Indehiscent pods with several seeds. Pods often leathery and hard – in *Inga* and *Pithecellobium* seeds are imbedded in a pulp	Extraction often difficult as it requires disintegration of the pods. Threshing or milling (e.g. hammer mill) is easiest after drying. Where pulp is fleshy/soft it may be removed by washing or high-pressure water
Leguminosae – Papilionaceae • *Dalbergia* • *Ormosia* • *Pterocarpus* • *Cordyla* • *Sophora* • *Tephrosia*	Indehiscent pods with fibrous or woody pericarp. The pods are flat and have usually developed an extension of a wing. In *Pterocarpus* there is a surrounding wing and sometimes spines	Seeds can be extracted from pods with fibrous pericarp by threshing or milling. Seeds are generally not extracted from fruits with woody pericarp, but wings are sometimes removed to reduce bulk
Leguminosae – Papilionaceae • *Derris* • *Cordyla* • *Sophora* • *Sesbania* • *Erythrina* • *Tephrosia*	Dehiscent pods with many seeds. Seeds usually release easily from the pods	Sun-drying until dehiscence. Depending on the strength of funicle attachment, shaking or tumbling is usually sufficient to release seeds
Leguminosae – Papilionaceae • *Milletia* • *Crotolaria* • *Derris* • *Pongamia*	Dehiscent woody pods. This group contains several species with very hard pods	Woody pods require a long time and strong drying to split open. If pods do not open, splitting can be performed by manually pounding them in a mortar or threshing in a hammer mill. Seeds usually detach themselves readily from the pods
Lycythidaceae • *Barringtonia*	1 seeded berry	Seeds extracted by wet extraction, e.g. soaking in water with subsequent washing under high water pressure

(*Continued*)

Table A.1. Seed processing table – species list—Cont'd.

Family/genus / species group	Prevailing fruit type – description	Extraction procedure
Lythraceae • *Lagerstroemia*	Dehiscent capsules with many seeds	Fruits dehisce upon drying and seeds fall out after gentle tumbling or other turning
Magnoliaceae • *Magnolia* • *Michelia* • *Manglietia* • *Elmerrillia*	Compound fruits consisting of a long axis with dehiscent follicles each containing 1 or more seeds surrounded by a fleshy aril	Seeds extracted from the follicles by drying until the fruits open, then removal of the seeds manually or, in some species, by tumbling/flailing or beating. Seeds often retain a strong funicle attachment to the fruit. The fleshy aril removed by washing or strong water pressure
Meliaceae • *Amoora* • *Cedrela* • *Chukrassia* • *Khaya* • *Swietenia* • *Entandophragme* • *Toona*	Dehiscent capsule withseeds attached to a central receptacle. The fruits are usually large, e.g. up to 20 cm in *Swietenia* and *Entandophragma*. The pericarp is shed shortly before dispersal	The pericarp will open and fall apart during drying. Seeds will fall off the receptacle with minimum impact, e.g. raking or shaking drying fruits. Large wings are occasionally broken off manually to reduce bulk
Meliaceae • *Melia* • *Azadirachta* • *Aglaia* • *Ekebergia* • *Sandoricum*	Drupe with fleshy mesocarp and hard endocarp. Usually several seed in the pyrene. *Aglaia* spp. have berry capsules	Fruit flesh removed by washing or water pressure. In *Aglaia* the exocarp is preliminarily removed manually
Moraceae • *Arthocarpus* • *Antiaris* • *Ficus* • *Morus* • *Bosqueia* • *Chlorophora*	Multiple fleshy fruits, many with edible pulp. Very variable in size from less than 1-cm diameter in *Ficus* spp. to more than 50-cm long in *Arthocarpus*	Extraction sometimes in connection with use of fruits for consumption. Otherwise depulping by soaking, mechanical depulping or water pressure. Sticky 'milk' can hamper mechanical depulping
Myrtaceae • *Eucalyptus* • *Melaleuca* • *Syzygium*	Capsules with various degrees of dehiscence. Opening by dentate operculum. *Syzygium* has 1–2 seeded fleshy fruit	Dehiscence by sun or kiln drying and extraction by subsequent tumbling. Floss or other mechanical constrictions can hamper extraction in species with an inferior ovary
Pinaceae • *Pinus* • *Abies* • *Tsuga*	Dehiscent cones with many seeds	Cone scales split open upon drying in most species – in *Abies* the cones disintegrate by dehiscence of the cone scales. In serotinous cones high temperature is required to melt the resin before dehiscence can occur

Table A.1. Seed processing table – species list—Cont'd.

Family/genus / species group	Prevailing fruit type – description	Extraction procedure
Podocarpaceae • *Podocarpus* • *Dacrycarpus* • *Dacrydium* • *Nageia*	Seed-bearing structure consists of 1 or more sterile cone scales upon which the seed with surrounding fleshy aril is borne	Seed with aril removed from branchlets by threshing or pulling the branches through a rake. Removal of the aril by wet depulping
Proteaceae • *Grevillea* • *Helicia* • *Macademia* • *Banksia* • *Hakea*	Follicles with 1 to several seeds. From thin and fragile in, e.g., *Grevillea* to very hard in some *Hakea* species. In *Banksia* individual fruits are united into a dense, woody multiple fruit	*Grevillea* seeds are easily extracted after drying, but sometimes the maturation period is short. Seeds of some *Hakea* and *Banksia* species can be extracted after strong drying,but many species require scorching, e.g. over a charcoal fire. Seeds must be rapidly cooled when extracted
Rhamnaceae • *Ziziphus* • *Maesopsis*	Drupe, often with thick endocarp	Wet depulping, e.g. by softening of pulp followed by high-pressure water
Rhizophoraceae • *Bruguiera* • *Kandelia* • *Ceriops* • *Rhizophora*	Fruits have one viviparous seed, which may grow up to 25 cm in *Bruguiera* and *Rhizophora*; significantly smaller in *Kandelia* and *Ceriops*	Seeds are not extracted. The viviparous seed is kept cool and moist and sown as soon as possible after collection
Rubiaceae • *Anthocephalus*	Multiple fruits consisting of many drupes in a globose multiple fruit. Many tiny seeds	Fruits are soaked in water until they get soft and can be split up by washing. Fruit pulp easiest to remove by flotation as seeds are very small
Rutaceae • *Teclea* • *Fagara* • *Zanthoxylum* • *Flindersia*	Variable, e.g. dehiscent capsule in *Fagara* and *Flindersia* and drupe in *Teclea* and *Zanthoxylum*. Often large fruits	Seeds from capsules extracted by dry extraction after drying, e.g. tumbling. Fruit pulp of drupes removed by washing after short softening treatment
Santalaceae • *Santalum*	Drupe	Fruit pulp removed by wet extraction
Salvadoraceae • *Salvadora* • *Dobera*	Berry or drupe with thin endocarp	Depulping by wet extraction. Species with very thin seed coat must be depulped gently, e.g. manually removing the exocarp and cleaning seeds under running water
Sapindaceae • *Pometia* • *Sapindus*	Mostly drupes with fleshy mesocarp and exocarp.	Wet extraction, e.g. high-pressure water or washing after softening of the pulp by soaking and fermentation

(*Continued*)

Table A.1. Seed processing table – species list—Cont'd.

Family/genus / species group	Prevailing fruit type – description	Extraction procedure
Sapotaceae • *Madhuca* • *Manilkara* • *Palaquium* • *Payena* • *Eberhardtia* • *Sideroxylon* • *Vitellaria*	Berries with thin or thick pericarp, containing 1–6 seeds	Depulping by wet extraction, e.g. softening by soaking and depulping by mechanical treatment or water pressure
Simaroubaceae • *Ailanthus*	Samara with large wing surrounding the seed	No extraction but fruit wing often removed to reduce bulk
Sterculiaceae • *Heritiera* • *Scaphium* • *Tarrietia*	Samara	Usually no extraction. Wings may be removed to reduce bulk
Sterculiaceae • *Brachychiton* • *Pterospermum* • *Sterculia*	Single follicles in *Brachychiton.* In *Sterculia* and *Pterospermum* the follicles are compressed into a starlike structure. Seeds large	The seeds fall out readily from the dehiscent fruits upon drying
Taxodiaceae • *Cunninghamia*	Dry dehiscent cones, morphologically similar to those of pines	Dry extraction as in pines
Theaceae • *Schima* • *Ternstroemia*	*Schima* has a woody capsule, *Ternstroemia* a berry capsule with arillate seed	Capsule opens upon drying. Fleshy aril removed by washing
Thymeleaceae • *Gonostylus* • *Aquillaria*	Round, woody 2–5-valved capsules with 1–5 seeds, often with arils	Capsules open at maturity by drying. Seed coats often thin and fragile and easily damaged by mechanical handling
Verbenaceae • *Avicennia* • *Gmelina* • *Vitex* • *Tectona* • *Peronema*	Most species with fleshy or juicy drupes with 1–4 seeds. *Tectona* and *Peronema* have dry drupes. In *Tectona* the pericarp is felty. In *Peronema* the fruits split into 4 parts exposing several pendulous seeds	Fleshy fruits are depulped by moist extraction or, as in some *Vitex* species, are dried without depulping. *Peronema* seeds are extracted by drying. *Tectona* is extracted by mechanical treatment which removes the enclosing involucre and felty pericarp

Appendix 2: Seed Testing Forms

Different laboratories use different forms as seed testing records. It is practical to use forms where all relevant data are filled in, e.g. weight of containers in moisture analysis. The standard test form has two parts, the first part pertaining to seed weight, moisture content and purity, and the second one pertaining to germination (Figs. A.1, A.2). Each sheet is indicated by seed lot number, seed source name and reference number, and the species name as appears from the test request submitted to the laboratory, and should follow the test result.

A test normally starts with a purity test, since the pure seed fraction can then be used for other seed tests. A purity test is normally carried out on two replicates of 5–10 g, depending on seed size. For large seeds, up to 50–100 g may be applicable. Once a clear 'pure seed definition' has been established, the two fractions are weighed separately and the percentage calculated. The weight of the container is not necessary for this calculation.

Determination of the 1,000-seed weight is carried out on pure seeds, e.g. those identified in the purity analysis can be used in order to save time. For most seeds, eight replicates of 100 seeds are used. The number may be reduced for very large seeds. The 1,000-seed weight is calculated as 10 times the average of the eight replicates. A statistical variation coefficient is calculated for the results: the smallest figure is subtracted from the largest one in order to calculate the range, which is used as a shortcut to calculate the standard deviation – the range is divided by 2.85, which is a table figure for $n = 8$. After the standard deviation has been calculated, the variation coefficient is easily found as the percentage of the average 100-seed weight. If this figure exceeds 4, the variation is too large (which could indicate a sample error), and the analysis should be redone.

Moisture content analysis is usually carried out on two replicates of 5–10 g, depending on seed size. Samples used for seed weight or purity analysis may be used again for moisture content. The weight of the empty container is indicated as this figure must be used the following day after oven-drying to calculate the loss of weight.

The final result of tests of purity, seed weight and moisture content is transferred to seed test form II.

SEED TEST FORM I: Weight, purity and moisture content sheet

Seed lot no. ____________ Seed source name: ____________________ Seed source ref. no: ________________ Analysis no. ____________

Species ____

Date of completed germination analysis

1000-grain weight (8x100) seed

Replicate	X
1	
2	
3	
4	
5	
6	
7	
8	
Sum	
Average X= (n=8)	
Weight of 1000 seed= gram	

Range (= Largest - smallest)=

Estimated standard deviation:
Range / 2.85* =

Var. Coeff: $\frac{100 \times std.}{X}$ =

Not to exceed 4.0

New analysis ☐ yes ☐ no

*Table value for n = 8

Purity

Replicate	Weight of total sample	Weight of impurities	Weight of pure seed	Percentage impurities	Percentage pure seed	Difference between A and B:
A	g	g	g	%	%	Max difference according to ISTA:
B	g	g	g	%	%	New analysis ☐ yes ☐ no
Average						

Moisture content

Replicate	Weight of empty container	Weight of fresh sample	Weight of oven-dry sample	Difference = weight of water	Moisture content	Difference between A and B. Moisture content = ____ % Max difference according to ISTA ____ New analysis ☐ yes ☐ no
A	g	g	g	g	%	
B	g	g	g	g	%	
C	g	g	g	g	%	
D	g	g	g	g	%	
Average					%	

Remarks (extra analysis, analysis errors, mechanical seed damage etc.)

Fig. A.1. Seed test form I: weight, purity and moisture content sheet

SEED TEST FORM II : Germination test sheet

Seed lot no. ______ Seed source name :______ Seed source ref. no:______ Analysis no. ______

Species ______

Date of completed germination analysis

Summary						
Purity (%)	1000 seed wgt (g)	No. of seed/kg	Moisture content (%)	Germ. Cap. (%)	Speed of germ. (%)	Germination criteria

Germination

Replica tion	Normal germination after days								Total norm. Germ. (a)	Fresh not germ. (b)	Abnormal germ. (c)	Dead seed		Total (a-e)	Rotten (mouldy seed) (f)	Polyem bryony (g)	Insect damaged seed (h)	Calculated germ. % (i)	Living seed (excl. Emp. Seed) (j)
	Start date	Date	Date	Date	Date	Date	Date	Date				Empty (d)	Full (e)						
A																			
B																			
C																			
D																			
Total																			
Av. %																			

Germination table / Cabinet type______
Germination temp.°C ______
Germination substrate*: TP, BP, PP, TS, S
Pretreatment:
Method ______
Time ______
Temp. (°C)______

Damage	Fungi	Insects	Other
None	☐	☐	☐
Small	☐	☐	☐
Average	☐	☐	☐
Large	☐	☐	☐

Difference in germination % A-B-C-D (%)______
Max. Difference (ISTA) (%) ______
New analysis ☐ yes ☐ no
New analysis start ______
New analysis number ______
Germination test done by______

*TP = Top of paper, BP = Between paper, PP = Pleated Paper, TS = Top of sand, S = In sand

Fig. A.2. Seed test form II: germination test sheet

The germination test is usually carried out on 50–100 seeds. Germination may be recorded once a week – for fast-germinating species more often and for slow-germinating species often with 1 or 2 weeks before the first count. Germination criteria should be established, e.g. whether there should be a full development of a seedling or whether radicle protrusion, e.g. equal to the length of the seed, is accepted as an indication of germination. The exact criteria are important for calculation of the speed of germination. Abnormal germination is counted separately and entered in the column indicated by 'c'. After the end of the test period, non-germinated seeds are examined and classified in various categories, b, d, e, f and g. Polyembryony (h) is indicated since this may give rise to more than one seedling per seed, usually in species with several morphological seeds within an endocarp (Box 7.4).

The speed of germination is calculated as the germination percentage at one third of the duration of the test.

At the bottom of the sheet there is room for information on germination conditions, viz. pretreatment, germination temperature (ambient or degrees centigrade) and substrate (top of paper, sand, top of sand, etc.)

Observations of damage by insects, fungi or other types are classified as 'none', 'small', 'average' or 'large'.

References

Adjers G, Srivastava PBL (1993) Nursery practices. In: Awang K, Taylor D (eds) *Acacia mangium*, growing and utilization. Winrock International, Bangkok

Agarwal VK, Sinclair JB (1987) Principles of seed pathology. Lewis, Boca Raton

Ahenda JO (1991) Processing of pulpy seeds. In: Proceedings of the 1st national tree seed workshop, Nairobi. Kenya Forestry Seed Centre/Kenya Forestry Research Institute, pp 114–126

Ahmed A el Houri (1986) Some aspects of dry land afforestation in the Sudan with special reference to *Acacia tortilis* (Forsk.) Hayne, *A senegal* Wild. and *Prosopis chilensis* (Molina) Stuntz. For Ecol Manage 16:209–221

Airi S, Rawal RS, Dhar U (2005) Presowing treatment effects on germination of *Cornus capitata* seeds. Seed Sci Technol 33:77–86

Albrecht J (1993) Forest seed handling. In: Pancel L (ed) Tropical forestry handbook, vol 1. Springer, Berlin Heidelberg New York, pp 383–462

Aldhous JR (1972) Nursery practice. Forestry Commission bulletin no 43. Her Majesty's Stationary Office, London

Amata-archachai P, Wasuwanich P (1986) Mechanical extraction and cleaning of nuts of some tropical species. Embryon 2:1–8

Ammer C, Mosandl R, El Kateb H (2002) Direct seeding of beech (*Fagus sylvatica* L.) in Norway spruce (*Picea abies* (L.) Karst.) stands – effect of canopy density and fine root biomass on germination. For Ecol Manage 159:59–72

Anderson EW, Spanos KA, Mullin TJ, Lindgren D (1998) Phenotypic selection compared to restricted combined index selection for many generations. Silva Fenn 32:111–120

AOSA (1983) Seed vigor testing handbook. Contribution no 32 to the handbook on seed testing. Association of Official Seed Analysts, Beltsville

AOSA (1997) Rules for testing seeds. Association of Official Seed Analysts, Beltsville

Arboricultural Association (2005) A guide to good climbing practice. Arboricultural Association, Romsey

ASEAN (1991) Standard germination test. Training course proceedings no 2. ASEAN–Canada Forest Tree Seed Centre Project, Muak Lek

Ashton DH (1975) Root and shoot development of *Eucalyptus regnans* F. Muell. Aust J Bot 23:867–887

Ashton DH (1976) The development of even-aged stands of *Eucalyptus regnans* in central Victoria. Aust J Bot 24:397–414

ATSC (1995) Operation manual. Australian Tree Seed Centre, CSIRO, Canberra

ATSC (1996) *Eucalyptus pilularis* seed germination project. Report prepared by ATSC for State Forest of New South Wales, northern region. Australian Tree Seed Centre, CSIRO, Canberra

Auld TD (1986) Population dynamics of the shrub *Acacia suaveolens* (Sm.) Willd., dispersal and dynamics of the soil seed bank. Aust J Ecol 11:235–254

Babeley GS, Kandya AK (1988) On finding out some suitable pretreatment for *Cassia fistula* Linn. seeds. J Trop For 4:147–154

Bachelard EP (1967) Role of seed-coat dormancy in *Eucalyptus pauciflora* and *E. delegatensis* seeds. Aust J Biol Sci 20:1237–1240

Bapat AR, Phulari MM (1995) Teak fruit treatment machine – a prototype II. Indian For 121:545–549

Barbour J (2006) Use of the Westrup brush machine. USDA Forest Service, Washington. http://www.nsl.fs.fed.us/brush_machine_user.pdf

Barner H, Olesen K (1983a) Climbing into the crown by way of the bole 1. Portable ladders. Technical note no 3. Danida Forest Seed Centre, Humlebaek

Barner H, Olesen K (1983b) Climbing into the crown by the way of the bole, 2. Technical note no 5. Danida Forest Seed Centre, Humlebaek

Barner H, Olesen K (1984a) Climbing within the crown 1, equipment and safety rules. Technical note no 10. Danida Forest Seed Centre, Humlebaek

Barner H, Olesen K (1984b) Climbing into the crown directly 2. Technical note no 8. Danida Forest Seed Centre, Humlebaek

Barner H, Willan RL (1985) Seed collection units 1. Seed zones. Technical note no 16. Danida Forest Seed Centre, Humlebaek

Barnett JP (1971) Flotation in ethanol reduced storability of Southern pine seeds. For Sci 17:50–54

Baskin JM, Baskin CC (2004) A classification system for seed dormancy. Seed Sci Res 14:1–16

Bebawi FF, Mohamed SM (1985) The pretreatment of seeds of six Sudanese acacias to improve their germination response. Seed Sci Technol 13:111–119

Bedell PE (1989) Preliminary observations on viability of teak in India. Indian For 115:72–81

Bergsten U, Sundberg M (1990) IDS sedimentation of *Cupressus lusitanica* seeds. In: Turnbull JW (ed) Tropical tree seed research. Proceedings of an international workshop held at the forestry training centre, Gympie, QLD, Australia, 1989, pp 99–102

Bergsten U, Wirklund K (1987) Some physical conditions for removal of mechanically damaged *Pinus sylvestris* L. seeds by using the PREVAC method. Scand J For Res 2:315–323

Berjak P, Pammenter NW (1996) Recalcitrant (desiccation-sensitive) seeds. In: Innovations in tropical tree seed technology. Proceedings of the IUFRO symposium of the project group P.2.04.00, 'Seed problems', Arusha, Tanzania, pp 14–29

Berjak P, Pammenter NW (2002) Orthodox and recalcitrant seed. In: Vozzo J (ed) Tropical tree seed manual. Agriculture handbook 721. USDA Forest Service, Washington, pp 137–147

Berjak P, Villiers TA (1972a) Ageing in plant embryos II. Age induced damage and its repair during early germination. New Phytol 71:135–144

Berjak P, Villiers TA (1972b) Ageing in plant embryos III. Acceleration of senescence following artificial ageing treatment. New Phytol 71:513–518

Berjak P, Villiers TA (1972c) Ageing in plant embryos IV. Loss of regulatory control in aged embryos. New Phytol 71:1069–1074

Berjak P, Villiers TA (1972d) Ageing in plant embryos V. Lyses of the cytoplasm in non-viable embryos. New Phytol 71:1076–1079

Bernald-Lugo I, Leopold AC (1998) The dynamics of seed mortality. J Exp Bot 49:1455–1461

Bewley JD (1997) Seed germination and dormancy. Plant Cell 9:1055–1066

Bewley JD, Black M (1982) Physiology and biochemistry of seeds – in relation to germination, vol II. Springer, Berlin Heidelberg New York, pp 1–59

Bewley JD, Black M (1994) Seeds. Physiology of development and germination. Plenum, New York

Bhardwaj SD, Chakraborty AK (1994) Studies on time of seed collection, sowing and presowing treatments of *Terminalia bellirica* Roxb. and *Terminalia chebula* Retz. Indian For 120:430–439

Bhodthipuks J, Pukittayacamee P, Saelim S, Wang BSP, Yu SL (eds) (1996a) Rapid viability testing of tropical tree seed. Training course proceedings no. 4. ASEAN Forest Tree Seed Centre Project, Muak Lek

Bhodthipuks J, Saelim S, Pukittayacamee P (1996b) Hydrogen peroxide (H_2O_2) testing for viability of tropical forest tree seed. In: Bhodthipuks J, Pukittayacamee P, Saelim S, Wang BSP, Yu SL (eds) Rapid viability testing of tropical tree seed. Training course proceedings no 4. ASEAN Forest Tree Seed Centre Project, Muak Lek, pp 11–15

Bird R, Lawrence J (1993) Direct seeding. In: Race D (ed) Agroforestry, trees for productive farming. Agmedia, Melbourne, pp 191–199

Blair DF (1995) Arborist equipment, a guide to the tools and equipment of tree maintenance and removal. International Society of Arboriculture, Champaign

Boland DJ, Brooker MIH, Turnbull JW, Kleinig DA (1990) Eucalyptus seed. CSIRO, Canberra

Boland DJ, Brooker MIH, Turnbull JW (1990) Eucalyptus seed. CSIRO, Australia

Bonner F (2003) Collection and care of acorns – a practical guide for seed collectors and nursery managers. http://www.nsl.fs.fed.us/nsl_fsstc.html

Bonner FT (1996) Commercial seed supply of recalcitrant and intermediate seed: present solutions to the storage problem. In: Ouédraogo AS, Poulsen K, Stubsgaard F (eds) Intermediate/recalcitrant tropical forest tree seeds. Proceedings of a workshop on improved methods for handling and storage of intermediate/recalcitrant tropical forest tree seeds, 1995, pp 27–33

Bonner FT, Vozzo JA, Elam WW, Land SB Jr (1994) Tree seed technology, training course. Instructors manual. General technical report SO-106. Southern Forest Experimental Station, USDA

Booth T (1996) Matching trees and sites. Proceedings of an international workshop held at Bangkok, Thailand, 27–30 March 1995. ACIAR proceedings no 63

Boshier DH (2000) Mating systems. In: Young A, Boshier D, Boyle T (eds) Forest conservation genetics principles and practices. CABI, Wallingford, pp 63–79

Boshier DH, Lamb AT (1997) Seed collection and storage. In: *Cordia alliodora.* Oxford forestry paper no 36. Oxford Forestry Institute, pp 67–71

Bowen MR (1980) Recording procedures for the collection, cleaning and drying (extraction), testing (certification), storage and issue of forest tree seeds. Working paper no 4, FAO/UNDP/MAL/78/009. Forest Research Centre, Sepilok

Bowen MR, Eusebio TV (1981) *Acacia mangium*: updated information on seed collection, handling and germination testing. Occasional technical and scientific notes: seed series no 5. Forest Research Centre, Sepilok

Bowen MR, Eusebio TV (1982) Seed handling practices: Four fast-growing hardwoods for humid sub-tropical plantations in the eighties. Malays For 45:534–547

Bradford KJ (2004) Seed technology and enhancement. In: Seed production and quality. University of California, Davis Extension, pp 112–117

Bradford KJ, Bewley JD (2002) Seeds: biology, technology and role in agriculture. In: Chrispeels MJ, Sadava DE (eds) Plants, genes and crop biotechnology. Jones and Bartlett, Sudbury, pp 210–239

Brahmam M (1996) Effect of pre-sowing treatments for hastening the germination of *Enterolobium cyclocarpum* (Jacq.) Griseb. and *Hemenaea courbaril* L. Indian For 122:740–745

Brandenburg NR (1983) A proposed severity index for seed handling. J Appl Seed Prod 1:8–11

Brant E, McKee GW, Cleveland RW (1971) Effect of chemical treatment on hard seed on Pengrift Crownvetch. Crop Sci 11:1–6

Brockwell J (1962) Studies on seed Pelleting as an aid to legume seed inoculation I: Coating materials, adhesives, and methods of inoculation. Aust J Agric Res 13:638–649

Brookman-Amissah J (1973) Seed problems as they affect forestry practices in Ghana. In: Seed problems. Proceedings of the international symposium on seed processing. Bergen, Norway, vol II. Working paper no 6. IUFRO working party S2.01.06

Brown NAC (1993) Promotion of germination of Fynbos seed by plant derived smoke. New Phytol 123:575–583

Brown NAC, van Stead J (1997) Smoke as a germination cue: a review. Plant Growth Regul 22:115–124

Brown R (1972) Germination. In: Steward FC (ed) Plant physiology – a treatise. Academic, New York, pp 3–48

Buker RS, Syvertsen JP, Burns JK, Roka FM, Miller WM, Salyani M, Brown GK (2004) Mechanical harvesting and tree health. University of Florida. http://edis.ifas.ufl.edu/HS199

Burger D (1972) Seedlings of some tropical trees and shrubs, mainly of South East Asia. PUDOC, Wageningen

CAF (1981) Studies ion the aerial seeding of *Pinus sinensis* (*P. tabulaeformis*) in the mountain areas of Hebei province. Linye Keji Tongxum 4:12–16 (in Chinese, reference based on abstract)

Campbell MW (1983) Importing seed into Nepal. Technical note no. 2/83. Nepal-Australia Forestry Project, Kathmandu

Casal JL, Sanches RA (1998) Phytochromes and seed germination. Seed Sci Res 8:317–329

Catalan LA, Macchiavelli RE (1991) Improving germination in *Prosopis flexuosa* D.C. and *P. alba* Griseb. with hot water treatments and scarification. Seed Sci Technol 19:253–262

CATIE (1997–1998) Catalogo de semillas forestales, Banco Latinoamericano de Semillas Forestales. Centro Agronomico Tropical de Investigacion y Ensenanza (CATIE), Turrialba

CATIE (1998) Proyecto de Semillas Forestales, PROSEFOR. Informe final, I fase: Setiembre 1992–Diciembre 1997. Preliminary report. CATIE, Turrialba

Chaichanasuwat O, Wang BSP, Wasuwanich P (1990) Evaluating seed quality of *Peltophorum pterocarpum* by X-radiography and germination. In: Turnbull J (ed) Tropical tree seed research. Proceedings of an international workshop held at the Forestry Training Centre, Gympie, QLD, Australia, 1989, pp 68–72

Chaisurisri K, Edwards DGW, El-Kassaby YA (1993) Accelerated aging of sitka spruce seeds. Silvae Genet 42:303–308

Chayiyasit L, Piewluang C, Wasuwanich P, Pukittayacamee P (1990) X-radiographic determination on seed germinability and early seedling survival in *Pterocarpus macrocarpus* Kurz. Embryon 3:1–5

Chaplin GE (1985) Seed cleaning with a winnowing chamber. Technical note no 25. Danida Forest Seed Centre, Humlebaek

Chen RZ, Fu JR (1984) Physiological studies on seed dormancy and germination of *Erythrophloeum fordii.* Sci Silvae Sin 20:35–41

Chien C-T, Lin T-P (1994). Mechanism of hydrogen peroxide in improving the germination of *Cinnamomum camphora* seed. Seed Sci Technol 22:231–236

Christensen CM (1973) Loss of viability in storage: microflora. Seed Sci Technol 1:547–562

Cipollini ML, Levey DJ (1997) Secondary metabolites of fleshy vertebrate-dispersed fruits: adaptive hypotheses and implications for seed dispersal. Am Nat 150:346–372

Close DC, Wilson SJ (2002) Provenance effects on pre-germination treatments for *Eucalyptus regnans* and *E. delegatensis* seed. For Ecol Manage 170:299–305

Cochard R, Jackes BR (2005) Seed ecology of the invasive tropical tree *Parkinsonia aculeate.* Plant Ecol 180:13–31

Coe M, Coe C (1987) Large herbivores, acacia trees and bruchid beetles. S Afr J Sci 83:624–635

Coffey M, Horlock J (1998) Hand direct sowing data sheet. Greening Australia Victoria, Heidelberg

Corbineau F, Come D (1988) Storage of recalcitrant seeds of four tropical species. Seed Sci Technol 16:97–103

Cornelius J (1994) The effectiveness of plus-tree selection for yield. For Ecol Manage 67:23–34

Correll MJ, Kiss JZ (2002) Interactions between gravitropism and phototropism in plants. J Plant Growth Regul 21:89–101

Costa ME, Sampaio DS, Paoli AAS, Leite SCAL (2004) Polyembryony and aspects of embryogenesis in *Tabebuia ochracea* (Chamisso) Standley. Rev Bras Bot 27:395–406

Craviotto RM, Arango MR, Salinas AR, Gibbons R, Bergmann R, Montero MS (2004) A device for automated digital x-ray imaging for seed analysis. Seed Sci Technol 32:867–871

Cremer KW (1990) Trees of rural Australia. Inkata, Melbourne, chap 5

Cresswell EG, Grime JP (1981) Induction of light requirement during seed development and its ecological consequences. Nature 291:583–585

Crocker W (1916) Mechanisms of dormancy in seeds. Am J Bot 3:99–120

Dale IR, Greenway PJ (1961) Kenya trees and shrubs. Buchanans Kenya Estates

Danusevicius D, Lindgren D (2002) Two-stage selection strategies in tree breeding considering gain, diversity, time and cost. For Genet 9:145–157

Danthu P, Gueye A, Bye A, Bauwens D, Sarr A (2000) Seed storage behaviour of four Sahelian and Sudanian species (*Boscia senegalensis, Butyrospermum parkii, Cordyla pinnata* and *Saba senegalensis*. Seed Sci Res 10:183–187

de Castilho CV, Magnusson WE, de Araujo RNO, da Costa Pereira E, de Souza SS (2006) The use of French spikes to collect botanical vouchers in permanent plots: Evaluation of potential impacts. Biotropica 38:555–557

Dell B (1980) Structure and function of the strophiolar plug in seeds of *Albizia lophantha*. Am J Bot 67:556–561

Delouche JC, Baskin CC (1973) Accelerated aging techniques for predicting the relative storability of seed lots. Seed Sci Technol 1:427–452

Delouche JC, Matthes RK, Dougherty GM, Boyd AH (1973) Storage of seed in subtropical and tropical regions. Seed Sci Technol 1:671–700

Desai BB, Kotecha PM, Salunkhe DK (1997) Seeds handbook, biology, production, processing and storage. Dekker, New York

Deubner CG (1932) Chemical treatment to shorten the rest period of red and black oak acorns. J For 30:672

Dickie JB, Smith RD (1995) Observation of the survival of seeds of *Agathis* spp. stored at low moisture contents and temperature. Seed Sci Res 5:5–14

Dixon KW, Roche S, Pate JS (1995) The promotive effect of smoke derived from burnt native vegetation on seed germination of Western Australian plants. Oecologia 101:1850–1920

Dobbs RC, Edwards DGW, Konishi J, Wallinger D (1976) Guidelines in collecting cones of British Columbia conifers. British Columbia Forestry Service and Canadian Forestry Service joint report no 3

Dogra PD (1967) Seed sterility and disturbances in embryogeny in conifers with particular reference to seed testing and tree breeding in Pinaceae. Stud For Suec 451–497

Doody P, O'Reilly C (2005) Effect of moist chilling and priming treatments on the germination of Douglas-fir and noble-fir seeds. Seed Sci Technol 33:63–73

Doran JC, Turnbull JW, Boland DJ, Gunn BV (1983) Handbook of seeds of dry zone acacias. A guide for collecting, extracting, cleaning and storing the seed and for treatment to promote germination of dry-zone acacias. FAO, Rome

DPI (1994) Direct seeding. Forest Service, tree note B3. Department of Primary Industries, Gympie

Dransfield J (2001) Taxonomy, biology and ecology of rattan. Unasylva 52:11–17

Duangpatra J (1991) Seed vigour testing. In: Standard germination test. Training course proceedings no 2. ASEAN–Canada Forest Tree Seed Centre Project, Muak Lek, pp 49–52

Duguma B, Kang BT, Okali DUU (1988) Factors affecting germination of Leucaena (*Leucaena leucocephala* (Lam.) de Wit.) seed. Seed Sci Technol 16:489–500

Dunlap JR, Barnett JP (1984) Influence on seed size on germination and early development of loblolly pine (*Pinus taeda* L.) germinants. Can J For Res 13:40–44

Eden Foundation (1992) Direct seeding by farmers – the natural solution to Sahel's problems. Reforestation by direct seeding. http://www.eden-foundation.org/project/articles.html

Edwards DGW (1981) Cone collection and processing, effect on seed quality and yield. In: High-quality collection and production of conifer seed. Proceedings of a workshop held on 14 November 1979 in Edmonton, Alberta. Northern Forestry Research Centre, Edmonton, pp 12–37

Edwards DGW, Wang BSP (1995) A training guide for laboratory analysis of forest tree seeds. Info report BC-X-356. Pacific Forestry Centre, Canadian Forest Service, Victoria

Elam WW, Blanche CA (1990) Accelerated aging: a potential vigour test for multipurpose tree seeds. In: Turnbull JW (ed) Tropical Tree Seed Research. Proceedings of an international workshop held at the Forestry Training Centre, Gympie, QLD, Australia, 1989, pp 63–67

El Atta HA (1993) The effect of *Caryedon serratus* Oliver (Col., Bruchidae) on viability and germination of seeds of *Acacia nilotica* (L.) Willd. ex Del. in the Sudan. For Ecol Manage 57:169–177

Eldridge K, Davidson J, Harwood C, van Wyk G (1993). Eucalypt domestication and breeding. Oxford Science, Oxford

Ellis JB (1986) Quantifying seed deterioration. In: McDonald MB Nelson CJ (eds) Physiology of seed deterioration. CSSA special publication no. 11. Crop Science Society of America, Madison, pp 101–123

Ellis JB (1988) The viability equation, seed viability nomographs and practical advice on seed storage. Seed Sci Technol 16:29–50

Ellis RH, Roberts EH (1980) Improved equations for the prediction of seed longevity. Ann Bot 45:13–30

Ellis RH, Hong TD, Roberts EH (1985) Handbook of seed testing for genebanks, vol 1. International Board for Plant Genetic Resources, Rome

Ellis RH, Hong TD, Roberts EH (1990) An intermediate category of seed storage behaviour? J Exp Bot 41:1167–1174

Emmanuel CJSK, Dharmaswamy SS (1991) Seed source variation in storage life of teak seeds (short note). Silvae Genet 40:249–250

Enescu V (1991) The tetrazolium test of viability. In: Tree and shrub seed handbook. International Seed Testing Association, Zurich, chap 9

Ezell AW (2004) Direct seeding, a forest regeneration alternative. Mississippi State University Extension Service

Fan HB, Huang YZ, Li YY, Lin DX (2005) Effects of simulated acid rain on seed germination and seedling growth of *Cunninghamia lanceolata*. Acta Agric Univ Jiangxiensis 27:875–879 (in Chinese with English abstract)

FAO (1967) Precleaner. FAO forestry equipment notes A.54.67. FAO, Rome

FAO (1984) Design and operation of cold stores in developing countries. FAO Agricultural Service bulletin 19/2. FAO, Rome, and International Institute of Refrigeration, Paris

Farrent JM, Pammenter NW, Berjak P (1988) Recalcitrance – a current assessment. Seed Sci Technol 16:155–166

Farrent JM, Pammenter NW, Berjak P, Walters C (1997) Sub-cellular organization and metabolic activity during development of seeds that attain different levels of desiccation tolerance. Seed Sci Res 7:135–144

Farnsworth E (2000) The ecology and physiology of viviparous and recalcitrant seed. Annu Rev Ecol Syst 31:107–138

Faulkner R (1975) Seed orchards. Forestry Commission bulletin no 54. Her Majesty's Stationary Office, London

Finkeldey R, Ziehe M (2004) Genetic implications of silvicultural regimes. For Ecol Manage 197:231–244
Flores EM (1992) Almendro de montana. Arboles Semillas Neotrop 1:1–22
Forestry Commission (1994) Seed manual. Procedures for seed collection, handling, storage, purchase and accounting. Forestry Commission, Hobart
Foster SG, Jones N, Kjaer ED (1995) Economics of tree improvement in development projects in the tropics. In: Shen S, Contreras-Hermosilla A (eds) Environmental and economic issues in forestry. Selected case studies in Asia. World Bank technical paper number 281. Asia Technical Department series
Fowels HA (1953) The effect of seed and stock sizes on survival and early growth of Ponderosa and Jeffry pine. J For 51:504–507
FRIM (1987) The silvicultural guide book of Malawi. Forest Research Institute of Malawi
Galop P, Webley DJ (1980) The use of plants and minerals as traditional protectants of stored products. Report of the Tropical Product Institute. FAO, Rome
Gamene CS, Eriksen EN (2004) Storage behaviour of *Khaya senegalensis* seeds from Burkina Faso. In: Sacande M, Jøker D, Dulloo ME, Thomsen KA (eds) Comparative storage – biology of tropical tree seeds. International Plant Genetic Resources Institute, Rome, pp 9–15
Gamene CS, Erdey D, Baxter D, Motete N, Berjak P (2004) Desiccation, germination and storage of *Sclerocarya birrea* seeds from Burkina Faso. In Sacande M, Jøker D, Dulloo ME, Thomsen KA (eds) Comparative storage – biology of tropical tree seeds. International Plant Genetic Resources Institute, Rome, pp 40–56
Gardner DE (1980) *Acacia koa* seedling wilt caused by *Fusarium oxysporum* f. sp. *koae*, f. sp. nov. Phytopathology 70:594–597
Ge X-J, Zhou X-L, Li Z-C, Hsu T-W, Schaal BA, Chiang T-Y (2005) Low genetic diversity and significant population structuring in the relict *Amentotaxus argotaenia* complex (Taxaceae) based on ISSR fingerprinting. J Plant Res 118:415–422
Gordon AG, Rowe DCF (1982) Seed manual for ornamental trees and shrubs. Forestry Commission bulletin no 59. Her Majesty's Stationary Office, London
Granhof J (1984) Seed extraction 2. Extraction of pine seeds by means of sun drying on elevated trays followed by tumbling. Technical note no 17. Danida Forest Seed Centre, Humlebaek
Graudal L, Lillesoe JPB (2006) Good practice for governance and provate sector development in tree seed supply in developing countries. Forest and Landscape Denmark, Hoersholm
Gray R (1990) Professional seed collection. In: Sowing the seeds, direct seeding and natural regeneration conference. Greening Australia, conference proceedings, pp 135–139
Gray R (2004) Professional seed collection. FloraBank, Yarralumla. http://www.florabank.org.au/support/articles/Professional%20Seed%20Collection.PDF
Greening Australia (2004) Hand direct sowing data sheet. http://www.greeningaustralia.org.au/GA/NAT/TipsAndTools/exchange/publications+library/Fact+sheet+and+case+studies
Green JW, Williams AV (1969) Collection of *Eucalyptus* branch specimens with the aid of a riffle. Aust For Res 4(2):19–30
Griffin AR (1972) The effects of seed size, germination time and sowing density on seedling development in radiata pine. Aust For Res 5(4):25–28

Griffin AR (1990) Effects of inbreeding on growth of forest trees and implications for management of seed supplies for plantation programmes. In: Bawa, KS, Hadley M (eds) Reproductive ecology of tropical forest plants. Man and the biosphere series, vol 7. Parthenon, Carnforth, pp 355–374

Groome JS, Lees HMN, Wigg LT (1957) A summary of information on *Pterocarpus angolensis*. For Abstr 18:153–162.

Gunn B (2001) Australian Tree Seed Centre operation manual. CSIRO Forestry and Forestry Products, Canberra

Gunn B, Agiwa A, Bosimbi D, Brammall B, Jarua L, Uwamariya A (2004) Seed handling and propagation of Papua New Guinea's tree species. CSIRO Forestry and Forest Products, Canberra

Haines RJ, Gould RJ (1983) Separation of seed of *Araucaria cunninghamii* Ait. ex D. Don (Hoop Pine) by flotation. Aust For Res 13:299–304

Halevy G (1974) Effects of gazelles and seed beetles (Bruchidae) on germination and establishment of *Acacia* species. Isr J Bot 23:120–126

Hallman RG (1981) Seed harvesting machine tested in the south. Tree Plant Notes 32:29–32

Hallman RG (1993) Loblolly tree seed collection system. Tree Plant Notes 44:101–104

Halloin JM (1986) Microorganisms and seed deterioration. In: McDonald MB Jr, Nelson CJ (eds) Physiology of seed deterioration. CSSA special publication no 11. Crop Science Society of America, Madison, pp 89–99

Hansen CP, Kjaer ED (1999) Appropriate planting material in tree plantings: Opportunities and critical factors. International expert meeting on the role of planted forests for sustainable forest development, Santiago, Chile, 6–9 April 1999

Harper JL (1977) Population biology of plants. Academic, London

Harrington JF (1972) Seed storage and longevity. In: Kozlowski TT (ed) Seed biology, vol. III. Academic, New York, pp 145–245

Hartmann HT, Kester DE (1983) Plant propagation, principles and practices, 5th edn. Prentice Hall, Englewood Cliffs

Hartmann HT, Kester DE, Davies FT Jr, Geneve RL (1997) Plant propagation, principles and practices, 6th edn. Prentice Hall, Englewood Cliffs

Hassen A, Rethman NFG, van Niekerk WA (2005) Effect of different treatment options on dormancy breaking, germination and emergence of *Ziziphus mucronata* (buffalo thorn) seed. Trop Grass 39:124–128

Hellum AK (1976) Grading seed by weight in white spruce. Tree Plant Notes 27:16–17, 23–24

Heydecker W (1972) Vigour. In: Roberts EH (ed) Viability of seeds. Syracuse University Press, New York, pp 209–252

Hilton GM, Packham JR (1997) A sixteen-year record of regional and temporal variation in the fruiting of beech (*Fagus sylvatica* L.) in England (1980–1995). Forestry 70:7–16

Hodgson TJ (1977) Effects of organic flotation media on the germination of *Pinus elliottii*, *P. patula* and *P. taeda* seed. S Afr For J 102:17–21

Holmes PM, McDonald IAW, Juritz J (1987) Effects on cleaning treatment on seed bank of alien invasive shrub *Acacia saligna* and *Acacia cyclops* in southern and south-western cape. S Afr J Appl Ecol 24:1045–1051

Holt C (1999) Direct seeding of native plants for revegetation. Farmnote 40/98. Agriculture Western Australia

Hong LT (1981) A note on some seed fungi of dipterocarps. Malays For 44:163–166
Hong TD, Ellis RH (1990) A comparison of maturation drying, germination and desiccation tolerance between developing seeds of *Acer pseudoplatanus* L. and *A. platanoides* L. New Phytol 116:589–596
Hong TD, Ellis RH (2002) Storage. In: Vozzo J (ed) Tropical tree seed manual. Agriculture handbook 721. USDA Forest Service, Washington, pp 125–136
Hor YL (1993) Seed testing for selected tropical trees in ASEAN region. Review paper no 2. ASEAN–Canada Forest Tree Seed Centre Project, Muak Lek
Hoskin M (1983) Forest nursery operations based on Malaysian experience with particular reference to Sabah and specific reference to Kemasul Nursery. In: Workshop on nursery and plantation practices in the ASEAN, Jakarta, Indonesia. 1983. ASEAN–New Zealand Afforestation Project Workshop
Howe HF (1990) Seed dispersal by birds and mammals: Implications for seedling demography. In: Bawa KS, Hadley M (eds) Reproductive ecology of tropical forest plants. Man and the biosphere series, vol. 7. Parthenon, Carnforth, pp 191–218
Howe RW (1972) Insects attacking seeds during storage. In: Kozlowski TT (ed) Seed biology, vol III. Academic, New York, pp 247–297
Hughes CE (1998) Leucaena – a genetic resource handbook. Tropical forestry paper no 37. Oxford Forestry Institute
Hughes CE, Styles BT (1984) Exploration and seed collection of multiple purpose dry zone trees in Central America. Int Tree Crops J 3:1–31
Hyde EOC (1954) The function of the hilum in some Papillionaceae in relation to ripening of the seed and the permeability of the testa. Ann Bot 18:241–256
ISTA (1986) ISTA handbook on seed sampling. International Seed Testing Association, Zurich
ISTA (1991) Tree and shrub seed handbook. International Seed Testing Association, Zurich
ISTA (1995) Handbook of vigour test methods. International Seed Testing Association, Zurich
ISTA (1996) International rules for seed testing, rules 1996. Seed science and technology 24 (supplement). International Seed Testing Association, Zurich
ISTA (1999) International rules for seed testing, rules 1999. Seed science and technology. International Seed Testing Association, Zurich
ISTA (2006) International rules for seed testing, rules 2006. Seed science and technology. International Seed Testing Association, Zurich
Ivory MH, Spreight MR (1993) Pest management. In: Pancel L (ed) Tropical forestry handbook. Springer, Berlin Heidelberg New York, pp 1142–1221
Jahufer MZZ, Borovoi VV (1992) The effect of mechanical damage to maize (*Zea mays* L.) seed germination, seed morphology and subsequent grain yield. J Appl Seed Prod 10:67–77
Janzen DH (1972) Escape in space by *Sterculia apetala* seeds from the bug *Dysdercus fasciatus* in a Costa Rican deciduous forest. Ecology 53:350–361
Jarvis BC (1975) The role of seed parts in the induction of dormancy of Hazel (*Corylus avellana* L.). New Phytol 75:491–494
Jasumback T (1994) Aerial lifts for working in tree tops. Tree Plant Notes 45:21–25
Javanovic T, Booth TH (2002) Improved species climatic profiles. A report for the RIRDC/L&W Australia/FWPRDC/MDC Joint Venture Agroforestry Program

Jensen KM (1987) Small scale cleaning machines. In: Madsen E, Langkilde NE (eds) ISTA handbook for cleaning of agricultural and horticultural seeds on small-scale machines part I. International Seed Testing Association, Zurich, pp 57–78

Johnson CD (1983) Handbook of seeds of *Prosopis* species. FAO, Rome

Johnson G, Chirco E (2005) Excised embryo tests of peach, apple and pear. In: Prochazkova Z, Gosling PG, Sutherland JR (eds) Proceedings of the ISTA Forest Tree and Shrub Seed Committee workshop, Prague, 20–22 October, 2003, pp 51–55

Johnson LPV (1946) Effect of chemical treatment on the germination of forest tree seeds. For Chron 22:17–24

Jones BJ (2003) Use and effectiveness of mobile elevating work platforms (MEWPS) to tree work. HSE Books, Sudbury

Kabay D, Lewis A (1987) Collection, handling and storage of Australian native plant seed. In: Longkamp P (ed) Germination of Australian native plants. Inkata, Melbourne, pp 20–30

Kamara AM, El-Lakani MH, Badran OA, Attia YG (1981) Seed pathology of *Araucaria* spp. I. A survey of seed-borne fungi associated with four *Araucaria* spp. Aust For Res 11:269–274

Kamondo BM, Kalanganire A (1996) Possibilities of mass propagation of *Melia volkensii.* In: Innovations in tropical tree seed technology. Proceedings of the IUFRO symposium of the project group P.02.04.00, 'Seed problems', Arusha, Tanzania, 1995, pp 95–101

Kannan CS, Sudhakara K, Augustine A, Ashokan PK (1996) Seed dormancy and pretreatments to enhance germination in selected *Albizia* species. J Trop For Sci 8:369–380

Karrfalt RP (1992) Increasing hardwood seed quality with brush machines. Tree Plant Notes 43:33–35

Karrfalt RP (2001) Seed testing. Reprint from: Woody plant seed manual. USDA, Washington, chap 5, pp 1–24

Kaul MLH (1979) The life-span of some indian forest seeds. Beitr Trop Lanwirtsch Veternarmed 17:283–286

Keiding H (1993) *Tectona grandis* Linn. f. Seed leaflet no 4. Danida Forest Seed Centre, Humlebaek

Keys MG, Dieters MJ, Walker SM, Huth JR (1996) Forest productivity gains from use of top quality seed and nursery plants: two case studies from Queensland, Australia. In: Yapa AC (ed) Proceedings of the international symposium on recent advances in tropical tree seed technology and planting stock production, ASEAN Forest Seed Centre, Muak-Lek, Saraburi, Thailand, pp 14–20

Kha LD, Son NH, Quang TH, Hunh NT (2004) Seed storage methods of star anise (*Illicium verum*), cinnamom (*Cinnamomum cassia*) and michelia (*Michelia mediocris*). In Sacande M, Jøker D, Dulloo ME, Thomsen KA (eds) Comparative storage – biology of tropical tree seeds. International Plant Genetic Resources Institute, Rome

Khasa PD (1992) Scarification of Limba seeds with hot water, bleach and acid. Tree Plant Notes 43:150–152

Kijkar S (1990) Prickling and transplanting. In: Plant stock production technology. Training Course proceedings no. 1. ASEAN–Canada Forest Tree Seed Centre Project, Muak Lek, pp 29–33

Kijkar S, Pong-anant K (1990) Selecting potting media. In: Planting stock production technology. Training course proceedings no 1. ASEAN–Canada Forest Tree Seed Centre Project, Muak Lek, pp 15–17

Kindt R, Muasya S, Kimontho J, Waruhiu A (1997) Tree seed suppliers directory: sources of seeds and microsymbionts. International Centre for Research in Agroforestry, Nairobi

King MW, Roberts EH (1979) The storage of recalcitrant seeds – achievements and possible approaches. International Board for Plant Genetic Resources Secretariat, Rome

Kitajima A, Sasagawa K, Hasegawa K (2003) Development of the abscission zone and morphological changes of abscission cells in the abscission process of Persimmon fruit. Acta Hortic 601:85–87

Kleinschmidt GJ (1989) Seeds, seed collection of eucalypts and ornamentals. Department of Primary Industries, Gympie

Kmecza NS (1979) Using tree shakers for pine cone collection in region 8. Tree Plant Notes 21:9–11

Kobmoo B, Hellum AK (1984) Hot water and acid improve the germination of *Cassia siamea* Britt. seeds. Embryon 1:27–33

Kommedahl T, Windels CE (1986) Treatment of maize seeds. In: Jeffs KA (ed) Seed treatment. British Crop Protection Council, Alton, pp 163–183

Koskela J, Kusipalo J, Sirikul W (1995) Natural regeneration of *Pinus merkusii* in northern Thailand. For Ecol Manage 77:169–177

Krishnapillay DB, Engelmann F (1996) Alternative methods for storage of recalcitrant and intermediate seeds: Slow growth and cryopreservation. In: Ouédraogo AS, Poulsen K, Stubsgaard F (eds) Intermediate/recalcitrant tropical forest tree seeds. Proceedings of a workshop on improved methods for handling and storage of intermediate/recalcitrant tropical forest tree seeds, 1995, pp 34–39

Kruse M (2004) ISTA handbook on seed sampling. International Seed Testing Association, Zurich

Lacey MJ, Line MA (1994) Influence of soil pH on the germination and survival of *Eucalyptus regnans* F. Muell. in southern Tasmania. Aust For 57:105–108

Laedem CL (1984) Quick tests for tree seed viability. British Columbia Ministry of Forests and Lands Research Branch, Victoria

Laedem CL, Bhodthipuks J, Clark JM (1995) Effects of stratification and temperature on the germination of *Dalbergia cochinchinensis*, *Pinus kesiya* and *Pinus merkusii*. J Trop For Sci 7:355–370

Lamprey HF, Halevy G, Makacha S (1974) Interaction between *Acacia*, bruchid beetles and large herbivores. East Afr Wildl J 12:81–85

Laurent N, Chamshama SAO (1987) Studies on the germination of *Erythrina abyssinica* and *Juniperus procera*. Int Tree Crops J 4:291–298

Lauridsen EB (1994) Seed documentation. Lecture note no 13. Danida Forest Seed Centre, Humlebaek

Lauridsen EB, Souvannavong S (1993) Neem provenance collection and seed handling. In: Genetic improvement of neem: strategies for the future. In: Proceedings of the international consultation on neem improvement, Bangkok, Thailand, 18–22 January 1993, pp 137–149

Lauridsen EB, Olesen K, Schøler E (1992) Packaging materials for tropical tree fruits and seeds. Technical note no 41. Danida Forest Seed Centre, Humlebaek

Leck MA, Parker VT, Simpson RL (1989) Ecology of seed banks. Academic, New York

Leon-Lobos P, Ellis RH (2005) Seed survival in Chilean *Nothofagus* in response to desiccation and storage. Seed Sci Res 15:113–123

Le Tam VT, Hong TD, Ellis RH, Ngoc Tam BT (2004) Seed storage of *Avicennia alba*. Seed Sci Technol 32:531–536

Liang SB, Yong WC (1985). Mechanical extraction and cleaning of *Gmelina arborea* nuts. Technical note no 20. Danida Forest Seed Centre, Humlebaek

Linington SH (2003) The design of seed banks. In: Smith RD, Dickie JB, Linington SH, Pritchard HW, Probert RJ (eds) Seed conservation – turning science into practice. The millennium seed bank project. Royal Botanic Gardens, Kew, pp 593–636

Lopez JH, Aviles RB (1988) The pretreatment of seeds of four Chilean prosopis to improve their germination response. Seed Sci Technol 16:239–246

Mahgoub S (1996) The influence of positioning of seed during sowing on seed germination of some tree species in relation to germination type and seed shape. In: Innovations in tropical tree Seed technology. Proceedings of the IUFRO symposium of the project group P.2.04.00, 'Seed problems', Arusha, Tanzania, 1995, pp 149–159

Maideen SK, Selvaraj JA, Vinaya Ray RS (1990) Presowing chemical treatments to hasten germination of *Casuarina equisetifolia*. Int Tree Crops J 6:173–181

Mangold RD, Bonner FT (2001) Certification of tree seeds and other woody plant materials. USDA Forest Service, Washington

Marcos-Filho J (1998) New approaches to seed vigor testing. Sci Agric 55:1–5

Marambe B, Gunasena HPM, Eeswara JP, Kumudinie KA (1998) Effect of stage of maturity of neem seeds (*Azadirachta indica* A. Juss.) on germination and physiological developments during cold storage. In: Proceedings of the final workshop: University of Peradenya Oxford Forestry Institute link project, pp 64–78

Martinez-Gomez P, Gradziel TM (2003) Sexual polyembryony in almond. Sex Plant Reprod 16:135–139

Martins-Loucao MA, Duarte PJ, Cruz C (1996) Phenological and physiological studies during carob (*Ceratonia siliqua* L.) seed germination. Seed Sci Technol 24:33–47

Martinussen JD (1999) Society, state and market. A guide to competing theories of development. Mellomfolkelig Samvirke, Copenhagen

Marx DH (1980) Ectomycorrhizal fungus inoculations: a tool for improving forestation practices. In: Mikola P (ed) Tropical mycorrhiza research. Oxford University Press, New York, pp 13–71

Marzalina M, Normah MN (2002) Cryopreservation techniques for the long-term storage of mahogany (*Swietenia macrophylla*) seeds. J Trop For Sci 14:525–535

Marzalina M, Baskaran K, Yap SK (1993) Collecting seed of tropical rain forest trees: problems and solutions. In: Proceedings of the international symposium on genetic conservation and production of tropical forest tree seed, ASEAN–Canada Forest Tree Seed Centre Project. Saraburi, Thailand, 1993, pp 63–67

Masamba C (1994) Presowing treatments on four African *Acacia* species: appropriate technology for use in forestry for rural development. For Ecol Manage 64:105–109

Masanga HP (1998) Seed germination of indigenous trees in Tanzania. Canadian Forest Service, Northern Forestry Centre

Masanga HP, Maghembe JA (1993) Germination of woodland mahogany (*Trichilia emetica*) following manual seed-coat scarification and potassium nitrate treatment. J Trop For Sci 5:518–527

Masano (1990) Sugar palm (Arenga pinnata) and its problems in Indonesia. In: Turnbull JW (ed) Tropical tree seed research. Proceedings of an international workshop held at the Forestry Training Centre, Gympie, QLD, Australia, 1989, pp 18–20

Masilamani P, Dharmalingam C, Annadurai K (2002) Effect of natural ageing on germination and seedling vigour of teak (*Tectona grandis* Linn. F) drupes. Indian J For 25:122–126

Maude RB (1996) Seed-borne diseases and their control, principles and practices. CABI, Wallingford

Mayer AM, Poljakoff-Mayber A (1982) The germination of seeds. Pergamon, Oxford

McKey D (1975) The ecology of coevolved seed dispersal systems. In: Gilbert, LE, Raven PH (eds) The coevolution of animals and plants. University of Texas Press, Austin, pp 159–191

Merritt DJ, Senaratna T, Touchell DH, Dixon KW, Sivasithamparan K (2003) Seed ageing of four Western Australian species in relation to storage environment and seed antioxidant activity. Seed Sci Res 13:155–165

Milimo PB (1986) The control of germination in *Melia volkensii* seeds. MSc thesis, University of Alberta, Edmonton

Mittal RK, Anderson RL, Mathur SB (1990) Microorganisms associated with tree seeds: world checklist 1990. Petawawa National Forestry Institute, Chalk River

Mohanan C, Sharma JK (1991) Seed pathology of forest tree species in India – present status, practical problems and future prospects. Commonw For Rev 70:133–151

Molina R, Trappe JM (1984) Mycorrhiza management in bareroot nurseries. In: Duryea ML, Landis TD (eds) Forest nursery manual, production of bareroot seedlings. USDA, Washington, pp 211–223

Morandini R (1962) Forest seed handling, equipment and procedures, I: Seed production, collection and extraction. Unasylva 15(4):1–15

Mori SA (1984) Use of Swiss tree grippers for making botanical collection of tropical trees. Biotropica 16:79–80

Mori SA (1995) Exploring for plant diversity in the canopy of a French Guianan forest. Selbyana 16:94–98

Moore RP (1972) Effects of mechanical injuries on viability. In: Roberts RP (ed) Viability of seeds. Syracuse University Press, New York, pp 94–113

Moore RP (1985) Handbook on tetrazolium testing. International Seed Testing Association, Zurich

Morrison RH (1999) Sampling in seed health testing. Phytopathology 89:1084–1087

Murthy BNS, Reddy YN (1989) Temperature dependence of seed germination and seedling growth in ber (*Ziziphus mauritiana* Lam.) and their modification by presowing treatments. Seed Sci Technol 18:621–627

Mycock DJ, Berjak P (1990) Fungal contaminants associated with several homoiohydrous (recalcitrant) seed species. Phytophylactica 22:413–418

Nair PKR (1993). An introduction to agroforestry. Kluwer, Dordrecht

Napier I, Robbins M (1989) Forest seed and nursery practices in Nepal. Nepal-UK Forestry Research Project, Kathmandu

Nan ZB, Hanson J, Yeshi WM (1998) Effects of sulphuric acid and hot water treatment on seedborne fungi and germination of *Stylosanches hamata*, *S. guianensis* and *S. Scabra*. Seed Sci Technol 26:33–43

Nathan I (2001) Organising tree seed programmes with a view to benefiting small farmers and ensuring sustainability: a case study of the National Tree Seed Programme (NTSP) in Tanzania. DFSC/NTSP

Nathan I, Shrestha KB, Jha PK, Suvedi S (2005) Commercial distribution of tree seed in small bags – results of a pilot and action research project in Nepal. Working paper no 9. Forest and Landscape, Denmark, Hoersholm

Neergaard P (1979) Seed pathology. Macmillan, New York

Negi AK, Todaria NP (1993) Research note: omprovement of germination of some Himalayan tree seeds by temperature treatment. Seed Sci Technol 21:675–678

Negi AK, Todaria NP (1995) Pretreatment methods to improve germination in *Terminalia tomentosa* Wight & Arn. Seed Sci Technol 23:245–248

Neya O, Golovina EA, Nijse J, Hoekstra FA (2004) Ageing increases the sensitivity of neem (*Azadirachta indica*) seeds to imbibitional stress. Seed Sci Res 14:205–217

Ng FSP (1991) Manual of forest fruits, seeds and seedlings. Malaysian forest record no 34. Forest Research Institute of Malaysia

Nikolaeva MG (1977) Factors affecting the seed dormancy pattern. In: Khan AA (ed): Physiology and biochemistry of seed dormancy and germination. North Holland, Amsterdam, pp 51–76

Nilson ME, Hjalten J (2002) Covering pine seeds immediately after seeding: effects on seedling emergence and on mortality through seed-predation. For Ecol Manage 176:446–457

Norton DA, Godley EJ, Heenan PB, Ladley JJ (2002) Germination of *Sophora* seeds after prolonged storage. N Z J Bot 40:389–396

Ochsner P (1984) Climbing within the crown 2, tree climbing using the prussic knot. Technical note no 11. Danida Forest Seed Centre, Humlebaek

Ogunnika CB, Kadeba O (1993) Effects of various methods of extraction and germination of *Gmelina arborea* seeds/fruits. J Trop For Sci 5:473–478

Oliveira PS, Galetti M, Pedron F, Morellato LPC (1995) Seed cleaning by *Mycocepurus goeldii* ants (Attini) facilitates germination in *Hymenaea courbaril* (Caesalpinaceae). Biotropica 27:518–522

Oloo WO, Campbell GK, Berjak P, Pammenter NW (1996) The effect of storage on genetic diversity of neem (*Azadirachta indica*) seed. In: Innovations in tropical tree seed technology. Proceedings of the IUFRO symposium of the project group P.2.04.00, 'Seed problems'. Arusha, Tanzania, 1995, pp 223–232

Omondi W (2004) Desiccation sensitivity of seeds of four tree species of economic importance in Kenya. In: Sacande M, Jøker D, Dulloo ME and Thomsen KA (eds) Comparative storage, – biology of tropical tree seeds. International Plant Genetic Resources Institute, Rome, pp 75–86

Omram TA, El-Lakani MH, Shehata MS (1989) Effect of storage on seed germination and subsequent seedling growth of *Casuarina*. Int Tree Crops J 5:247–256

Osborne DJ (1989) Abscission. Crit Rev Plant Sci 8:103–129

Otsamo R, Ådjers G, Kuusipalo J, Otsamo A, Susilo N, Tuomela K (1996) Effect of nursery practices on seed germination of selected dipterocarp species. J Trop For Sci 9:23–34

Owour BO, Gudo S, Niang A (2001) Direct seeding of *Sesbania sesban* for green manure in agroforestry systems – a short connumication. Agrofor Syst 52:23–25

Palani M, Dasrhagir MG, Kumaran K (1995) Effect of pre-sowing chemical treatment on germination and seedling growth in *Acacia nilotica*. Int Tree Crops J 8:189–192

Palmberg C (1985) Sampling in seed collection. In: Forest tree improvement. FAO forestry paper no 20. FAO, Rome, pp 41–45

Pammenter NW, Berjak P (1999) A review of recalcitrant seed physiology in relation to desiccation tolerance mechanisms. Seed Sci Res 9:13–37

Pammenter NW, Berjak P (2000) Evolutionary and ecological aspects of recalcitrant seed biology. Seed Sci Res 10:301–306

Panochit J, Wasuwanich P, Hellum AK (1984) Collection, germination and storage of *Shorea siamensis* Miq. seeds. Embryon 1:1–13

Patterson B, Vaillancourt RE, Potts BM (2001) Eucalypt seed collectors: beware of sampling seedlots from low in the canopy. Aust For 64:139–142

Peran R, Pammenter NW, Naiker J, Berjak P (2004) The influence of rehydration technique on the response of recalcitrant seed embryos to desiccation. Seed Sci Res 14:179–184

Perry DA (1981) Handbook of vigour test methods. International Seed Testing Association, Zurich

Perry DR (1978) A method of access into the crowns of emergent canopy trees. Biotropica 10:155–157

Perry DR, Williams J (1981) The tropical rain forest canopy: A method providing total access. Biotropica 13:283–285

Perry DR, Williams J (1985) Climbing into the crown directly 3. Technical note no 23. Danida Forest Seed Centre, Humlebaek

Peterson J (1987) Seed testing procedures for native plants. In: Langkamp P (ed) Germination of Australian native plant seed. Inkata, Melbourne, pp 31–45.

Phartyal SS, Thapliyal RC, Nayal JS, Rawat MMS (2002) Late maturation changes in Sal (Shorea robusta) seed and their evaluation as indices for proper timing of seed collection. J Trop For Sci 14:191–197

Piotto B, Piccini C (1998) Influence of pretreatment and temperature on the germination of *Fraxinus angustifolia* seeds. Seed Sci Technol 26:799–812

Ponnuswamy AS, Chellapilla KL, Vihaya-Ray RS, Surendran C (1991) Effect of collection date and hydration – dehydration treatment on seed viability and vigour of *Ailanthus excelsa* Roxb. Seed Sci Technol 19:591–595

Poulsen K, Stubsgaard F (1995) Three methods for mechanical scarification of hard-coated seed. Technical note no 27. Danida Forest Seed Centre, Humlebaek

Poulsen KM (1994) Seed testing. Lecture note no. C-8. Danida Forest Seed Centre, Humlebaek

Poulsen KM (1996) Ways and recommendations on how to test seed quality of tropical tree seed species. In: Innovations of tropical tree seed technology. Proceedings of the IUFRO symposium of the project group P.2.04.00, 'Seed problems', Arusha, Tanzania, pp 233–241

Poulsen KM, Parratt MJ, Gosling PG (eds) (1998) ISTA tropical and sub-tropical tree and shrub seed handbook. International Seed Testing Association, Zurich

Priestley DA (1986) Seed ageing, implications for seed storage and persistence in the soil. Comstock, Ithaca

Pritchard HW, Sacande M, Berjak P (2004) Biological aspects of tropical tree seed desiccation and storage responses. In: Sacande M, Jøker D, Dulloo ME, Thomsen KA (eds) Comparative storage – biology of tropical tree seeds. International Plant Genetic Resources Institute, Rome

Pukittayacamee P (1990) Pretreatment of seeds. In: Planting stock production technology. Training course proceedings no 1. ASEAN–Canada Forest Tree Seed Centre, Muak Lek, pp 4–11
Puntarulo S, Sanchez RA, Beveris A (1988) Hydrogen peroxide metabolism in soybean embryonic axes at the onset of germination. Plant Physiol 86:626–630
Purohit M, Jamaluddin, Mishra GP (1996) Testing of seeds of some tropical tree species for germination and mycoflora. Indian For 122: 492–495
Raae K, Christensen SJC (1997) Marketing at a tree seed programme. Technical note no 49. Danida Forest Seed Centre, Humlebaek
Rao PB, Singh SP (1985) Response breadths on environmental gradients on germination and seedling growth in two dominant tree species of Central Himalaya. Ann Bot 56:783–794
Rasmussen P (1992) A guide to the NTSP seed documentation system, forms, routines, files source and accession numbers. National Tree Seed Project, Morogoro
Razanamandranto S, Tigabu M, Sawadogo L, Oden PC (2005) Seed germination of eight savannah-woodland species from West Africa in response to different cold smoke treatments. Seed Sci Technol 33:315–328
Rees AA, Phillips DH (1986) Detection, presence and control of seedborne pest and diseases of trees with special reference to seeds of tropical and subtropical pines. Technical note no 28. Danida Forest Seed Centre, Humlebaek
Richards D, Beardsell D (1987) Seed dormancy. In: Langkamp P (ed) Germination of Australian native plant seed. Inkata, Melbourne, pp 1–13
Richardson MS (1990) ISTA handbook on seed health testing. International Seed Testing Association, Zurich
Riley JD, Craft IW, Rimmer DL, Smith RS (2004) Restoring of magnesian limestone grassland: optimising the time for seed collection by vacuum harvesting. Restor Ecol 12:311–317
Rizzini CT (1977) Nota sobre um embriao dormente em leguminosa esclerodermica (A dormant embryo in a leguminous plant with a hard seed-coat). Rodriguesia 29:33–39 (in Portuguese with English abstract)
Robbins AJ (1983a) *Pinus caribaea* Morelet (mimeo). Seed leaflet no 2. Danida Forest Seed Centre, Humlebaek
Robbins AMJ (1983b) Climbing into the crown by way of the bole 3. The use of spurs for tree climbing. Technical note no 6. Danida Forest Seed Centre, Humlebaek
Robbins AMJ (1984) Tools for harvesting tree fruits. Technical note no 13. Danida Forest Seed Centre, Humlebaek
Roberts EH (1972) Cytological, genetical and metabolic changes associated with loss of viability. In: Roberts EH (ed) Viability of seeds. Syracuse University Press, New York, pp 253–306
Roberts EH (1973a) Predicting storage life of seeds. Seed Sci Technol 1:499–514
Roberts EH (1973b) Loss of seed viability: chromosomal and genetic aspects. Seed Sci Technol 1:515–527
Roberts EH (1973c) Loss of seed viability: Ultrastructural and physiological aspects. Seed Sci Technol 1:529–545
Rungu C (1996) Pretreatment with water and sulphuric acid of different seed sources of *Acacia nilotica*. In: Innovations in tropical tree seed technology. Proceedings of the

IUFRO symposium of the project group P.2.04.00, 'Seed problems', Arusha, Tanzania, pp 242–255
Sacande M, Jøker D, Dulloo ME, Thomsen KA (eds) (2004) Comparative storage – biology of tropical tree seeds. International Plant Genetic Resources Institute, Rome
Saelim S, Pukittayacamee P, Bhodthpuks J, Wang BSP (1996) X-radiography testing for viability of tropical forest seed. In: Bhodthipuks J, Pukittayacamee P, Saelim S, Wang BSP, Yu SL (eds) Rapid viability testing of tropical tree seed. Training course proceedings no 4. ASEAN Forest Tree Seed Centre Project, Muak Lek, pp 17–31
Sajeevukumar B, Sudhakara K, Ashokan PK, Gopikumar K (1995) Seed dormancy and germination in *Albizia falcataria* and *Albizia procera*. J Trop For Sci 7:371–382
Salomao AN (2004) Desiccation, storage and germination of Genipa Americana seeds. In: Sacande M, Jøker D, Dulloo ME, Thomsen KA (eds) (2004) Comparative storage – biology of tropical tree seeds. International Plant Genetic Resources Institute, Rome, pp 263–269
Sandeep-Sharma, Bhardwaj SD, Pankaj-Panwar (2003) Standardizing seed collection time of *Acacia catechu* using leachate conductivity. Seed Res 31:107–109
Sandiford M (1988) Burnt offerings, an evaluation of the hot wire seed scarifier. Commonw For Rev 67:285–292
Sary H, Yameogo CS, Stubsgaard F (1993) The CO_2 method to control insect infestation in tree seed. Technical note no 42. Danida Forest Seed Centre, Humlebaek
Sasaki S, Asakawa S (1974) Effect of monochromatic lights on the germination of pine seeds processed in darkness. J Jpn For Soc 56:361–365
Schaefer C (1990a) Processing, storage and germination of *Prunus africana* seeds. Technical note no 10. Kenya Forestry Research Institute
Schaefer C (1990b) Seed testing research on species indigenous to Kenya. In: Turnbull JW (ed) Tropical tree seed research. Proceedings of an international workshop held at the Forestry Training Centre, Gympie, QLD, Australia, 1989, pp 132–139
Schaefer C (1991) Storage of tree seeds in Kenya, recommendations and problems. In: Proceedings of the 1st national tree seed workshop, Nairobi, 1–5 July 1991. Kenya Forestry Seed Centre/Kenya Forestry Research Institute, pp 99–113
Schmidt LH (2000) A guide to handling of tropical and subtropical forest tree seed. Danida Forest Seed Centre, Humlebaek
Sedgley D, Griffin AR (1989) Sexual reproduction of tree crops. Academic, New York
Seeber G (1976) Nursery techniques. In: Weidelt HJ (ed) Manual of reforestation and erosion control for the Philippines. GTZ, Eschborn, pp 229–389
Seeber G, Agpaoa A (1976) Forest tree seeds. In: Agpaoa A, Endangan D, Festin S, Gumayagay J, Hoenninger T, Seeber G, Unkel K, Weidelt HJ (eds) Manual of reforestation and erosion control for the Philippines. GTZ, Eschborn, pp 473–535
Silc T, Winston DA (1979) Recommendation for calibrating mechanical seeders. Tree Plant Notes 30(4):22–24
Silen R, Osterhaus C (1979) Reduction of genetic base by sizing of bulked Douglas-fir seed lots. Tree Plant Notes 30(1):24–30
Simak M (1973) Separation of forest seed through flotation. In: Seed problems. International symposium on seed processing, vol I, Bergen, paper no 16
Simak M (1991) Testing of forest tree and shrub seeds by X-radiography. In: Gordon AG, Gosling P, Wang BSP (eds)Tree and shrub seed handbook. International Seed Testing Association, Zurich, chap 14

Simons T (1997) The importance of germplasm policies in tree domestication. Advance proceedings of the international workshop on policy aspects of tree germplasm demand and supply organised at ICRAF, Nairobi, Kenya, 6–8 October 1997

Singh P, Bhandari RS (1987) Insect pests of *Acacia tortilis* in India. Indian For 113:734–743

Singh P, Bhandari RS (1988) Insect pests of leguminous forest tree seed and their control. Indian For 114:844–853

Singh S, Verma VPS, Suri RK (1979) Protection of sal seeds in storage against moulds. Indian For 105:811–814

Sirikul W (1990) Shoot growth and flower development in tropical pines, studies on genetic and environmental variation. Tropical forestry reports no 6. University of Helsinki

Sirikul W (1994) Seed production of pines in Thailand. In: Proceedings of the international symposium on genetic conservation and production of tropical forest seed, Thailand. ASEAN–Canada Forest Tree Seed Centre Project, Muak Lek, pp 229–232

Smith OE, Yen WW, Lyons JM (1968) The effects of kinetin in overcoming high temperature dormancy in lettuce seeds. Proc Am Soc Hortic Sci 93:444–453

Smith RD, Dickie JB, Linington SH, Pritchard HW, Probert RJ (eds) (2003) Seed conservation – turning science into practice. The millennium seed bank project. Royal Botanic Gardens, Kew

Somasegaran P, Hoben HJ (1994) Handbook for rhizobia. Methods in legume–rhizobium technology. Springer, Berlin Heidelberg New York

Some LM, Verwey H, Sacande M (1990) Methodology and costs for treatment of *Parkia biglosa* fruits. In: Tropical tree seed research. ACIAR proceedings no 28. IUFRO/ACIAR, Canberra

Song X, Chen Q, Wang D, Yang J (1984) A study on the principal storage conditions of *Hopea hainanensis* seeds. Sci Silvae Sin 20:225–236

Soon LG, Bottrell DG (1994) Neem pesticides in rice – potential and limitations. International Rice Research Institute, Laguna

Sorensen FC, Campbell RK (1993) Seed weight – seedling size correlation in coastal Douglas-fir: genetic and environmental components. Can J For Res 23:275–285

Southgate BJ (1983) Handbook on seed insects of Acacia species. FAO, Rome

Srimathi P, Ramanadane T, Malarkodi K, Natarajan K (2003) Seed extraction in Jamun (*Syzygium cuminii* Skeels). Prog Hortic 35:221–223

Stein WI, Slabaugh PE, Plummer AP (1974) Harvesting, processing and storage of fruits and seeds. In: Schopmeyer CS (ed) Seeds of woody plants in the United States. Agricultural handbook no 450. USDA Forest Service, Washington, pp 98–125

Stubsgaard F (1987) Climbing into the crown directly, 4. Technical note no 31. Danida Forest Seed Centre, Humlebaek

Stubsgaard F (1992) Seed storage. Lecture note no C-9. Danida Forest Seed Centre, Humlebaek

Stubsgaard F (1997) Tree climbing for seed collection. Technical note no 44. Danida Forest Seed Centre, Humlebaek

Stubsgaard F, Moestrup S (1991) Seed processing. Lecture note no C-7. Danida Forest Seed Centre, Humlebaek

Stubsgaard F, Poulsen KM (1995) Seed moisture and drying principles. Lecture note no C-5. Danida Forest Seed Centre, Humlebaek

Sun WQ, Liang Y (2001) Discrete levels of desiccation sensitivity in various sees as determined by the equilibrium dehydration method. Seed Sci Res 11:317–323
Sur K, Lahiri AK, Basu RN (1988) Hydration-dehydration treatment for improved storability of bamboo (*Dendrocalamus strictus* L.). Indian For 114:560–563
Suszka B, Tylkowski T (1982) Storage of acorn of the northern red oak (*Quercus bore alis* Mich. – *Q. rubra* L.) over 1–5 winters. Arbor Kornickie 26:253-306
Swaminathan C, Vinaya Rai RS, Suresh KK, Sivaganam K (1993) Improving seed germination of *Derris indica* by vertical sowing. J Trop For Sci 6:152–158
Sweeney JD, El-Kassaby YA, Taylor DW, Edwards DGW, Miller GE (1991) Applying the IDS method to remove seeds infested with the seed chalcid, *Megastigmus spremotropus* Wachtl. in Douglas fir, *Pseudotsuga menziesii* (Mirb.) Franco. New For 5:327–334
Taiz L, Zeiger E (1991) Plant physiology. Benjamin Cummings, San Francisco
Tanaka Y (1984) Assuring seed quality for seedling production: cone collection and seed processing, testing, storage and stratification. In: Duryea ML, Landis TD (eds) Forest nursery manual: production of bareroot seedlings. Nijhoff and Junk, Dordrecht, chap 4, pp 27–39
Teketay D (1996a) The effect of different pre-sowing treatments, temperature and light on the germination of five senna species from Ethiopia. New For 11:155–171
Teketay D (1996b) Germination ecology of twelve indigenous and eight exotic multipurpose leguminous species from Ethiopia. For Ecol Manage 80:209–223
TeKrony DM (2003) Effect of maturation on seed germination of *Dalbergia cochinchinensis* Pierre. Seed Technol 25:124–127
TeKrony DM (2005) Accelerated aging test: Principles and procedures. Seed Technol 27:135–146
Thapliyal RC, Naithani KC (1996) Inhibition of germination in *Nyctanthes arbor-tristis* (Oleaceae) by pericarp. Seed Sci Technol 24:67–73
Thapliyal RC, Sood OP, Rawat MMS (1991) Effect of moisture content and storage temperature on the viability of *Bambusa tulda* seed. Int Tree Crops J 7 67–75
Thomas TH (1992) Some reflections on the relationship between endogenous hormones and light-mediated seed dormancy. Plant Growth Regul 11:239–248
Todd-Bockarie AH, Duryea ML, West SH, White TL (1993) Pretreatment to overcome seed-coat dormancy in *Cassia sieberiana.* Seed Sci Technol 21:383–398
Tombesi A, Guelfi P, Nottiani G (1998) Ottimizzazione della raccolta delle olive e meccanizzazione (Optimization and mechanization of olive harvesting). Inf Agrar 54(46):79–84 (in Italian with English abstract)
Tompsett PB (1992) A review of the literature on storage of dipterocarp seeds. Seed Sci Technol 20:251–267
Traveset A, Verdu M (2002) A meta-analysis of the effect of gut treatment on seed germination. In: Levey DJ, Galetti M (eds) Seed dispersal and frugivory: ecology, evolution and conservation. CABI, Wallington, pp 339–350
Tschinkel H (1967) La madurez y el almacemaniento de las semillas de Cordia alliodora (Ruiz and Pav.) Cham. Turrialba 17:89–90
Turakka A, Luukkanen O, Bhummibhamon S (1982) Notes on *Pinus kesiya* and *P. merkusii* and their natural regeneration in watershed areas of northern Thailand. Acta For Fenn 178
Turnbull J, Doran J (1987) Seed development and germination in the Myrtaceae. In: Langkamp P (ed) Germination of Australian native plant seed. Inkata, Melbourne, pp 46–57

Turnbull JW, Martens PN (1983) Seed production, collection and germination in Casuarinaceae. In: Midgley SJ, Turnbull JW, Johnston RD (eds) Casuarina, ecology, management and utilization. Proceedings of an international workshop, Canberra, Australia, 1981, pp 126–132

Tweddle JC, Dickie JB, Baskin CC, Baskin JM (2003) Ecological aspects of seed desiccation sensitivity. J Ecol 91:294–304

Uniyal RC, Nautiyal AR (1999) Seed germination and seedling extension growth in *Ougenia dalbergioides* Benth. Under water and salinity stress. New For 16:265–272

Valera FP, Kageyama PY (1991) *Pinus patula.* Seed leaflet no 8a. Danida Forest Seed Centre, Humlebaek

van der Pijl L (1981) Principles of dispersal in higher plants. Springer, Berlin Heidelberg New York

Vargas-Maciel R (2003) Se avanza en el dominio de la siembre direckta con simillas peletizadas. Forestal XXI 6(5):23–24

Vasista HB, Soni P (1988) Germination capacity of *Trema politoria* Planch., an economic plant species. Indian For 114:102–104

Vasquez-Yanes C (1982) Phytochrome control of seed germination in the tropical rain forest pioneer trees *Cecropia obtusifolia* and *Piper auritum* and its ecological significance. New Phytol 92:477–485

Vasquez-Yanes C, Orozco-Segovia A (1996) Comparative longevity of seeds of five tropical rain forest woody species stored under different moisture conditions. Can J Bot 74:1635–1639

Vasquez-Yanes C, Rojas-Arechiga M, Sanches-Coronado ME, Orozco-Segovia A (1996) Comparison of light-regulated seed germination in *Ficus* spp. and *Cecropia obtusifolia*: ecological implications. Tree Physiol 16:871–875

Veira CP, Veira RD, Paschoalick JHN (1994) Effects of mechanical damage during soybean seed processing on physiological seed quality and storage potential. Seed Sci Technol 22:581–589

Vertucci CW, Crane J, Vance NC (1996) Physiological aspects of *Taxus brevifolia* seeds in relation to seed storage characteristics. Physiol Plant 98:1–12

Vijayan AK, Rehill PS (1990) Effect of culture filtrates of some seed borne fungi of *Dalbergia sissoo* Roxb. on seed germination and seedling growth. Indian For 116:559–563

Villiers TA (1972) Seed dormancy. In: Kozlowski TT (ed) Seed biology, vol. II. Academic, New York, pp 220–281

Vleeshouwers M, Bouwmeester HJ, Karssen CM (1995) Redefining seed dormancy: an attempt to integrate physiology and ecology. J Ecol 83:1031–1037

Vogel EF (1980) Seedlings of dicotyledons. PUDOC, Wageningen

Vozzo JA, McConnel L, Edwards JL, Milodragovich E (1988) Seed collection from Loblolly pine cones after tree shaking. Tree Plant Notes 39(2):19–21

Walters C, Pammenter NW, Berjak P, Crane J (2001) Desiccation damage, accelerated ageing and respiration in desiccation tolerant and sensitive seeds. Seed Sci Res 11:135–148

Wang BSP (1975) Tree and shrub seeds. In: Bradnock WT (ed) Advances in research and technology of seeds, part 1. PUDOC, Wageningen, pp 34–43

Wang BSP, Downic B, Weyzel S, Palamarek D, Hamilton R (1992) Effect of cone scorching on germinability and vigour of lodgepole pine (*Pinus contorta* var. *latifolia* seeds in Alberta. Seed Sci Technol 20:409–419

Wang ZF, Xu LK (1985) A study on the technology of aerial sowing for afforestation. Sci Silvae Sin 21:1–9 (in Chinese, reference based on abstract)

Wasuwanich P (1984) Collection and handling of *Gmelina arborea* Roxb. seeds. Embryon 1:14–20

Were J, Munjuga N, Daws MI, Monete N, Erdey D, Baxter D, Berjak P, Pritchard HW, Harris C, Howard CA (2004) Desiccation and storage of *Prunus Africana* seeds (Hook.f.) Kalkm. In: Sacande M, Joker D, Dulloo ME, Thomsen KA (eds) Comparative storage – biology of tropical tree seeds. International Plant Genetic Resources Institute, Rome, pp 67–74

Werker E (1980) Review: seed dormancy as explained by the anatomy of the embryo envelopes. Isr J Bot 29:22–44

Whitehead M (1981) Tree climbing with the Prussik loop system. J Arboric 7(5):123–126

Whitmore TC (1984) Tropical rain forest of the Far East. Clarendon, Oxford

Willan RL (1985) A guide to forest seed handling. FAO forestry paper no 20/2. FAO, Rome

Willan RL (1993) Phytosanitary problems in seed procurement. Lecture note no. C-12. Danida Forest Seed Centre, Humlebaek

Willan RL (1995) International transfer of forest seed. Technical note no 35. Danida Forest Seed Centre, Humlebaek

Willan RL, Barner H (1993) Control measures in transfer of forest seed. Lecture note no C-11. Danida Forest Seed Centre, Humlebaek

Williams CG, Bridgewater FE, Nelson CD (2004) A funder project: making the domestication baseline for forest trees. For Genet Resour 31:20–24

Wilson DO Jr, McDonald MB (1986) The lipid peroxidation model of seed ageing. Seed Sci Technol 14:269–300

Winer N (1983) Germination of pretreated seeds of Mesquite (*Prosopis chilensis*) under arid conditions in northern Sudan. For Ecol Manage 5:307–312

Yadav JP (1992) Pretreatment of teak seed to enhance germination. Indian For 118:260–264

Yang JC, Kuo SR, Lin TP (2005) Storability and treatments to overcome dormancy of sweetheart tree (*Euscaphis japonica* (Thunb.) Kanitz) seeds. Taiwan J For Sci 20: 179–192 (in Chinese with English abstract)

Yeatman CW, Nieman TC (1978) Safe tree climbing in forest management. Forestry technical report no 24. Canadian Forest Service. Ottawa. (Re-issued as: Safe tree climbing in forest management. Technical note no. 24. Danida Forest Seed Centre, Humlebaek)

Yu SL, Wang BSP (1996) Tetrazolium testing for viability of tree seed. In: Bhodthipuks J. Pukittayacamee P, Saelim S, Wang BSP, Yu SL (eds) Rapid viability testing of tropical tree seed. Training course proceedings no 4. ASEAN Forest Tree Seed Centre Project, Muak Lek, pp 33–58

Zobel B, Talbert J (1984) Applied forest tree improvement. Wiley, New York

Subject Index

Printing: Krips bv, Meppel
Binding: Stürtz, Würzburg